# Plant Breeding Technology: Future Trends and Challenges

## CABI Biotechnology Series

Biotechnology, in particular the use of transgenic organisms, has a wide range of applications including agriculture, forestry, food and health. There is evidence that it could make a major impact in producing plants and animals that are able to resist stresses and diseases, thereby increasing food security. There is also potential to produce pharmaceuticals in plants through biotechnology, and provide foods that are nutritionally enhanced. Genetically modified organisms can also be used in cleaning up pollution and contamination. However, the application of biotechnology has raised concerns about biosafety, and it is vital to ensure that genetically modified organisms do not pose new risks to the environment or health. To understand the full potential of biotechnology and the issues that relate to it, scientists need access to information that not only provides an overview of and background to the field, but also keeps them up to date with the latest research findings.

This series, which extends the scope of CABI's successful "Biotechnology in Agriculture" series, addresses all topics relating to biotechnology including transgenic organisms, molecular analysis techniques, molecular pharming, *in vitro* culture, public opinion, economics, development and biosafety. Aimed at researchers, upper-level students and policy makers, titles in the series provide international coverage of topics related to biotechnology, including both a synthesis of facts and discussions of future research perspectives and possible solutions.

## Titles Available

1. Animal Nutrition with Transgenic Plants
   Edited by G. Flachowsky

2. Plant-derived Pharmaceuticals: Principles and Applications for Developing Countries
   Edited by K.L. Hefferon

3. Transgenic Insects: Techniques and Applications
   Edited by M.Q. Benedict

4. Bt Resistance: Characterization and Strategies for GM Crops Producing *Bacillus thuringiensis* Toxins
   Edited by Mario Soberón, Yulin Gao and Alejandra Bravo

5. Plant Gene Silencing: Mechanisms and Applications
   Edited by Tamas Dalmay

6. Ethical Tensions from New Technology: The Case of Agricultural Biotechnology
   Edited by Harvey James

7. GM Food Systems and their Economic Impact
   Tatjana Brankov and Koviljko Lovre

8. Endophyte Biotechnology: Potential for Agriculture and Pharmacology
   Edited by Alexander Schouten

9. Forest Genomics and Biotechnology
   Richard Meilan and Matias Kirst

10. Transgenic Insects: Techniques and Applications, 2nd Edition
    Edited by M.Q. Benedict and Maxwell J. Scott

11. Plant Omics: Advances in Big Data Biology
    Edited by Hajime Ohyanagi, Eiji Yamamoto, Ai Kitazumi and Kentaro Yano

12. Next-generation Sequencing and Agriculture
    Edited by Phillip Bayer and Dave Edwards

13. Aquaculture and Fisheries Biotechnology: Genetic Approaches
    Rex A. Dunham

14. Algal Biotechnology
    Edited by Qiang Wang

15. Genetic Engineering of Vegetable Crops
    Edited by Pritam Kalia

16. Plant Breeding Technology: Future Trends and Challenges
    Edited by Zeba Khan

# Plant Breeding Technology: Future Trends and Challenges

*Edited by*

**Zeba Khan**

*Department of Botany, Raja Mahendra Pratap Singh State University, Aligarh, India*

CABI

**CABI is a trading name of CAB International**

CABI
Nosworthy Way
Wallingford
Oxfordshire OX10 8DE
UK

Tel: +44 (0)1491 832111
E-mail: info@cabi.org
Website: www.cabi.org

CABI
200 Portland Street
Boston
MA 02114
USA

T: +1 (617)682-9015
E-mail: cabi-nao@cabi.org



A catalogue record for this book is available from the British Library, London, UK.

ISBN-13: 9781800626614 (hardback)
9781800626621 (ePDF)
9781800626638 (ePub)

DOI: 10.1079/9781800626638.0000

Commissioning Editor: Ward Cooper
Editorial Assistant: Emma McCann
Production Editor: Shankari Wilford

Typeset by Exeter Premedia Services Pvt Ltd, Chennai, India
**Printed and bound by CPI Group (UK) Ltd, Croydon, CR0 4YY**

# Contents

# Contributors

**Anees, Arshi**, Multidisciplinary Centre for Advanced Research and Studies (MCARS), Jamia Millia Islamia, New Delhi – 110025, India. Email: arshianees744@gmail.com

**Ansari, Mohd Yunus Khalil**, Department of Botany, Aligarh Muslim University, Aligarh – 202001, India. Email: mykansari1@gmail.com

**Bolouri, Parisa**, Department of Genetics and Bioengineering, Yeditepe University, Istanbul, Turkey. Email: bolouri88@gmail.com

**Debnath, Sadhan**, National Institute of Plant Genome Research (NIPGR), Aruna Asaf Ali Marg, New Delhi, India; College of Post Graduate Studies in Agricultural Sciences (CPGSAS), Central Agricultural University (Imphal), Umiam, Meghalaya, India. Email: debnathsadhan1989@yahoo.com

**Devi, Kirti**, National Agri-food Biotechnology Institute (NABI), Mohali, Punjab, India. Email: kirtirana999@gmail.com

**Dikmen, Sevval**, Department of Molecular Biology and Genetics, Biruni University, Istanbul, Turkey. Email: 190401025@st.biruni.edu.tr

**Dinkar, Vishal**, ICAR-Central Institute of Temperate Horticulture, Srinagar – 191132, J&K, India. Email: vishaldinkar1991@gmail.com

**Firdaus, Aliya**, Biocontrol Research Laboratory, Integral Information and Research Centre-4 (IIRC-4), Integral University, Lucknow, UP – 226026, India. Email: aliyafir@student.iul.ac.in

**Izhar, Syed Khalida**, Biocontrol Research Laboratory, Integral Information and Research Centre-4 (IIRC-4), Integral University, Lucknow, UP – 226026, India. Email: syedkhali@student.iul.ac.in

**Jain, Shikha**, ICAR-Indian Agricultural Research Institute, New Delhi – 110012, Delhi, India. Email: jain64235@gmail.com

**Karki, Shanta**, National Centre for Fruit Development, Kirtipur, Nepal. Email: shantakyoto@gmail.com

**Khan, Zeba**, Department of Botany, Raja Mahendra Pratap Singh State University, Aligarh – 202002, India. Email:khanzeba02@gmail.com

**Kumar, Girjesh**, Department of Botany, University of Allahabad – 211002, India. Email: kumar_girjesh@yahoo.com

**Majumder, Sujan**, ICAR-Indian Institute of Vegetable Research (IIVR), Varanasi, India. Email: sujaniari@gmail.com

**Meetei, Ng. Tombisana**, College of Post Graduate Studies in Agricultural Sciences (CPGSAS), Central Agricultural University (Imphal), Umiam, Meghalaya, India. Email: ntmeetei@gmail.com

**Mhetre, Vishal B.**, Defence Institute of High Altitude Research, Leh, UT of Ladakh – 194101, India. Email: vishalmhetre8@gmail.com

**Mohanty, Ayesha**, AINP on Soil Biodiversity-Biofertilizers, Odisha University of Agriculture and Technology, Bhubaneswar, Odisha, India. Email: ayeshamohanty89@gmail.com

**Mohapatra, Sourav Ranjan**, Department of Forest Biology and Tree Improvement, Odisha University of Agriculture and Technology, Bhubaneswar, Odisha, India. Email: mohapatrasourav23@gmail.com; srmohapatra.forestry@ouat.ac.in

**Padhy, Jyoti Sankar**, Department of Genetics and Plant Breeding, Bidhan Chandra Krishi Vishwavidalaya, Mohanpur, West Bengal, India. Email: padhy.jyotisankar@bckv.edu.in

**Pandey, Asha**, Department of Botany, University of Allahabad – 211002, India. Email: asha2124shankar@gmail.com

**Pandey, Saroj**, Department of Botany, Iswar Saran Degree College, University of Allahabad, Prayagraj – 211004, India. Email: saroj.sri@gmail.com

**Panigrahi, Kaushik Kumar**, AICRN on Potential Crops, Odisha University of Agriculture and Technology, Bhubaneswar, Odisha, India. Email: kaushikouat@gmail.com

**Patil, Manjunath S.**, Zonal Agriculture Research Station (MPKV, Rahuri), Solapur – 413002, Maharashtra, India. Email: patilmsflori@gmail.com

**Rizal, Govinda**, Institute of Agriculture and Animal Science, Tribhuvan University, Kirtipur, Nepal. Email: grizal@gmail.com

**Sahin, Fikrettin**, Department of Genetics and Bioengineering, Yeditepe University, Istanbul, Turkey. Email: fsahin@yeditepe.edu.tr

**Sarkar, Ritabrata**, Department of Genetics and Plant Breeding, ICAR-Indian Institute of Agricultural Biotechnology, Ranchi, India. Email: ritabratablg98@gmail.com

**Sirasapalli, Bhargav Kiran**, Centurion University of Technology and Management, R. Sitapur – 761211, Odisha, India. Email: bhargavsirasapalli@gmail.com

**Singh, Rajani**, Department of Botany, Iswar Saran Degree College, University of Allahabad, Prayagraj – 211004, India. Email: singh.rajani1995@gmail.com

**T.S., Chaithra**, ICAR-Indian Institute of Soil and Water Conservation, Research Centre, Kota – 324002, Rajasthan, India. Email: chaithra.571996@gmail.com

**Vittal, Hatkari**, ICAR-Indian Agricultural Research Institute, New Delhi – 110012, Delhi, India. Email: vittalhatkari@gmail.com

**Yunus, Mohd Hadi**, Department of Horticultural Bioscience, Pusan National University, Miryang 50463, Republic of Korea. Email: mhadiyunus@gmail.com

# 1 Harnessing Next-Generation Genetics and Genomics for Enhanced Crop Improvement: Strategies, Progress and Potential

**Rajani Singh[1]*, Saroj Pandey[1], Asha Pandey[2] and Girjesh Kumar[2]**
*[1]Department of Botany, Iswar Saran Degree College, University of Allahabad, Prayagraj – 211004, India; [2]Department of Botany, University of Allahabad – 211002, India*

## Abstract

One of the foremost challenges confronting humanity pertains to the provision of food, clothing and nourishment to an increasingly expanding global populace, amidst adverse climatic conditions, limited water supplies and the imperative for sustainable energy alternatives, and the necessity for environmental stewardship. In the coming decades, enhancing crop improvement through next-generation genetics and genomics or next-generation sequencing (NGS) represents a promising frontier in agriculture. More accurate and precise phenotypic strategies are necessary to empower high-resolution linkage mapping and genome-wide association studies and for training genomic selection models in plant improvement. Genome selection (GS) may be achieved using high-throughput genotyping, transcriptome sequencing, metagenomics, whole-genome sequencing (WGS) and NGS. These transformative technologies may also be used for disease detection, identification and eradication, diversity analysis, and the genetic and epigenetic characterization of germplasm.

**Keywords:** conventional breeding, genome selection, next-generation sequencing, metagenomics, whole-genome sequencing

## Abbreviations

BC – Backcross, BV – Breeding value, CRISPR-Cas – clustered regularly interspaced short palindromic repeats/CRISPR-associated protein, GBS – Genotyping-by-sequencing, GEBV – Genomic estimated breeding values, GO – Gene Ontology, GS – Genomic selection, GWS – Genome-wide selection, $h^2$ – Narrow-sense heritability, $H^2$ – Broad-sense heritability, KEGG – Kyoto Encyclopedia of Genes and Genomes, MAB – Marker-assisted breeding, MABC – Marker-assisted backcrossing, MAGIC – Multi-parent advanced-generation inter-cross, MARS – Marker-assisted recurrent selection, MAS – Marker-assisted selection, NGS – Next-generation sequencing, QTL – Quantitative trait locus, RAD-seq – Restriction-site-associated DNA sequencing, RIL – Recombinant inbred lines, SNP – Single-nucleotide polymorphism, SNV – Single-nucleotide

*Corresponding author: singh.rajani1995@gmail.com


DOI: 10.1079/9781800626638.0001

variant, SSR – Simple sequence repeat, STR – Short tandem repeat, SV – Structural variation, WGS – Whole-genome sequencing.

## 1.1 Introduction

Recent remarkable technological advancements have expanded our knowledge from the genome to the gene level. It has made it easier to comprehend how gene networks in many model, or important crop, species relate to plant development and agronomy. Functional genomics, transcriptome/genome sequencing, physical mapping, trait mapping and molecular markers are some of these technologies. More intriguingly, genomics techniques and tools are transforming breeding methods through directed mutagenesis and molecular breeding, which greatly increase the efficiency of breeding to enhance agronomic characteristics. This process is known as 'genomics-assisted breeding'. Plant biotechnology is further accelerated by genomics, which provides more native target genes. Together with the previously listed methods, other cutting-edge technologies have arisen in the past 5 years. These technologies, which include high-throughput marker genotyping and next-generation sequencing (NGS), are thought to have a stronger influence on plant genetics research and breeding initiatives. Similarly, using natural variation is proposed to be achieved by a number of contemporary genetic techniques. In addition to the widely used marker-assisted selection (MAS), additional techniques that are becoming increasingly well liked are genome-wide selection (GWS), marker-assisted recurrent selection (MARS), and marker-assisted backcrossing (MABC). As part of an international scientific research effort known as the Human Genome Project (Buniello *et al.*, 2019), the first draft of the human genome sequence, made up of approximately 3 billion nucleotides, was released in 2001. Scientists from around the world sequenced sections of the human genome in rooms crammed with capillary sequencers as part of this enormous project, which came at a cost of approximately 3 billion US dollars. In 2003, the project was formally deemed finished. With the sequencing methods available today, a single lab would be able to finish the same project quickly and affordably within a few hours. NGS is a DNA sequencing technique that is highly parallel and efficient. Whole-genome sequencing (WGS) has become increasingly affordable, with costs as low as US$600 per genome, while sequencing technology and bioinformatics technologies have advanced to record speeds, from budget short reads to whole genome sequencing at accelerated speeds. Industry leaders anticipate that NGS will cost US$100 per genome (Preston *et al.*, 2021). This means that any lab will soon have access to the potential of personalized medicine. This chapter gives a synopsis of the key technologies that define the various generations of NGS, as well as a brief history of the technology.

### 1.1.1 Next-generation sequencing: a synopsis

It is critical to improve our capacity to comprehensively examine the nucleic acid-based information within the cell for genomics-assisted breeding. In this regard, it has been demonstrated that molecular markers are highly helpful genetic tools for creating physical and genetic maps, as well as trait mapping, in a number of crop species (Varshney and Tuberosa, 2007). In fact, these markers have been employed in breeding programmes to produce enhanced varieties and/or superior lines in a number of temperate cereal species. Simple sequence repeat (SSR) or microsatellite markers have been found to be the most widely used molecular marker technology for breeding reasons. Nevertheless, high-throughput genotyping technologies (Varshney, 2009) and next-generation sequencing technologies (Varshney, 2009) have emerged recently, and it is anticipated that single-nucleotide polymorphism (SNP) markers may eventually replace SSR markers over the next 5 years or so. Library preparation, clonal amplification, sequencing and analysis are steps in the fundamental NGS procedure. NGS technologies can be utilized to provide parental genotype sequence data for mapping populations, which may then be used for extensive SNP mining. Aligning human NGS data with reference genome sequence data (if available), or transcript sequence data obtained from expressed sequence tag (EST) sequencing projects, is quite simple for model plant species and major crop species. After that, millions of template molecules are created for the sequencing procedure by amplifying the template DNA using droplet, or solid phase polymerase chain reaction (PCR), which involves DNA amplification occurring on a solid surface rather than in solution. Every nucleotide location on the template molecule produces an optical (like fluorescence) or chemical (like pH) signal during the sequencing reaction in response to a process (like nucleotide addition on a developing complementary strand). To ascertain the sequence, this signal is captured and subjected to computer analysis. The four main steps of NGS are as follows:

1. Library preparation: To prepare for amplification, DNA is broken apart and ligated to certain adaptors.
2. Clonal amplification: The template is amplified using bridge PCR (solid surface) or emulsion PCR (on beads).
3. Sequencing: The NGS instrument is used for sequencing, and the output signal is recorded.
4. Data analysis: To ascertain the sequence, an optical or chemical signal is analysed, using, for example, the whole-genome genotyping Infinium assay, which (WGIA) This assay is based on comparative genomic hybridization. It facilitates measurement of signal intensity variation and changes in allele composition simultaneously. This assay includes whole genome amplification to increase the amount of DNA, followed by fragmentation and capture on to a bead array through an SNP-specific primer.

### 1.1.2 Trait mapping and QTL analysis

A quantitative trait locus (QTL) is a section of DNA linked to a particular phenotype or feature that fluctuates within a population. Qualitative trait loci (QTLs) are generally linked to continuous features, like skin colour or height, as opposed to discrete traits, like eye or hair colour. A statistical technique called QTL mapping is used to determine which molecular markers cause a certain trait's quantitative variation. Imputation or whole-genome sequencing is a crucial requirement for QTL mapping, to precisely identify the relevant molecular marker, as a single locus may have several variations. Variants that act at various stages along the genotype-to-phenotype continuum have been added to QTLs.

### 1.1.3 Multi-parent advanced generation inter-cross (MAGIC): an overview

With the assistance of several parents, the MAGIC population is a mapping resource for crop growth that is in its second generation. This population mapping strategy maps genetic variation in populations that are segregating for several QTLs using linkage and association approaches. This approach was initially used on mice, where a large number of parent recombinant inbred lines (RILs) were used to map a large number of QTLs controlling complex characteristics (Yalcin *et al.*, 2005). This approach is known as the MAGIC populace in the plant community (Cavanagh *et al.*, 2008). A previous study in mice has demonstrated a certain degree of subcentromere range mapping resolution using the eight-parent RIL population with 1000 progeny (MAGIC population) (Valdar *et al.*, 2006).

The goal of molecular biology in the area of functional genomics is to comprehend the intricate link that exists between an organism's observable qualities, or phenotype, and its genetic makeup, or genotype, which is the result of interactions between genes and the environment. Though functional genomics studies how genes are activated and function in a dynamic, context-specific, cooperative manner to answer biological problems, while genetics and genomics research genes and an organism's total genetic material. This is achieved by employing diverse genomes and associated 'omics' datasets. The primary objectives since the completion of the sequencing of the human genome have been to comprehend the role of genes, assess the patterns of their expression, investigate the significance of proteins and investigate the influence of genetic diversity on diverse biological processes.

### 1.1.4 Marker-assisted selection (MAS) and marker-assisted backcrossing (MAB)

Breeding programmes include MABC backcrossing and MAS. The next stage is to incorporate these markers into the breeding programme once the markers associated with a particular trait have been identified by techniques like transcriptomics, advanced backcross QTL analysis, association mapping and linkage mapping (Utomo and Linscombe, 2009). Within this framework, the process of choosing one or a small number of genes (known as QTLs) through the use of molecular markers via backcrossing is an exceptionally effective strategy (Collard and Mackill, 2008). MABC consists of three levels:

1. The introduction of foreground selection entails using molecular markers to identify the target gene or QTL. Recessive alleles can also be chosen at this stage for backcrossing, as they require one generation of selfing to manifest. Finding and choosing the progeny of a cross that carry the desired gene and have experienced recombination events between the target locus and nearby markers is known as 'recombinant selection'.
2. Reducing the detrimental effect brought on by the presence of markers surrounding the target gene is the aim of this selection procedure. A major obstacle to selecting using traditional breeding methods is the existence of linkage drag. Furthermore, two backcross (BC) generations are generally used to carry out this recombination selection process.
3. Furthermore, background selection comprises the utilization of markers that are not linked to the target locus to select BC progeny with the highest proportion of the recurrent parent (RP).

To summarize, the MABC technique utilizes linked markers to identify the desired gene or QTL from the donor parent, and unlinked tags to reintroduce the recurrent parent. The conventional methods for recovering the RP genome need up to six BC generations, while the utilization of markers allows achievement of the same outcome, even as early as BC2. The MABC technique has gained significant popularity among the molecular breeding community (Hospital *et al.*, 1997; Holland, 2004). Although MABC techniques are commonly used in many multinational enterprises, they have also been effectively employed in certain breeding initiatives within the public sector. Success stories of MABC have been studied in both legumes (Varshney *et al.*, 2010) and cereals (Varshney *et al.*, 2009), where it has proven instrumental in the development of improved genotypes and varieties. The MABC method has recently become more well liked in emerging nations, such as Bangladesh, Thailand, China and India. In these countries, the public sector has been collaborating with international agriculture research centres to develop improved varieties that are suitable for their specific environments.

### 1.1.5 Marker-assisted recurrent selection (MARS)

This approach is applied in cases where tiny individual effects of several genes, collectively referred to as QTLs, influence quantitative variance. But when it comes to introducing these genes, there are problems with MARS, in connection with the ideal quality. Furthermore, biparental mapping populations are often employed in MABC to determine the markers associated with a certain characteristic. Examining allelic variation and genetic background – two crucial components of crop-breeding programmes – becomes more challenging as a result. One limitation of MAS is the lack of comprehensive statistical methods to investigate polygenic variables affected by several loci with small individual effects. Although MAS has increased the effectiveness of crossing and selection, it remains limited for crops with complex genetics (Shahwar *et al.*, 2023).

Furthermore, there is a variable influence of small QTLs on the characteristic. Although the influence of these minor QTLs is constant, the requirement for a higher number of progeny background – which can occasionally be an unmanageable background – makes incorporating them into a MABC strategy extremely difficult. The number of QTLs involved determines whether lines in MABC are appropriate.

MARS can be used in the previously indicated situations to merge several genes/QTLs (with little effect) into a single genotype (Hospital *et al.*, 1997; Bernardo and Charcosset, 2006).

### 1.1.6 Genome-wide selection

Genome-wide selection (GWS) is an additional tactic that may be utilized in conjunction with MARS to promote advantageous alleles for QTLs with small effects at the whole-genome level. (Meuwissen *et al.*, 2001).

Unlike MABC or MARS, the GWS technique accounts for all the diversity in phenotype by computing the effects of markers throughout the whole genome. Using the marker loci or haplotype data derived from the progeny lines, the genomic estimated breeding values (GEBVs) are computed (Meuwissen *et al.*, 2001). It is essential to stress that genotyping data and a model trained on people with both phenotyping and genotyping data are used to compute the GEBVs for individual subjects. These GEBVs are then used to choose the offspring lines that will advance through the breeding process. In summary, GWS provides a way to choose an individual without visible traits by predicting the individual's genetic value using a model. None the less, it is essential to carefully choose a training population that appropriately replicates the candidates for selection in the breeding programme, where GWS will be applied in order to maximize the accuracy of GEBVs in GWS. There is not, as far as the authors are aware, any published information about the use of GWS in crop breeding. Nonetheless, it is now feasible to gather marker data throughout the whole genome and use GWS in breeding programmes thanks to the availability of high-throughput genotyping instruments in a number of crops (Buniello *et al.*, 2019). It is anticipated that GWS will soon be implemented in at least a few breeding projects for important crops including maize, rice and soybeans.

### 1.1.7 Plant phenomics

Plant phenomics is a high-throughput, innovative field that satisfies every need for the quick, precise and comprehensive collection of phenotypic data. Plant phenomics is a method for identifying complicated features that are suitable for selection and offers pertinent details about the reasons behind why a specific genotype might be more prominent in a given environment. Plant phenotyping is an approach that may be used with management methods at various stages, from the gene to the entire plant in a specific environment. In this chapter, we discuss recent advances in plant phenomics and provide a number of restricted and field-based high-throughput techniques for applications in forward and reverse genetics. These plant phenomics approaches enable the identification of stress, the investigation of physiological processes, rapid and efficient screening, dissection, and validation, to understand the genetic basis of different phenotypes, genes and features. High-throughput phenomics technologies are crucial for verifying gene or trait functional analysis, phenotyping huge germplasm populations, and avoiding human error and time consumption. NGS technologies have become crucial in the field of plant research. They are used for assembling crop reference genomes, sequencing the transcriptome to study gene expression, developing molecular markers for the entire genome, and identifying markers in genes with known functions (Fig. 1.1). Some of these have proven valuable in the cultivation of different crops. We will explore these techniques in detail in the subsequent sections of this chapter.

## 1.2 Next-Generation Gene Sequencing

The development of next-generation sequencing (NGS) techniques was prompted by the need to efficiently sequence a greater number of samples at a reduced cost. The subsequent Sanger-based approaches have been inadequate for handling large quantities of samples, as well as for doing extensive parallel signature sequencing (Brenner *et al.*, 2000). In the plant research area, next-generation sequencing technologies have

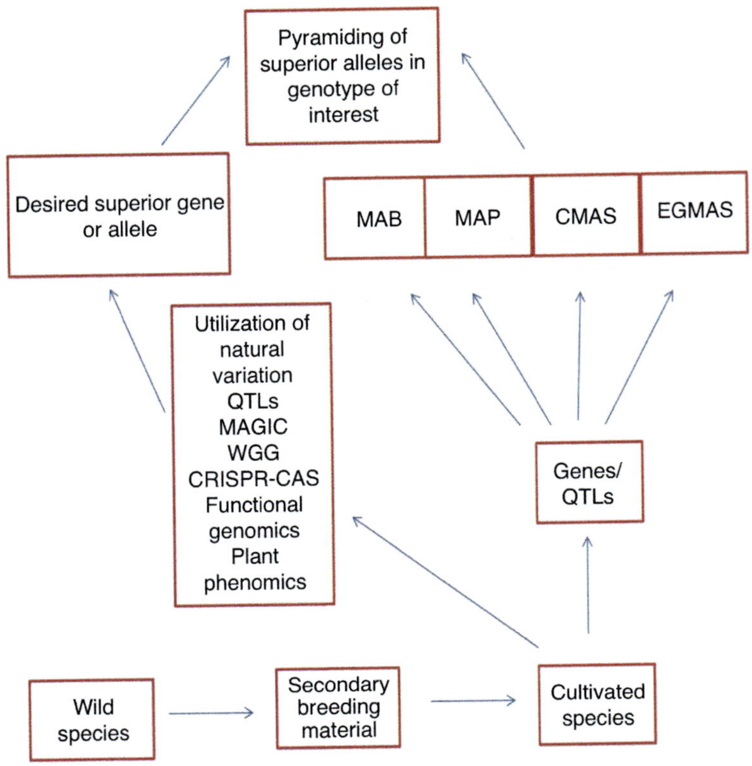


**Fig. 1.1.** A linear chart showing the use of novel genomics technologies together with modern breeding methodologies in an integrated way. CMAS – Combined MAS, CRIPSR-Cas – Clustered regularly interspaced short palindromic repeats/CRISPR-associated protein, EGMAS – Early-generation MAS, MJAP – Marker-assisted pyramiding, WGG – Whole-genome genotyping.

become important tools for the construction of crop reference genomes, and transcriptome sequencing is used to investigate gene expression, while whole-genome molecular marker development is employed to identify markers in genes with established functions. Several of them have proven to be valuable in the cultivation of different crops. Genotyping in large plant breeding populations can be a time-consuming process, typically requiring several months, depending on the marker system used. This extended duration of genotyping adds to the overall expense of the process. In order to advance plant breeding techniques, it is necessary to develop more efficient methods that can quickly and cost-effectively genotype huge populations of plants. The accessibility of WGS has transformed the methodology for detecting DNA markers. The focus has switched from fragment-based polymorphism identification to sequence-based SNP identification. The purpose of this modification is to enhance the efficiency of the marker identification procedure and augment the quantity of informative markers. Nevertheless, the WGS technologies that depend on Sanger sequencing are not efficient in terms of both time and money, and they just provide information on the specific individual under investigation. Consequently, their utilization in gene discovery is restricted.

The utilization of this technique in extensive breeding populations is restricted due to time and financial constraints. The emergence of next-generation sequencing (NGS) technologies and advanced computational pipelines has significantly decreased the expense of whole genome sequencing, enabling the identification, sequencing, and genotyping of several markers in a single process (Stapley *et al.*, 2010). NGS has become a highly effective method for identifying a wide range of DNA sequence variations in a short period of time. This technology, as shown in Fig. 1.2, has rapidly gained popularity as a valuable tool in the field of plant breeding for future generations. NGS-based marker development begins with the construction of a library before sequencing. Various focused marker discovery methods have been developed using Next-Generation Sequencing (NGS) technologies. These methods require representing only a portion of the genome and can be used even without prior knowledge of WGS. There are two main types of partial genome representation libraries: complexity reduced-representation libraries, which are created using restriction enzymes, and sequence-capture libraries, which do not require restriction digestion. The first category includes a range of techniques, such as reduced-representation libraries (Gore *et al.*, 2009), complexity reduction of polymorphic sequences (Mammadov *et al.*, 2010), restriction-site associated DNA sequencing (RAD-seq) (Pfender *et al.*, 2011), sequence-based polymorphic marker technology (Sahu *et al.*, 2012), multiplexed shotgun genotyping (Andolfatto *et al.*, 2011), and genotyping-by-sequencing (GBS) (Elshire *et al.*, 2011).

The second category consists of techniques such as molecular inversion probes (Porreca *et al.*, 2007), solution hybrid selection (Gnirke *et al.*, 2009), and microarray-based genomic selection (GS) (Albert *et al.*, 2007). Sequence capture can be performed to selectively target either wide-ranging or precise sections within the genome, such as exome sequencing (Teer and Mullikin, 2010) or sequencing of the genomic region associated with a certain trait (Teer and

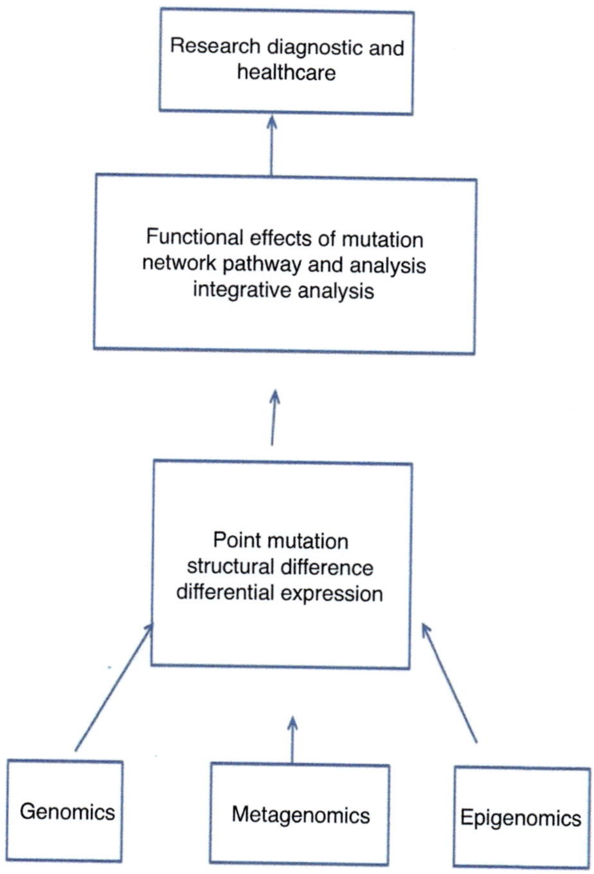


**Fig. 1.2.** Various approaches used for genome analysis and NGS application. (Author's own image.).

Mullikin, 2010). NGS technologies are gaining widespread acceptance in the realm of crop breeding. Several Next-Generation Sequencing (NGS) methods can efficiently detect and categorize SNPs, hence accelerating the overall process (Fig. 1.3). Moreover, the extensive accessibility of gene and transcript sequence data in the public domain facilitates the development of genetic molecular markers or functional markers. Previous studies have shown that RAD-seq and GBS, two separate NGS methods, have been effective in the field of next-generation plant breeding (Yang *et al.*, 2012; Glaubitz *et al.*, 2014).

### 1.2.1 Restriction site-associated DNA sequencing (RAD-seq)

RAD-seq is a method for analysing bulked segregants that is mostly focused on SNPs. In this technique, the genomic DNA is fragmented using a specific restriction enzyme. Then, a barcoded adaptor with a molecular identification is attached to the fragments through ligation. This method was described by Pfender *et al.* (2011) and Yang *et al.* (2012). Subsequently, the DNA sample that has undergone processing from numerous individuals (about 20) is combined and then randomly fragmented, ensuring that only a portion of the resulting fragments have a barcoded adaptor (Poethig, 2001).

The fragments are linked to another divergent adaptor in preparation for PCR. The amplification of fragments with both adaptors is ensured by the divergent adaptor. An Illumina platform is used to sequence the amplicons that are produced. Finally, samples with unique identifiers are split, and a standard bioinformatics process is used to identify SNPs (Fig. 1.3). Prior knowledge of the genome sequence is not necessary for this technique to work. Single nucleotide polymorphisms generated from RAD-seq were employed to generate a genetic map in aubergine and identify quantitative trait loci (QTLs) linked to the fruit's anthocyanin pigmentation (Barchi *et al.*, 2012). Furthermore, Yang *et al.* (2012) used RAD-seq tagged SNPs to find a resistance gene against anthracnose disease in lupin.

### 1.2.2 Genotyping-by-sequencing (GBS)

The progress made in NGS has significantly reduced the expenses associated with DNA sequencing. As a result, GBS has become a viable option for studying species with high genetic diversity and huge genomes (Elshire *et al.*, 2011). GBS is a streamlined and efficient method for creating libraries with reduced representation for the Illumina NGS platform; it was created by the Buckler group (Elshire *et al.*, 2011). The study conducted by Beissinger *et al.* (2013) produced a substantial amount of SNPs that can be utilized in genetic studies and genotyping. The main features of this approach consist of its affordability, minimized sample manipulation, decreased number of PCR and purification stages, absence of size separation,

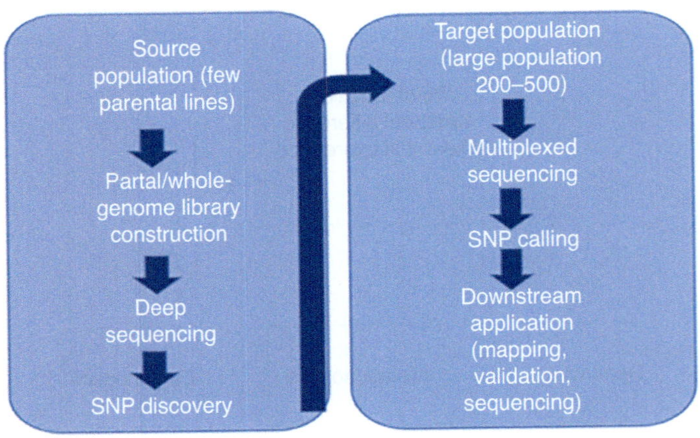


**Fig. 1.3.** RAD sequencing application.

absence of constraints on reference sequences, effective barcoding and ease of scalability (Davey *et al.*, 2011). Genotyping-by-sequencing is gaining significance as an economical and distinctive method for genomics-guided breeding in several plant species. The method was initially designed for conducting high-resolution association studies in maize and has subsequently been adapted for use in several species with intricate genomes, similar to RAD. Unlike other genotyping methods that are primarily used for well-known 'reference' genomes, GBS provides a cost-effective method for identifying and genotyping SNPs in various crop species and populations. A technologically uncomplicated and highly versatile method, GBS is well suited for population studies, germplasm characterization, plant genetics and breeding in various crops. It has been extensively utilized in numerous large crop genomes to densely populate mapping and breeding populations with tens to hundreds of thousands of SNP markers (Poland *et al.*, 2012).

The construction of GBS libraries relies on the reduction of genomic complexity using restriction enzymes (Elshire *et al.*, 2011). This method is straightforward, efficient, exceptionally precise, highly replicable and has the potential to target crucial genomic areas that cannot be accessed by sequence-capture methods. By selecting suitable regular expressions, it is possible to prevent the occurrence of repetitive sections in genomes and enhance the efficiency of targeting lower-copy regions by two to three times (Gore *et al.*, 2007). This greatly simplifies computationally complex alignment issues in species with significant genetic variation. The GBS approach was illustrated using maize and barley recombinant inbred populations, with approximately 200,000 and 25,000 sequence tags mapped, respectively (Elshire *et al.*, 2011).

## 1.3 CRISPR-Cas Gene Editing

Clustered regularly interspaced short palindromic repeats/CRISPR-associated protein (CRISPR-Cas) technology enables the addition, removal or alteration of genetic material at specific locations in the genome by replacing an existing sequence with a customized DNA segment (Shahwar *et al.*, 2023). This technology consists of a synthetic short guide RNA (sgRNA) and a programmable Cas9 enzyme. The guide RNA gives DNA its target specificity. Because of this, the CRISPR-Cas9 system is, comparatively, simpler to construct than zinc finger or transcription activator-like nucleases (TALENs). Cong *et al.* (2013) stated that its affordability, ease of use, effectiveness and ability to target many genes simultaneously set it apart. The CRISPR/Cas9 gene-editing techniques include the deliberate creation of double-strand breaks (DSBs) in targeted portions of the genome, which are subsequently repaired by biological mechanisms. The cell machinery repairs the double-strand break (DSB) via either homology-directed repair (HDR) or non-homologous end joining (NHEJ), depending on the cell state and the availability of a repair template (Martinez-Lage *et al.*, 2018). The precision of CRISPR/Cas9 editing is derived from the sgRNAs (Fig. 1.4), which form interactions with target sequences through Watson–Crick base pairing (Hsu *et al.*, 2014). Overcoming the challenges of delivering CRISPR-Cas9 editing tools to specific cells in the body and minimizing unexpected off-target effects are crucial for the successful use of this technology in medical treatments (Chen *et al.*, 2019). Therefore, novel gene-editing techniques, such as prime editors, cytosine base editors and CRISPRon have begun to appear. These systems exhibit enhanced precision and reduced off-target effects compared to the usual CRISPR/Cas9 method (Anzalone *et al.*, 2019; Doman *et al.*, 2020).

CRISPR-based target enrichment techniques serve as a supplementary tool to enhance the analytical performance of NGS, thus indirectly aiding in the detection of nucleic acids. The utilization of the CRISPR-Cas system for targeted DNA cleavage enables the extraction of intact, sizable segments from disease-associated genomic areas. The CRISPR-Cas system can specifically identify and separate the desired region without the need for amplification. Following this, long-read sequencing technologies were also utilized (Khan and Shahwar, 2024). Applications for CRISPR-Cas9 technology are numerous. It has been used in medical studies on pathogen detection, cancer, viral infections, and genetic issues. Research by Long *et al.* (2016) and Kang *et al.* (2015) has demonstrated that the method has been effectively applied in mice to fix mutations in monogenic diseases, such as Duchenne

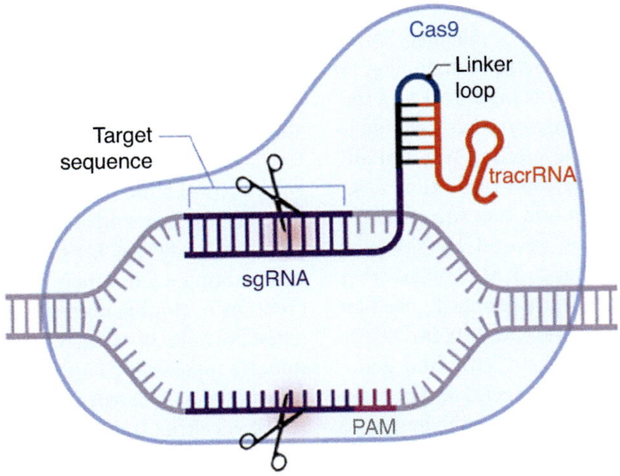


**Fig. 1.4.** CRISPR/Cas9 editing.

muscular dystrophy. Using CRISPR-Cas9, interference has been observed with human papillomaviruses (Kennedy *et al.*, 2014), hepatitis B virus (Kennedy and Cullen, 2015), and HIV-1 provirus (Ebina *et al.*, 2013). Moreover, CRISPR-Cas9 has demonstrated remarkable potential in the treatment of Hutchinson-Gilford progeria disease (Beyret *et al.*, 2019) and cancer (Chen *et al.*, 2019). Additionally, it has been applied to treat persons with certain genetic liver problems (Yang *et al.*, 2016; Yin *et al.*, 2016).

Structural variations (SVs) have a significant role in both genetic diversity and the development of genetic disorders. SVs refer to alterations in specific sections of the genome that involve changes in the number of copies (deletion or duplication) or the location of the genome. These alterations can also include expansions of repetitive sequences, mobile elements or combinations of these components. There are still a considerable number of structural variations that have not been identified due to the constraints of current technologies.

Over the past ten years, PCR and hybridization have been employed to detect and uncover SVs. However, the detection of a significant quantity of SVs persists as a result of constraints in current technology. Short-read massive parallel sequencing exhibits great precision in finding single-nucleotide variations (SNVs), although its capacity to identify regions with high guanine-cytosine nucleotide content and major structural changes is limited due to intrinsic PCR constraints. NGS platforms utilizing short reads have a high false detection rate (up to 85%) and a low sensitivity (ranging from 30–70%) when it comes to detecting SVs (De Coster and Van Broeckhoven, 2019). Recently, a number of long-read platforms have been created to accurately identify various types of SVs, while minimizing false positives and maximizing sensitivity (Sedlazeck *et al.*, 2018). Significantly, these techniques enable the detection of SVs within repetitive sequences and segmental duplications, which were previously inaccessible using short reads. Oxford Nanopore Technologies and PacBio commercialized these platforms utilizing long-read single-molecule sequencing or single-molecule real-time sequencing. The expensive nature and limited efficiency of current sequencing technologies prevent the examination of SVs on a genomic level for therapeutic purposes (Roberts *et al.*, 2013; Jain *et al.*, 2016). The applications of CRISPR include target enrichment for NGS and the detection of structural variations. Our focus is on CRISPR-based technologies specifically designed for the identification of SVs.

## 1.4 Trait Mapping and QTL Analysis

Genetic linkage analysis enables the identification of genes or QTLs that regulate characteristics. This study relies on the notion of genetic recombination occurring during meiosis (Tanksley, 1993). This allows for the creation of linkage maps consisting of genetic markers for a particular population. Frequently, populations such as F2, F3, or BC populations are employed for segregation. Nevertheless, it is more advantageous to work with populations that can be sustained and generated indefinitely, such as recombinant inbreds and doubled haploids, as they enable the execution of duplicated and repeated studies. These populations may not be suitable for breeding cereals that undergo outbreeding, as inbreeding depression can lead to non-random alterations in gene frequency and a decline in the viguor of the lines. By employing statistical techniques like single-marker analysis or interval mapping, it is possible to identify the presence of genes or QTLs in connection to a linkage map, by examining the relationships between DNA markers and phenotypic data (Kearsey, 1998). The utilization of DNA markers for the identification of QTLs represented a significant advancement in the characterization of quantitative traits (Paterson *et al.*, 1988).

There have been multiple reports of DNA markers that are associated with genes or QTLs (Francia *et al.*, 2005). Figure 1.5 provides a comprehensive summary of marker development. Previously, it was believed that the majority of markers linked to QTLs from initial mapping studies were directly applicable in MAS. Nevertheless, there is now a widespread consensus that the process of confirming, validating, and finely mapping QTLs may be necessary (Langridge *et al.*, 2001). While there are instances of exceptionally precise initial QTL mapping data, as evidenced by subsequent QTL mapping studies (Price, 2006), it is advisable to conduct a confirmation phase. This is because the accuracy of QTL placements and effects can be compromised by variables including sampling bias (Melchinger *et al.*, 1998). QTL validation often involves confirming the efficacy of a QTL in several genetic contexts (Langridge *et al.*, 2001). Further marker-testing procedures may include the identification of a set of markers within a 10 cM range that encompasses and surrounds a QTL (due to the limited genetic variation of individual markers in various genotypes) and the transformation of markers into a format that necessitates less complex detection methods.

Once markers that are strongly associated with a specific characteristic phenotype have been found, they can be utilized for MAS. The primary benefits of MAS compared to traditional phenotypic selection are as follows:

1. Phenotypic screening can be more complex compared to other methods, potentially leading to time, resource and effort savings. Challenging and arduous features to quantify include the resistance of wheat to cereal cyst nematode and root lesion nematode (Eagles *et al.*, 2001; Zwart *et al.*, 2004). Additional instances include quality characteristics that typically necessitate costly screening methods.
2. Seedling stage is suitable for conducting selection. This can be beneficial for numerous characteristics, particularly those that manifest during later stages of development. Consequently, unfavourable plant genotypes can be swiftly eradicated. This could yield significant advantages in rice breeding, as conventional rice production methods entail planting pre-germinated seeds and transferring seedlings to rice paddies, allowing for the selective transplantation of chosen seedlings to the main field.
3. Individual plants can be chosen. Conventional screening procedures for many qualities involve growing plant families or plots, as single-plant selection is unreliable due to environmental influences. Using MAS, specific plants can be chosen based on their genotype. Conventional phenotypic screening is unable to differentiate between homozygous and heterozygous plants for most characteristics.

Breeders can utilize these benefits to expedite the breeding process (Ribaut and Hoisington, 1998; Morris *et al.*, 2003). By targeting specific genotypes, the selection process can be optimized, allowing for the accelerated growth and release of desired features. Markers can serve as an alternative to phenotyping, enabling selection in off-season nurseries and increasing the cost-effectiveness of growing many generations each year (Ribaut and Hoisington, 1998). Using MAS also offers the advantage of reducing the overall number of lines that require testing. After the process of MAS in a breeding programme,

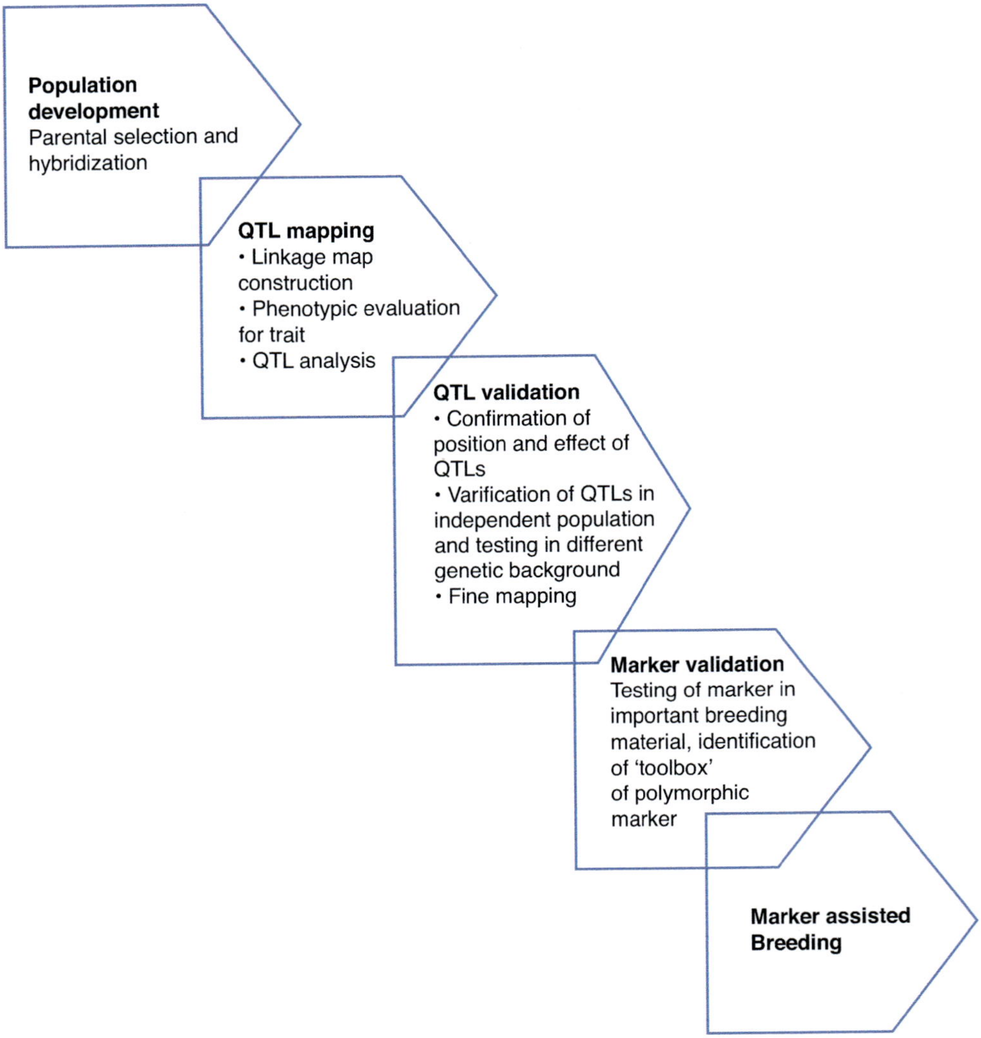


**Fig. 1.5.** Diagram showing QTL analysis.

it becomes possible to exclude many lines. This allows for a more efficient utilization of limited glasshouse and/or field space, as only the crucial breeding material is retained.

## 1.5 Multi-Parent Advanced Generation Inter-Cross (MAGIC)

Since it enables both coarse and fine mapping of RILs derived from multiple parents by sampling seed of any generation with larger genetic variety, the MAGIC approach can be regarded as superior to association and linkage mapping. It has been shown that epistatic interactions can be used to study the complex architecture of several variables linked to crop production and quality when a large number of RILs are generated (Cavanagh *et al.*, 2008). Nonetheless, it is crucial to remember that a big sample size is crucial for making it easier to identify and screen the genes causing complex features. Additionally, the MAGIC lines might exhibit strong plant developmental trait segregation, such as height and maturity, which might restrict the application of the MAGIC population in the analysis of complex features. The work to develop MAGIC populations in some crop species, such as rice (Hei Leung, personal communication), sorghum (Tom Hash, personal communication), and chickpea (Pooran Gaur and Hari Upadhyay, personal communication), is quite advanced, even though there are no published reports on the subject. It is predicted that MAGIC populations will provide a significant resource for the identification, isolation and transfer of critical genes to support agricultural advancement.

## 1.6 Functional Genomics

Functional genomics is a branch of molecular biology that focuses on comprehending the intricate connection between genotype (an organism's genetic makeup) and phenotype (the observable characteristics resulting from the interaction of genes with the environment). Genetics and genomics focus on the examination of genes and an organism's complete genetic material, while functional genomics seeks to address biological inquiries regarding the activation and functioning of genes in a dynamic, context-specific and cooperative manner. This is accomplished by utilizing various genomics and related omics datasets (Krassowski *et al.*, 2020; Kumar *et al.*, 2021). Utilizing integrative methodologies involving mathematical or relational models is essential to effectively incorporate multiple types of omics data, in order to comprehend molecular interactions and physiological mechanisms at a whole-organism scale.

Following the completion of the human genome sequencing, the primary objective has been to elucidate the function of genes, comprehend their expression patterns, investigate the significance of proteins, and assess the influence of genetic diversity on these biological processes (Kibe *et al.*, 2020). This initial policy briefing provides an overview of the concept of functional genomics and the methodologies employed within this discipline. The data integration encompasses information obtained from several biological processes associated with DNA sequencing, gene expression and protein function, including coding and non-coding transcription, protein translation, and protein–DNA, protein–RNA and protein–protein interactions. These data are collectively utilized to simulate interconnected and evolving networks that control the expression of genes, the specialization of cells and the advancement of the cell cycle. The many techniques involved in functional genomics are as follows:

- **Gene function annotation**
  This involves a combination of experimental approaches, such as gene knockouts (disabling a gene to observe the effects) and computational methods, including sequence analysis and comparisons to known gene functions.
- **Gene expression profiling**
  Microarray technology and RNA sequencing (RNA-seq) are commonly used to profile gene expression on a genome-wide scale. These methods provide insights into how genes are regulated and how their expression changes under different conditions.
- **Proteomics**
  Mass spectrometry and other proteomic techniques are used to identify and quantify proteins, providing insights into post-translational modifications, protein–protein interactions and subcellular localization.
- **Metabolomics**
  Analysing the complete set of small molecules (metabolites) within a biological system. Mass spectrometry and nuclear magnetic resonance (NMR) spectroscopy are employed to identify and quantify metabolites. This helps in understanding the metabolic pathways and how they respond to genetic and environmental changes.
- **Functional genomics databases**
  Databases such as Gene Ontology (GO), Kyoto Encyclopedia of Genes and Genomes (KEGG) and others provide curated information on gene functions, pathways and interactions.

In short, functional genomics is essential for enhancing our comprehension of the molecular underpinnings of biological processes, development and illnesses. The knowledge acquired from functional genomics investigations has a significant impact on multiple domains such as health, agriculture and biotechnology. It serves as a basis for the creation of new treatment approaches, enhanced crops and revolutionary biotechnological advancements.

## 1.7 Marker-Assisted Breeding

Marker-assisted selection is mostly employed in plant breeding to preserve a certain trait. The technique entails the identification of particular genes by the utilization of molecular or genetic markers, which are sequences of nucleic acid comprising a segment of DNA. The markers are positioned in close proximity to the DNA sequence of the target gene, and the transfer of traits occurs according to the principles of heredity. Due to the close proximity of the markers and genes on the same chromosome, the desired characteristics are consistently passed on to subsequent generations of plants. This procedure facilitates plant breeding by ensuring the preservation of superior quality and disease-resistant characteristics in the resulting variety.

### 1.7.1 Marker-assisted backcrossing

Markers are utilized in three stages of backcross breeding: frontal area, recombinant and foundation determinations. During the primary stage of cutting-edge selection, markers are employed to choose the desired characteristic. The use of markers in this context is robust due to the presence of certain traits that undergo consistent phenotypic selection processes or contain passive alleles whose influence may have been repressed by dominant genes. The recombinant selection is the second phase, wherein backcross offspring exhibiting the desired trait and closely linked flanking markers are chosen, thereby reducing the occurrence of linkage drag (Fig. 1.6).

Using background markers, backcross descendants who have already been selected for their desirable traits are picked in the third stage, which is called background selection. The background markers should be pure rice markers rather than ones that are affixed to

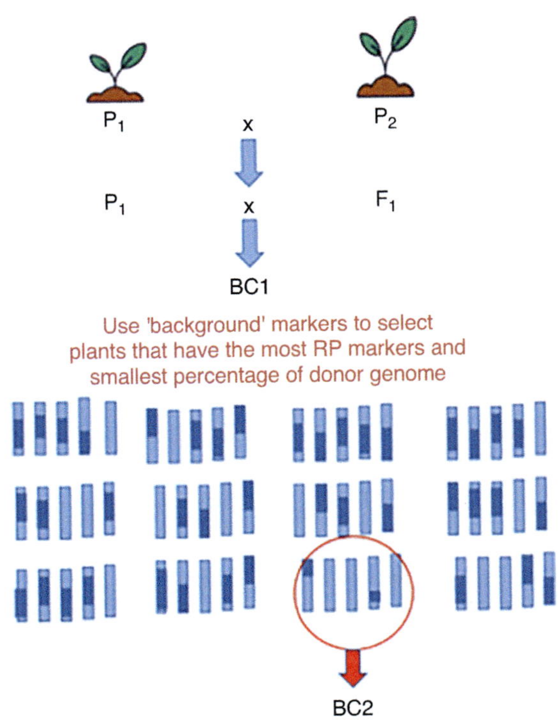


**Fig. 1.6.** Marker-assisted backcrossing.

the markers permanently (Ribaut and Betrán, 1999). By using markers to target the donor genome specifically, the genetic contribution of the donor parent can be decreased, and the quantity of the intermittent parent genome can be increased (Koebner and Summers, 2003).

### 1.7.2 Marker-assisted pyramiding

This refers to the process of combining multiple desirable traits in a plant by the use of molecular markers. Pyramiding refers to the process of incorporating many genes or quantitative trait loci into a single genotype simultaneously (Singh *et al.*, 2001). The pyramiding of characteristics is crucial for achieving a wide spectrum of disease resistance against bacterial leaf blight (BLB) in rice, as well as for enhancing the persistence of resistance. Molecular markers facilitate determination as they are non-invasive and allow for the examination of multiple desirable genes with a single DNA test, without the need for phenotyping. The most widespread application of pyramiding in plant breeding has been the incorporation of multiple disease resistance genes or QTLs to enhance durability against diseases (Castro *et al.*, 2003). While traditional breeding methods can still be used, it becomes challenging or even impossible in the initial stages due to the requirement of phenotypically evaluating each plant for all the qualities being assessed. Assessing plants from specific populations, such as F2, or determining their quality by destructive bioassays (Servin *et al.*, 2004), becomes exceedingly challenging.

### 1.7.3 Early-generation marker-assisted selection

While markers can be employed at any point in a conventional plant breeding process, MAS offers a significant advantage in the initial generations as it allows for the removal of plants with unfavourable gene combinations. This enables breeders to concentrate their attention on a smaller number of top-priority lines in future generations (Fig. 1.7).

The maximum effectiveness of MAS occurs in the initial generations when the connection between the marker and the targeted QTL is not strongly linked. This is because there is a higher likelihood of recombination between the marker and QTL. The primary drawback of implementing MAS in early generations is the financial burden associated with genotyping a larger population of plants.

Ribaut and Betrán (1999) presented an approach for early-generation MAS dubbed single large-scale MAS (SLS-MAS). The authors suggested that a solitary MAS procedure may be conducted on F2 or F3 generations obtained from superior progenitors. This method employed adjacent genetic markers (within a distance of less than 5 cM) on both sides of a specific genetic locus to identify and analyse up to three QTLs simultaneously in a single MAS procedure. Optimally, these QTLs should explain most of the phenotypic variation and exhibit stability across diverse settings. Due to the intense selection pressure, the population numbers are likely to decrease significantly, creating a favourable condition for genetic drift to take place at non-target loci. Therefore, it is advisable to utilize large population sizes, as suggested by Ribaut and Betrán (1999). To mitigate this issue, one can opt for F3 populations instead of F2 populations. This is because the proportion of selected individuals in an F3 population is greater than that in an F2 population. Specifically, for a single target locus, 38% of the F3 population will be selected, but only 25% of the F2 population would be picked. Ribaut and Betrán (1999) also suggested that, in theory, linkage drag may be reduced by employing extra flanking markers around the target QTLs, similar to the approach used in MAB.

### 1.7.4 Combined marker-assisted selection

Phenotypic screening can be carefully integrated with MAS in certain situations. Initially, the concept of 'combined MAS' (introduced by Moreau *et al.*, 2004) may offer benefits compared to phenotypic screening or MAS alone, with the aim of maximizing genetic improvement

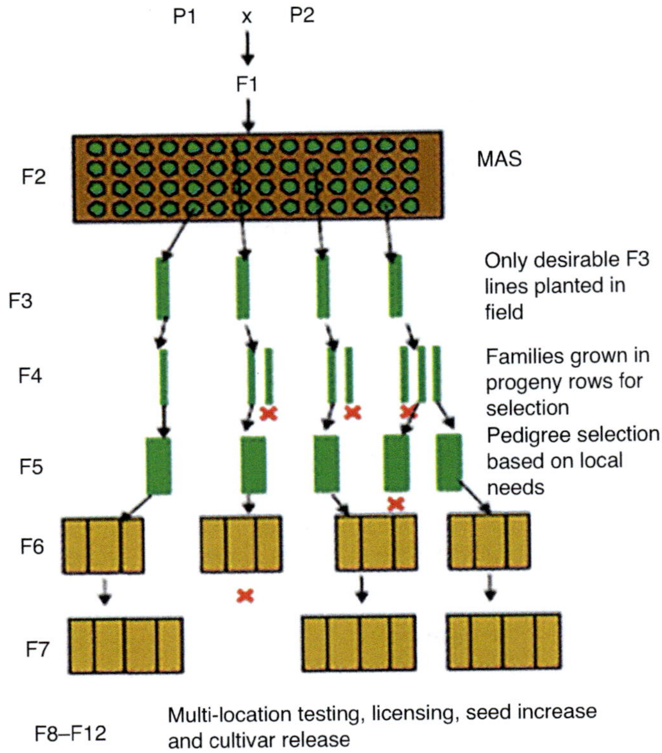


**Fig. 1.7.** Early-generation marker-assisted selection.

(Lande and Thompson, 1990). This strategy can be employed in situations when there are still unidentified QTLs that govern a particular trait, or when there is a need to manipulate a large number of QTLs (Fig. 1.8). Simulation experiments have demonstrated that this method is more effective than relying just on phenotypic screening, particularly when dealing with huge population sizes and low trait heritability (Hospital *et al.*, 1997). In their study, Bohn *et al.* (2001) examined the potential of MAS in enhancing insect resistance in tropical maize. They discovered that MAS, when used alone, was not as effective as conventional phenotypic selection. Nevertheless, the combination of MAS and phenotypic screening resulted in a modest improvement in relative efficiency. When studying wheat, researchers found that the combination of MAS and phenotypic screening was more successful than using phenotypic screening alone to identify a significant QTL on chromosome 3BS that is associated with resistance to *Fusarium* head blight (Zhou *et al.*, 2003). Practically, all MAS schemes will be implemented within the framework of the entire breeding programme, which includes the application of phenotypic selection at different phases. It is essential to validate the findings of MAS and identify features or genes for which map location is currently unidentified. In certain circumstances, there may be a limited degree of recombination between a marker and a QTL, unless markers that surround the QTL are employed (Sanchez *et al.*, 2000; Sharp *et al.*, 2001). Put simply, a marker assay cannot guarantee a phenotype prediction with absolute certainty. Nevertheless, the utilization of these markers for plant selection remains advantageous for breeders as it enables the identification of a subset of plants, hence reducing the number of plants requiring phenotypic evaluation. This can be especially beneficial when the expense of marker genotyping is lower than that of phenotypic screening, especially for quality traits (Han

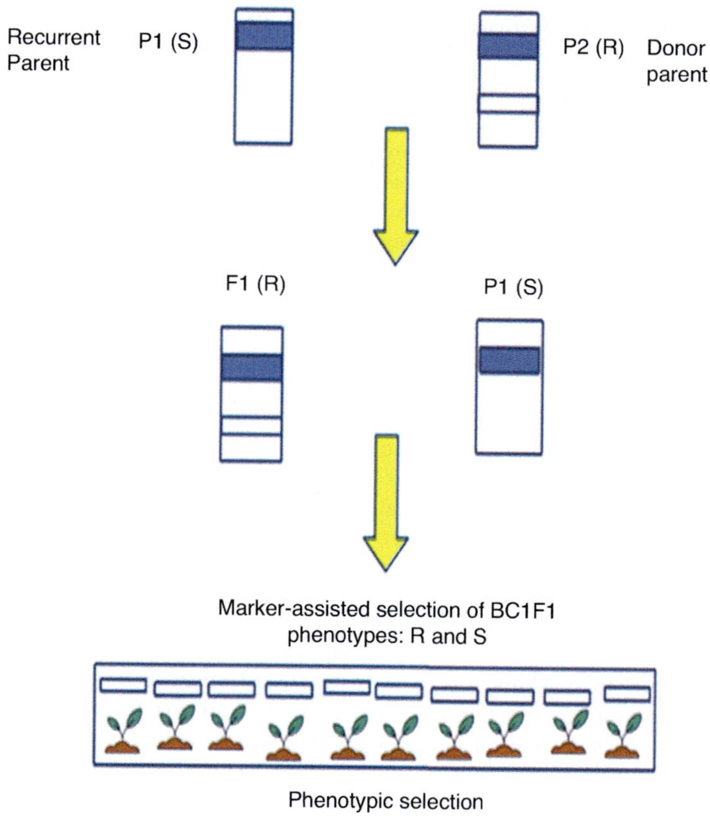


**Fig. 1.8.** Combined marker-assisted selection. R – resistant, S – sensitive.

*et al.*, 1997). The term 'tandem selection' was coined by Han *et al.* (1997), although Langridge and Chalmers (2004) referred to it as 'stepwise selection'. Mapping QTLs for significant qualities not only complements traditional breeding techniques but also offers an indirect advantage in a conventional breeding effort. Often, this happens when qualities that were previously believed to be regulated by a complex set of genes are discovered to be influenced by one or a few significant QTLs. For instance, the study conducted by Jones *et al.* (1995) revealed that the resistance to pearl millet downy mildew is determined by genes with significant influence. Similarly, the ability of rice to withstand submergence was determined to be governed by the primary QTL Sub1, which facilitated the streamlining of breeding efforts for this characteristic (Mackill, 2007).

Breeders can utilize these benefits to expedite the breeding process (Ribaut and Hoisington, 1998; Morris *et al.*, 2003). By targeting certain genotypes, the selection process can be optimized, allowing for the expedited development and marketing of new varieties by accelerating the expression of desired features. Markers can serve as an alternative to phenotyping, enabling selection in off-season nurseries and increasing the cost-effectiveness of growing many generations each year (Ribaut and Hoisington, 1998). Another advantage of utilizing MAS is the potential reduction in the overall number of lines requiring testing. Discarding numerous lines after MAS at an early stage in a breeding programme allows for a more efficient utilization of limited glasshouse and/or field area. This is because only the essential breeding material is retained.

## 1.8 Genomic Selection

Genomic selection, also called genome-wide selection, is a method that uses genetic data from the entire genome to accurately predict any trait, making it possible to identify a desirable individual (Meuwissen *et al.*, 2001). Based on a genomic estimation of breeding values, the most suitable individual is chosen. In the animal breeding industry, breeding values are a common and widely used statistic. Breeding values are computed by adding the weighted total of the estimated breed impact and the estimated genetic deviation. These estimations are based on the additive infinitesimal model (Fig. 1.9) applied to phenotypic data acquired from family pedigrees. Quantitative genetics and the infinitesimal genetic model were used to improve the efficacy of animal breeding selection, specifically in cattle and pigs. In animal breeding, the Bayesian framework and best linear unbiased prediction are frequently employed in the assessment of breeding value. A novel method known as GEBV was developed for the aim of selection following the introduction of genome sequencing in numerous model animals (Meuwissen *et al.*, 2001). The principles of genomic selection are covered in this chapter, along with how it can be used to speed up plant breeding.

### 1.8.1 Statistical techniques employed in genomic selection

The process of identifying a cause-and-effect relationship between a genetic element and a certain attribute is a widely utilized approach in pedigree-based phenotypic selections, traditional marker-assisted selection and genomic selection. Conventional QTL mapping and MAS methods tend to either exaggerate or disregard the impact of markers. The development of genomic selection was driven by the aspiration to employ high-density genotyping technology for the accurate prediction of complex traits. Meuwissen *et al.* (2001) stated that minimizing bias during effect estimation and genetic value computations can be achieved by avoiding marker selection. Marker selection yields a larger predictor impact (P) compared to the number of data (lower n). Insufficient degrees of freedom cause the ordinary least squares estimator to be ineffective in estimating the effects of all predictors, leading to the creation of an over-fitted model. Consequently, the prediction ability of ordinary least squares is negatively affected. In order to tackle this problem, several genomic selection models have been created. Models such as shrinkage methods, variable selection models, kernel

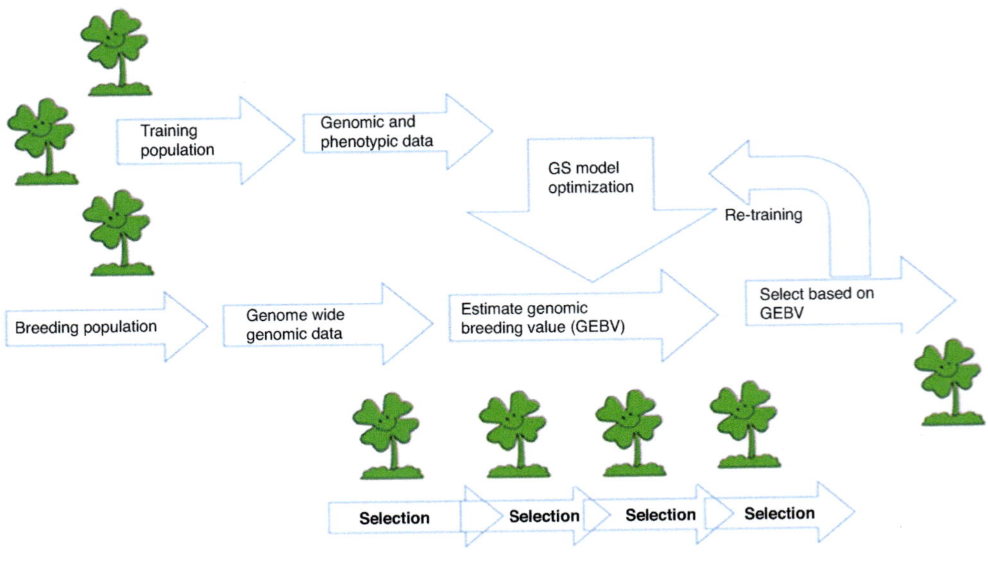


**Fig. 1.9.** Phases of genome-wide GS. Genotypic and phenotypic information from the training population allows GS model optimization and breeding value estimation.

approaches and dimension reduction methods can be cited.

The fundamental genetic model that links an individual's phenotype (P) to the cumulative genetic values (G) is based on the assumption that only the impacts of genetic factors are inherited by subsequent generations. The genetic values encompass genetic, dominance and epistatic influences, as well as the residual environmental effect (E). The mathematical notation for this is:

$$P = G + E \tag{1.1}$$

When there is no interaction between G and E, the covariance (COV) between G and E is reduced to zero. Thus, the phenotypic variance V (P) can be mathematically represented as:

$$(P) = V(G) + V(E) + 2COV(G, E)$$

$$V(P) = V(G) + V(E) + 0 = V(G) + V(E) \tag{1.2}$$

The GEBV is typically equivalent to G.

### 1.8.2 The degree to which traits are influenced by genetic factors

Heritability is the term used to describe the proportion of phenotypic variation (V(P)) that may be attributed to variation in genetic value (V(G)). It evaluates the extent to which a population's phenotypic traits are inherited by the next generation. Heritability can be explained using two approaches: the wide sense and the limited sense. Broad-sense heritability ($H^2$) quantifies the proportion of phenotypic variance that can be attributed to genetic value. The focus is on all genetic factors, encompassing additive, dominance and epistatic effects. Thus, it can be expressed numerically as:

$$H^2 = V(G) / V(P) \tag{1.3}$$

The narrow-sense heritability ($h^2$) specifically measures the amount of genetic variation resulting from additive genetic effects (V(A)) and the residual effect variance represented as $V(\varepsilon)$. It is represented by Eqn 1.4, as follows:

$$h^2 = V(G) / V(P) \tag{1.4}$$

Thus, the genetic model for $h^2$ can be restated as:

$$V(P) = V(A) + V(\varepsilon) \tag{1.5}$$

This depicts the total phenotypic variance (V(P)) as the sum of the additive genetic variance (V(A)) and the residual variance (V(V)). The residual variance, V(V), accounts for effects that are not accounted for by the additive genetic effect, such as dominant and epistatic effects.

The narrow-sense heritability is crucial in plant selection as it encompasses almost all the genetic diversity that influences the response to selection, approaching 100%. Meuwissen *et al.* (2001) proposed that the genetic variation (V(A)) can be subdivided into different DNA markers, such as V(A1), V(A2), V(A3), and so forth. This facilitated the computation of the plant's breeding value by utilizing markers that spanned the entire genome.

### 1.8.3 The genetic worth passed on to offspring through breeding

The possibility of bias can be divided into two main concerns: the possibility of bias in the data utilized to create the algorithms and bias in the algorithms themselves. Extensive datasets are utilized to train machine learning systems to recognize patterns in the data, and the accuracy of the resulting algorithm's performance is heavily reliant on the quality of the training and validation datasets. If the dataset does not accurately represent the population it is being used for, then the outcomes produced by the machine learning algorithm will be influenced by bias. This raises worries that the conclusions drawn by machine learning systems may occasionally mirror and perpetuate prejudices present in society. The issue of bias is especially worrisome when machine learning is applied to genetic sequencing due to the historical absence of racial and ethnic diversity in human genetic and genomic research (Knerr *et al.*, 2011).

An individual's breeding value is the cumulative sum of the average effects of all the alleles that the individual possesses. If an A allele is assigned a value of +5 and a B allele is assigned a value of −2, an AB heterozygote would have a breeding value of 3. It is the cumulative sum of an individual's genetic worth. Breeding value (BV) can be defined as the difference between the average phenotype of offspring and the average phenotype of the population, based on the idea of narrow-sense heritability ($h^2$). Mathematically, this can be represented as:

$$\begin{aligned} BV &= (\overline{m_0}) + h^2(yi - m_0) \\ &= m_0 + (yi - m_0)\, V(A)V(P) \end{aligned} \qquad (1.6)$$

which represents the calculation of BV, where $y_i$ represents the phenotypic value of individual i (where i ranges from 1 to n) and $m_0$ represents the mean phenotypic value of the population. An estimate of breeding value (EBV) refers to the estimation of a breeding value that is derived from hereditary factors. Genomic selection utilizes markers across the entire genome to assess the impact of genotypes and breeding value, leading to the determination of GEBVs (Nakaya and Isobe, 2012).

Prior to generating phenotypes, conventional cultivar development in plant breeding requires a comprehensive understanding of biological function. Genomic selection allows for cost-effective breeding by bypassing the need for gene/QTL mapping and characterization, while still obtaining valuable functional data. In crop breeding programmes, GS can significantly increase the rate of genetic improvement. This is achieved by improving the accuracy of GEBVs through the use of dense markers, larger training populations, higher trait heritability, effective GEBV prediction models and advanced imputation techniques.

## 1.9 Plant Phenomics

This is a cutting-edge, high-throughput field that satisfies all criteria for the quick, precise and comprehensive collecting of phenotypic data. Plant phenomics is a strategy that helps identify complicated features that are suitable for selection and offers pertinent details on why a specific genotype can be more prominent in a given environment. Plant phenotyping is a technique that may be applied at many levels, ranging from the gene to the complete plant in a particular environment with management strategies. We explore the latest developments in plant phenomics in this chapter, emphasizing various field and enclosed high-throughput technologies for use in forward and reverse genetics. These plant phenomics techniques are highly significant for understanding physiological processes, identifying stress, and quickly and effectively screening, dissecting, and verifying.

To improve a phenotype (such as plant disease resistance, plant cell-wall modification for biomass conversion properties, or plant tolerance to abiotic stresses), strategies for understanding the processes underlying the phenotype and its genetic variation must be developed. Rapid detection of variance in natural (various germplasm collections, association genetic panels) and derived (recombinant inbred, mutant, broad introgression lines, etc.) genetic populations will be facilitated by high throughput phenomic techniques (Fig. 1.10).

The creation and modification of robust screens to enable high-throughput, trustworthy, and significant comparisons will provide the challenge for the effective implementation of phenomic techniques. The advent of high-throughput sequencing and phenotyping platforms is expected to deepen our genetic understanding and improve the accuracy of phenotypic evaluations (Shahwar *et al.*, 2024). Conventional phenotyping techniques are often labour intensive and low throughput. However, because next-generation genotyping and phenotyping technologies are quick, precise, and assist in speeding up crop breeding programmes, they have an edge over conventional breeding programmes.

Under the vertical intensification scenario, increasing yields will need cutting-edge molecular techniques, and plant breeding will increase output while improving resource efficiency. Rapid and accurate phenotypic evaluation can help in the production of innovative and superior cultivars under a variety of climatic circumstances. This requires evaluating thousands of breeding lines.

## 1.10 Challenges and Ethical Consideration

Next-generation sequencing has immense potential in both academic and clinical settings, the reduction in the amount of DNA needed for NGS library preparation, it is believed that virtually every sample can be sequenced, hence maximizing the amount of information that can be extracted from each biological sample (Matheson, 2016). Law enforcement agencies have intensified their attempts to exchange DNA information, leading to an expansion in international databases analysing genetic markers. Given the numerous studies that have shown how easily de-identified genetic data can be linked back to individuals (Gymrek *et al.*, 2013), the utilization of de-identified data does not alleviate concerns regarding privacy. Lin *et al.* (2004) provided evidence that an individual's unique identification is possible by utilizing SNPs from that individual. The proliferation of genomic data on the internet, facilitated by genomic data repositories and databases, along with advancements in data-analysis methods, pose a growing challenge in preventing the re-identification of genomic data (Heeney *et al.*, 2011). The probability of re-identification increases as more information about mutations in an individual's DNA becomes available (Masca *et al.*, 2011). As NGS becomes more often used in clinical settings, it is important for patients and institutions to address the issue of informing patients about the potential privacy hazards associated with re-identification of their data and the sharing of clinical genomic data for research reasons. There are some people who argue that genomic data should not be seen as information that has had its identity removed.

The issue of bias is especially worrisome when machine learning is applied to genetic sequencing due to the historical absence of racial and ethnic diversity in human genetic and genomic research (Knerr *et al.*, 2011).

There are two main problems with bias: bias in the data used to create algorithms and prejudice in the algorithms themselves. Extensive datasets are utilized to train machine learning systems to recognize patterns in the data. The accuracy of the resulting algorithm's performance is heavily influenced by the quality of the training and validation datasets. If the dataset is not a true representation of the population it is used for, bias will affect the results produced by the machine learning algorithm. There are concerns that the results of machine learning systems can reflect and perpetuate prejudices present in society. Bias is a significant issue in the use of machine learning for genetic sequencing due to the limited representation of racial and ethnic diversity in historical human genetic and genomic studies (Knerr *et al.*, 2011).

## 1.11 Conclusion

The exploration of next-generation genetics and genomics in the realm of crop improvement presents a promising avenue for addressing the challenges of food security and sustainable agriculture. The strategies discussed, ranging from marker-assisted breeding to CRISPR-based gene editing, showcase the diverse tools available for researchers and breeders. The progress made in decoding plant genomes, understanding gene functions, and uncovering molecular pathways has laid a solid foundation for precision breeding. As evidenced by various success stories and ongoing research initiatives, the potential for enhanced crop improvement is vast. Increased yields, improved resistance to pests and diseases, and optimized nutritional profiles are within reach. However, it is essential to navigate ethical considerations and regulatory frameworks associated with genetic modifications to ensure responsible and safe deployment. The collaboration between traditional breeding methods and cutting-edge genomic technologies is proving to be a dynamic and effective approach. Furthermore, the accessibility of these technologies to a wider range of researchers and farmers will be crucial in democratizing the benefits of enhanced crop varieties. The integration of data-driven approaches and artificial intelligence in analysing vast genomic datasets opens up new frontiers for accelerating crop improvement programmes.

In moving forward, continued investment in research, technology and education will be pivotal. Additionally, fostering international collaborations and sharing knowledge across borders can expedite progress and contribute to global food security. As we stand at the intersection of innovation and agriculture, the potential impact of harnessing next-generation genetics and genomics for crop improvement is nothing

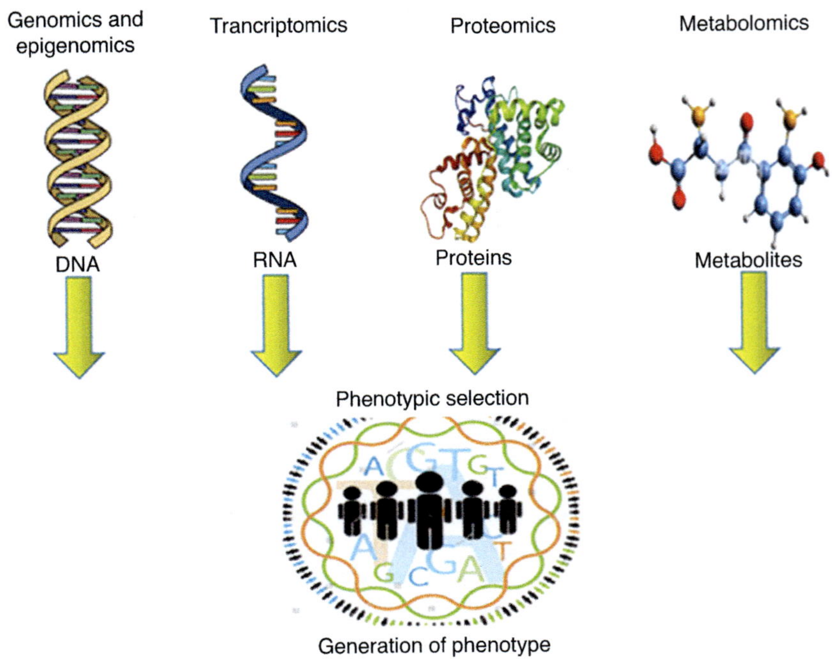


**Fig. 1.10.** Flow diagram of plant phenomics.

to strengthen our knowledge of health and illness, and to enhance treatment methods. Simultaneously, specific aspects of the research and clinical environment for NGS provide difficulties for ethical frameworks governing research and clinical practices. The pursuit of enhanced research collaboration and data sharing, together with the utilization of large-scale datasets and big-data methodologies for genetic research, pose significant privacy concerns.

What are the possible hazards that a person may face when their genomic data is disclosed to third parties? Genomic data has the potential to be utilized for drawing conclusions about an individual's present and future well-being. Such information has the potential to adversely affect an individual in terms of insurance, employment or legal matters. NGS data not only presents privacy hazards for individuals, but also raises concerns regarding groups and family members. Conducting research on a subset of a specific group may result in making assumptions about the entire population, which can lead to over-generalizations or even stigmatization (Mello and Wolf, 2010). Some individuals within the group may have a sense of violation when they become aware that others believe they possess knowledge about the group's members (Alpert, 2000).

The application of NGS in forensic investigations also poses concerns and potential threats in relation to privacy. DNA analysis is a crucial instrument in forensic science, utilized in child custody proceedings and the identification of individuals for legal and criminal justice objectives. Most contemporary DNA forensic techniques examine the diversity in short tandem repeat (STR) markers (Butler, 2015). NGS enhances existing techniques by enabling researchers to identify multiple STR locations on both autosomes and sex chromosomes. This allows for the collection of data on various locations throughout the genome, enabling the investigation of a broader variety of inquiries in a single analysis. Additionally, NGS has the potential to generate physical descriptions of probable suspects. An inherent issue with existing forensic tests is the potential limitation in the quantity or quality of the given sample. With short of transformative. With responsible stewardship, this scientific frontier holds the key to addressing the challenges of a growing population and changing environmental conditions, ensuring a sustainable and resilient future for agriculture.

## References

Albert, T.J., Molla, M.N., Muzny, D.M., Nazareth, L., Wheeler, D. *et al.* (2007) Direct selection of human genomic loci by microarray hybridization. *Nature Methods* 4(11), 903–905. DOI: 10.1038/nmeth1111.

Alpert, S. (2000) Privacy and the analysis of stored tissues. In: *Research Involving Human Biological Materials: Ethical Issues and Policy Guidance. Ethical issues and Policy Guidance. Volume II, Commissioned Papers*. National Bioethics Advisory Commission, Rockville, Maryland.

Andolfatto, P., Davison, D., Erezyilmaz, D., Hu, T.T., Mast, J. *et al.* (2011) Multiplexed shotgun genotyping for rapid and efficient genetic mapping. *Genome Research* 21(4), 610–617. DOI: 10.1101/gr.115402.110.

Anzalone, A.V., Randolph, P.B., Davis, J.R., Sousa, A.A., Koblan, L.W. *et al.* (2019) Search-and-replace genome editing without double-strand breaks or donor DNA. *Nature* 576(7785), 149–157. DOI: 10.1038/s41586-019-1711-4.

Barchi, L., Lanteri, S., Portis, E., Valè, G., Volante, A. *et al.* (2012) A RAD tag derived marker based eggplant linkage map and the location of QTLs determining anthocyanin pigmentation. *PLOS ONE* 7(8), e43740. DOI: 10.1371/journal.pone.0043740.

Beissinger, T.M., Hirsch, C.N., Sekhon, R.S., Foerster, J.M., Johnson, J.M. *et al.* (2013) Marker density and read depth for genotyping populations using genotyping-by-sequencing. *Genetics* 193(4), 1073–1081. DOI: 10.1534/genetics.112.147710.

Bernardo, R. and Charcosset, A. (2006) Usefulness of gene information in marker-assisted recurrent selection: A simulation appraisal. *Crop Science* 46(2), 614–621. DOI: 10.2135/cropsci2005.05-0088.

Beyret, E., Liao, H.-K., Yamamoto, M., Hernandez-Benitez, R., Fu, Y. *et al.* (2019) Single-dose CRISPR-Cas9 therapy extends lifespan of mice with Hutchinson–Gilford progeria syndrome. *Nature Medicine* 25(3), 419–422. DOI: 10.1038/s41591-019-0343-4.

Bohn, M., Groh, S., Khairallah, M.M., Hoisington, D.A., Utz, H.F. *et al.* (2001) Re-evaluation of the prospects of marker-assisted selection for improving insect resistance against *Diatraea* spp. in tropical maize by cross validation and independent validation. *Theoretical and Applied Genetics* 103(6–7), 1059–1067. DOI: 10.1007/s001220100708.

Brenner, S., Johnson, M., Bridgham, J., Golda, G., Lloyd, D.H. *et al.* (2000) Gene expression analysis by massively parallel signature sequencing (MPSS) on microbead arrays. *Nature Biotechnology* 18(6), 630–634. DOI: 10.1038/76469.

Buniello, A., MacArthur, J.A.L., Cerezo, M., Harris, L.W., Hayhurst, J. *et al.* (2019) The NHGRI-EBI GWAS catalog of published genome-wide association studies, targeted arrays and summary statistics 2019. *Nucleic Acids Research* 47(D1), D1005–D1012. DOI: 10.1093/nar/gky1120.

Butler, J.M. (2015) The future of forensic DNA analysis. *Philosophical Transactions of the Royal Society B: Biological Sciences* 370(1674), 20140252. DOI: 10.1098/rstb.2014.0252.

Castro, A.J., Capettini, F., Corey, A.E., Filichkina, T., Hayes, P.M. *et al.* (2003) Mapping and pyramiding of qualitative and quantitative resistance to stripe rust in barley. *Theoretical and Applied Genetics* 107(5), 922–930. DOI: 10.1007/s00122-003-1329-6.

Cavanagh, C., Morell, M., Mackay, I. and Powell, W. (2008) From mutations to MAGIC: resources for gene discovery, validation and delivery in crop plants. *Current Opinion in Plant Biology* 11(2), 215–221. DOI: 10.1016/j.pbi.2008.01.002.

Chen, M., Mao, A., Xu, M., Weng, Q., Mao, J. *et al.* (2019) CRISPR-Cas9 for cancer therapy: opportunities and challenges. *Cancer Letters* 447, 48–55. DOI: 10.1016/j.canlet.2019.01.017.

Collard, B.C.Y. and Mackill, D.J. (2008) Marker-assisted selection: an approach for precision plant breeding in the twenty-first century. *Philosophical Transactions of the Royal Society B: Biological Sciences* 363(1491), 557–572. DOI: 10.1098/rstb.2007.2170.

Cong, L., Ran, F.A., Cox, D., Lin, S., Barretto, R. *et al.* (2013) Multiplex genome engineering using CRISPR/Cas systems. *Science* 339(6121), 819–823. DOI: 10.1126/science.1231143.

Davey, J.W., Hohenlohe, P.A., Etter, P.D., Boone, J.Q., Catchen, J.M. *et al.* (2011) Genome-wide genetic marker discovery and genotyping using next-generation sequencing. *Nature Reviews Genetics* 12(7), 499–510. DOI: 10.1038/nrg3012.

De Coster, W. and Van Broeckhoven, C. (2019) Newest methods for detecting structural variations. *Trends in Biotechnology* 37(9), 973–982. DOI: 10.1016/j.tibtech.2019.02.003.

Doman, J.L., Raguram, A., Newby, G.A. and Liu, D.R. (2020) Evaluation and minimization of Cas9-independent off-target DNA editing by cytosine base editors. *Nature Biotechnology* 38(5), 620–628. DOI: 10.1038/s41587-020-0414-6.

Eagles, H.A., Bariana, H.S., Ogbonnaya, F.C., Rebetzke, G.J., Hollamby, G.J. *et al.* (2001) Implementation of markers in Australian wheat breeding. *Australian Journal of Agricultural Research* 52(12), 1349–1356. DOI: 10.1071/AR01067.

Ebina, H., Misawa, N., Kanemura, Y. and Koyanagi, Y. (2013) Harnessing the CRISPR/Cas9 system to disrupt latent HIV-1 provirus. *Scientific Reports* 3(1), 2510. DOI: 10.1038/srep02510.

Elshire, R.J., Glaubitz, J.C., Sun, Q., Poland, J.A., Kawamoto, K. *et al.* (2011) A robust, simple genotyping-by-sequencing (GBS) approach for high diversity species. *PLOS ONE* 6(5), e19379. DOI: 10.1371/journal.pone.0019379.

Francia, E., Tacconi, G., Crosatti, C., Barabaschi, D., Bulgarelli, D. *et al.* (2005) Marker assisted selection in crop plants. *Plant Cell, Tissue and Organ Culture* 82(3), 317–342. DOI: 10.1007/s11240-005-2387-z.

Glaubitz, J.C., Casstevens, T.M., Lu, F., Harriman, J., Elshire, R.J. *et al.* (2014) TASSEL-GBS: a high capacity genotyping by sequencing analysis pipeline. *PLOS ONE* 9(2), e90346. DOI: 10.1371/journal.pone.0090346.

Gnirke, A., Melnikov, A., Maguire, J., Rogov, P., LeProust, E.M. *et al.* (2009) Solution hybrid selection with ultra-long oligonucleotides for massively parallel targeted sequencing. *Nature Biotechnology* 27(2), 182–189. DOI: 10.1038/nbt.1523.

Gore, A.V., Cheong, A., Gilligan, P.C. and Sampath, K. (2007) Gore et al. reply. *Nature* 450(7167), E2–E4. DOI: 10.1038/nature06315.

Gore, M.A., Chia, J.-M., Elshire, R.J., Sun, Q., Ersoz, E.S. *et al.* (2009) A first-generation haplotype map of maize. *Science* 326(5956), 1115–1117. DOI: 10.1126/science.1177837.

Gymrek, M., McGuire, A.L., Golan, D., Halperin, E. and Erlich, Y. (2013) Identifying personal genomes by surname inference. *Science* 339(6117), 321–324. DOI: 10.1126/science.1229566.

Han, F., Romagosa, I., Ullrich, S.E., Jones, B.L., Hayes, P.M. *et al.* (1997) Molecular marker-assisted selection for malting quality traits in barley. *Molecular Breeding* 3(6), 427–437. DOI: 10.1023/A:1009608312385.

Heeney, C., Hawkins, N., de Vries, J., Boddington, P. and Kaye, J. (2011) Assessing the privacy risks of data sharing in genomics. *Public Health Genomics* 14(1), 17–25. DOI: 10.1159/000294150.

Holland, J.B. (2004) Implementation of molecular markers for quantitative traits in breeding programs – challenges and opportunities. In: *Proceedings of the 4th International Crop Science Congress*, 26 September–1 October 2004, Brisbane, Australia, Vol. 26, pp. 1–13.

Hospital, F., Moreau, L., Lacoudre, F., Charcosset, A. and Gallais, A. (1997) More on the efficiency of marker-assisted selection. *Theoretical and Applied Genetics* 95, 1181–1189. DOI: 10.1007/s001220050679.

Hsu, P.D., Lander, E.S. and Zhang, F. (2014) Development and applications of CRISPR-Cas9 for genome engineering. *Cell* 157(6), 1262–1278. DOI: 10.1016/j.cell.2014.05.010.

Jain, M., Olsen, H.E., Paten, B. and Akeson, M. (2016) The Oxford Nanopore MinION: delivery of nanopore sequencing to the genomics community. *Genome Biology* 17(1), 1–11. DOI: 10.1186/s13059-016-1103-0.

Jones, E.S., Liu, C.J., Gale, M.D., Hash, C.T. and Witcombe, J.R. (1995) Mapping quantitative trait loci for downy mildew resistance in pearl millet. *Theoretical and Applied Genetics.* 91(3), 448–456. DOI: 10.1007/BF00222972.

Kang, H., Minder, P., Park, M.A., Mesquitta, W.T., Torbett, B.E. *et al.* (2015) CCR5 disruption in induced pluripotent stem cells using CRISPR/Cas9 provides selective resistance of immune cells to CCR5-tropic HIV-1 virus. *Molecular Therapy: Nucleic Acids* 4, e268. DOI: 10.1038/mtna.2015.42.

Kearsey, M.J. (1998) The principles of QTL analysis (a minimal mathematics approach). *Journal of Experimental Botany* 49(327), 1619–1623. DOI: 10.1093/jxb/49.327.1619.

Kennedy, E.M. and Cullen, B.R. (2015) Bacterial CRISPR/Cas DNA endonucleases: A revolutionary technology that could dramatically impact viral research and treatment. *Virology* 479–480, 213–220. DOI: 10.1016/j.virol.2015.02.024.

Kennedy, E.M., Kornepati, A.V.R., Goldstein, M., Bogerd, H.P., Poling, B.C. *et al.* (2014) Inactivation of the human papillomavirus E6 or E7 gene in cervical carcinoma cells by using a bacterial CRISPR/Cas RNA-guided endonuclease. *Journal of Virology* 88(20), 11965–11972. DOI: 10.1128/JVI.01879-14.

Khan, Z. and Shahwar, D. (2024) Genome editing in plants via CRISPR/Cas9: A genomic scissor borrowed from bacterial immune system. In: *Genome Editing and Global Food Security*. Routledge, pp. 1–16. DOI: 10.4324/9781003382102-1.

Kibe, M., Nyaga, C., Nair, S.K., Beyene, Y., Das, B. *et al.* (2020) Combination of linkage mapping, GWAS, and GP to dissect the genetic basis of common rust resistance in tropical maize germplasm. *International Journal of Molecular Sciences* 21(18), 6518. DOI: 10.3390/ijms21186518.

Knerr, S., Wayman, D. and Bonham, V.L. (2011) Inclusion of racial and ethnic minorities in genetic research: advance the spirit by changing the rules? *Journal of Law, Medicine & Ethics* 39(3), 502–512. DOI: 10.1111/j.1748-720X.2011.00617.x.

Koebner, R.M.D. and Summers, R.W. (2003) 21st century wheat breeding: plot selection or plate detection? *Trends in Biotechnology* 21(2), 59–63. DOI: 10.1016/S0167-7799(02)00036-7.

Krassowski, M., Das, V., Sahu, S.K. and Misra, B.B. (2020) State of the field in multi-omics research: from computational needs to data mining and sharing. *Frontiers in Genetics* 11, 610798. DOI: 10.3389/fgene.2020.610798.

Kumar, R., Sharma, V., Suresh, S., Ramrao, D.P., Veershetty, A. *et al.* (2021) Understanding omics driven plant improvement and *de novo* crop domestication: some examples. *Frontiers in Genetics* 12, 637141. DOI: 10.3389/fgene.2021.637141.

Lande, R. and Thompson, R. (1990) Efficiency of marker-assisted selection in the improvement of quantitative traits. *Genetics* 124(3), 743–756. DOI: 10.1093/genetics/124.3.743.

Langridge, P. and Chalmers, K. (2004) The principle: identification and application of molecular markers. In: *Molecular Marker Systems in Plant Breeding and Crop Improvement*. Springer, Berlin/Heidelberg, pp. 3–22. DOI: 10.1007/3-540-26538-4_1.

Langridge, P., Lagudah, E.S., Holton, T.A., Appels, R., Sharp, P.J. *et al.* (2001) Trends in genetic and genome analyses in wheat: a review. *Australian Journal of Agricultural Research* 52(12), 1043–1077. DOI: 10.1071/AR01082.

Lin, Z., Owen, A.B. and Altman, R.B. (2004) Genetics. genomic research and human subject privacy. *Science* 305(5681), 183. DOI: 10.1126/science.1095019.

Long, C., Amoasii, L., Mireault, A.A., McAnally, J.R., Li, H. *et al.* (2016) Postnatal genome editing partially restores dystrophin expression in a mouse model of muscular dystrophy. *Science* 351(6271), 400–403. DOI: 10.1126/science.aad5725.

Mackill, D.J. (2007) Molecular markers and marker-assisted selection in rice. In: *Genomics-Assisted Crop Improvement: Vol. 2: Genomics Applications in Crops*. Springer, Dordrecht, The Netherlands, pp. 147–168. DOI: 10.1007/978-1-4020-6297-1_7.

Mammadov, J.A., Chen, W., Ren, R., Pai, R., Marchione, W. *et al.* (2010) Development of highly polymorphic SNP markers from the complexity reduced portion of maize [*Zea mays* L.] genome for use in marker-assisted breeding. *Theoretical and Applied Genetics* 121(3), 577–588. DOI: 10.1007/s00122-010-1331-8.

Martinez-Lage, M., Puig-Serra, P., Menendez, P., Torres-Ruiz, R. and Rodriguez-Perales, S. (2018) CRISPR/Cas9 for cancer therapy: hopes and challenges. *Biomedicines* 6(4), 105. DOI: 10.3390/biomedicines6040105.

Masca, N., Burton, P.R. and Sheehan, N.A. (2011) Participant identification in genetic association studies: improved methods and practical implications. *International Journal of Epidemiology* 40(6), 1629–1642. DOI: 10.1093/ije/dyr149.

Matheson, S. (2016) DNA phenotyping: snapshot of a criminal. *Cell* 166(5), 1061–1064. DOI: 10.1016/j.cell.2016.08.016.

Melchinger, A.E., Utz, H.F. and Schön, C.C. (1998) Quantitative trait locus (QTL) mapping using different testers and independent population samples in maize reveals low power of QTL detection and large bias in estimates of QTL effects. *Genetics* 149(1), 383–403. DOI: 10.1093/genetics/149.1.383.

Mello, M.M. and Wolf, L.E. (2010) The Havasupai Indian tribe case – lessons for research involving stored biologic samples. *New England Journal of Medicine* 363(3), 204–207. DOI: 10.1056/NEJMp1005203.

Meuwissen, T.H., Hayes, B.J. and Goddard, M.E. (2001) Prediction of total genetic value using genome-wide dense marker maps. *Genetics* 157(4), 1819–1829. DOI: 10.1093/genetics/157.4.1819.

Moreau, L., Charcosset, A. and Gallais, A. (2004) Experimental evaluation of several cycles of marker-assisted selection in maize. *Euphytica* 137(1), 111–118. DOI: 10.1023/B:EUPH.0000040508.01402.21.

Morris, M., Dreher, K., Ribaut, J.M. and Khairallah, M. (2003) Money matters (II): costs of maize inbred line conversion schemes at CIMMYT using conventional and marker-assisted selection. *Molecular Breeding* 11(3), 235–247. DOI: 10.1023/A:1022872604743.

Nakaya, A. and Isobe, S.N. (2012) Will genomic selection be a practical method for plant breeding? *Annals of Botany* 110(6), 1303–1316. DOI: 10.1093/aob/mcs109.

Paterson, A.H., Lander, E.S., Hewitt, J.D., Peterson, S., Lincoln, S.E. *et al.* (1988) Resolution of quantitative traits into Mendelian factors by using a complete linkage map of restriction fragment length polymorphisms. *Nature* 335(6192), 721–726. DOI: 10.1038/335721a0.

Pfender, W.F., Saha, M.C., Johnson, E.A. and Slabaugh, M.B. (2011) Mapping with RAD (restriction-site associated DNA) markers to rapidly identify QTL for stem rust resistance in *Lolium perenne. Theoretical and Applied Genetics.* 122(8), 1467–1480. DOI: 10.1007/s00122-011-1546-3.

Poethig, R.S. (2001) Life with 25,000 genes. *Genome Research* 11(3), 313–316. DOI: 10.1101/gr.180001.

Poland, J., Endelman, J., Dawson, J., Rutkoski, J., Wu, S. *et al.* (2012) Genomic selection in wheat breeding using genotyping-by-sequencing. *Plant Genome* 5(3). DOI: 10.3835/plantgenome2012.06.0006.

Porreca, G.J., Zhang, K., Li, J.B., Xie, B., Austin, D. *et al.* (2007) Multiplex amplification of large sets of human exons. *Nature Methods* 4(11), 931–936. DOI: 10.1038/nmeth1110.

Preston, J., VanZeeland, A. and Peiffer, D.A. (2021) Innovation at lillumina: the road to the $600 human genome. *Nat. Portf.*

Price, A.H. (2006) Believe it or not, QTLs are accurate! *Trends in Plant Science* 11(5), 213–216. DOI: 10.1016/j.tplants.2006.03.006.

Ribaut, J.M. and Betrán, J. (1999) Single large-scale marker-assisted selection (SLS-MAS). *Molecular Breeding* 5(6), 531–541. DOI: 10.1023/A:1009631718036.

Ribaut, J.M. and Hoisington, D. (1998) Marker-assisted selection: new tools and strategies. *Trends in Plant Science* 3(6), 236–239. DOI: 10.1016/S1360-1385(98)01240-0.

Roberts, R.J., Carneiro, M.O. and Schatz, M.C. (2013) The advantages of SMRT sequencing. *Genome Biology* 14(7), 405. DOI: 10.1186/gb-2013-14-6-405.

Sahu, B.B., Sumit, R., Srivastava, S.K. and Bhattacharyya, M.K. (2012) Sequence based polymorphic (SBP) marker technology for targeted genomic regions: its application in generating a molecular map of the *Arabidopsis thaliana* genome. *BMC Genomics* 13(1), 20. DOI: 10.1186/1471-2164-13-20.

Sanchez, A.C., Brar, D.S., Huang, N., Li, Z. and Khush, G.S. (2000) Sequence tagged site marker-assisted selection for three bacterial blight resistance genes in rice. *Crop Science* 40(3), 792–797. DOI: 10.2135/cropsci2000.403792x.

Sedlazeck, F.J., Lee, H., Darby, C.A. and Schatz, M.C. (2018) Piercing the dark matter: bioinformatics of long-range sequencing and mapping. *Nature Reviews Genetics* 19(6), 329–346. DOI: 10.1038/s41576-018-0003-4.

Servin, B., Martin, O.C., Mézard, M. and Hospital, F. (2004) Toward a theory of marker-assisted gene pyramiding. *Genetics* 168(1), 513–523. DOI: 10.1534/genetics.103.023358.

Shahwar, D., Khan, Z. and Park, Y. (2023) Molecular marker-assisted mapping, candidate gene identification, and breeding in melon (*Cucumis melo* L.): A review. *International Journal of Molecular Sciences* 24(20), 15490. DOI: 10.3390/ijms242015490.

Shahwar, D., Khan, Z. and Park, Y. (2024) Molecular markers for marker-assisted breeding for biotic and abiotic stress in melon (*Cucumis melo* L.): A review. *International Journal of Molecular Sciences* 25(12), 6307. DOI: 10.3390/ijms25126307.

Sharp, P.J., Johnston, S., Brown, G., McIntosh, R.A., Pallotta, M. *et al.* (2001) Validation of molecular markers for wheat breeding. *Australian Journal of Agricultural Research* 52(12), 1357–1366. DOI: 10.1071/AR01052.

Singh, S., Sidhu, J.S., Huang, N., Vikal, Y., Li, Z. *et al.* (2001) Pyramiding three bacterial blight resistance genes (*xa5, xa13 and xa21*) using marker-assisted selection into indica rice cultivar PR106. *Theoretical and Applied Genetics* 102(6–7), 1011–1015. DOI: 10.1007/s001220000495.

Stapley, J., Reger, J., Feulner, P.G.D., Smadja, C., Galindo, J. *et al.* (2010) Adaptation genomics: the next generation. *Trends in Ecology & Evolution* 25(12), 705–712. DOI: 10.1016/j.tree.2010.09.002.

Tanksley, S.D. (1993) Mapping polygenes. *Annual Review of Genetics* 27(1), 205–233. DOI: 10.1146/annurev.ge.27.120193.001225.

Teer, J.K. and Mullikin, J.C. (2010) Exome sequencing: the sweet spot before whole genomes. *Human Molecular Genetics* 19(R2), R145–151. DOI: 10.1093/hmg/ddq333.

Utomo, H.S. and Linscombe, S.D. (2009) Current patents and future development underlying marker-assisted breeding in major grain crops. *Recent Patents on DNA & Gene Sequences* 3(1), 53–62. DOI: 10.2174/187221509787236174

Valdar, W., Flint, J. and Mott, R. (2006) Simulating the collaborative cross: power of quantitative trait loci detection and mapping resolution in large sets of recombinant inbred strains of mice. *Genetics* 172(3), 1783–1797. DOI: 10.1534/genetics.104.039313.

Varshney, R.K. (2009) Gene-based marker systems in plants: high throughput approaches for marker discovery and genotyping. In: Jain, S. and Brar, D. (eds) *Molecular Techniques in Crop Improvement*, 2nd edn. Springer, Dordrecht, pp. 119–142. DOI: 10.1007/978-90-481-2967-6_5.

Varshney, R.K. and Tuberosa, R. (eds) (2007) *Genomics-Assisted Crop Improvement, Vol. 2*. Genomics Applications in Crops. Springer. DOI: 10.1007/978-1-4020-6297-1.

Varshney, R.K., Nayak, S.N., May, G.D. and Jackson, S.A. (2009) Next-generation sequencing technologies and their implications for crop genetics and breeding. *Trends in Biotechnology* 27(9):522–530. DOI: 10.1016/j.tibtech.2009.05.006.

Varshney, R.K., Thudi, M., May, G.D. and Jackson, S.A. (2010) Legume genomics and breeding. *Plant Breeding Reviews* 33, 257–304. DOI: 10.1002/9780470535486.

Yalcin, B., Flint, J. and Mott, R. (2005) Using progenitor strain information to identify quantitative trait nucleotides in outbred mice. *Genetics* 171(2), 673–681. DOI: 10.1534/genetics.104.028902.

Yang, H., Tao, Y., Zheng, Z., Li, C. *et al.* (2012) Application of next-generation sequencing for rapid marker development in molecular plant breeding: a case study on anthracnose disease resistance in *Lupinus angustifolius* L. *BMC Genomics* 13(1), 318. DOI: 10.1186/1471-2164-13-318.

Yang, Y., Wang, L., Bell, P., McMenamin, D., He, Z. *et al.* (2016) A dual AAV system enables the Cas9-mediated correction of a metabolic liver disease in newborn mice. *Nature Biotechnology* 34(3), 334–338. DOI: 10.1038/nbt.3469.

Yin, H., Song, C.-Q., Dorkin, J.R., Zhu, L.J., Li, Y. *et al.* (2016) Therapeutic genome editing by combined viral and non-viral delivery of CRISPR system components *in vivo*. *Nature Biotechnology* 34(3), 328–333. DOI: 10.1038/nbt.3471.

Zhou, W.C., Kolb, F.L., Bai, G.H., Domier, L.L., Boze, L.K. *et al.* (2003) Validation of a major QTL for scab resistance with SSR markers and use of marker-assisted selection in wheat. *Plant Breeding* 122(1), 40–46. DOI: 10.1046/j.1439-0523.2003.00802.x.

Zwart, R.S., Thompson, J.P. and Godwin, I.D. (2004) Genetic analysis of resistance to root-lesion nematode (*Pratylenchus thornei*) in wheat. *Plant Breeding* 123(3), 209–212. DOI: 10.1111/j.1439-0523.2004.00986.x.

# 2 Advancements and Challenges in RNA Interference (RNAi) Technology: from Molecular Mechanisms to Agricultural Applications

**Arshi Anees[1] and Zeba Khan[2*]**

*[1]Multidisciplinary Centre for Advanced Research and Studies (MCARS), Jamia Millia Islamia, New Delhi – 110025, India; [2]Department of Botany, Raja Mahendra Pratap Singh State University, Aligarh – 202002, India*

**Abstract**

RNA interference (RNAi) is a biological process that suppresses gene expression after the transcription of RNA molecules, and it is triggered by the presence of double-stranded RNA. It has been extensively utilized as a knock-down technique for analysing gene function in many organisms. Prior to its discovery in worms, comparable phenomena such as post-transcriptional gene silencing and coat protein-mediated protection from viral infection had also been identified in plants. RNAi is commonly employed in plants by introducing transgenes that generate hairpin RNA. RNAi has superior efficiency and stability compared to antisense-mediated gene silencing and co-suppression, making it advantageous for genetic enhancement of crop plants. Additionally, it provides benefits compared to mutation-based reverse genetics by effectively controlling the expression of transgenes in a regulated fashion within multigene families.

**Keywords:** RNA interference, short RNA, post-transcriptional gene silencing, small interfering RNA, double-stranded RNA, gene silencing, small interfering RNA, microRNA, Piwi-interacting RNA, ribonuclease family, RISC, untranslated region, CRISPR-Cas9, ribonucleoprotein, Argonaute, Dicer, nematodes, virus

**Abbreviations**

ABA – Abscisic acid, ACMV – African cassava mosaic virus, Ago – Argonaute, BCH – β-Carotene hydroxylase, bp – Base pairs, BR – Brassinosteroid, CaMV – Cauliflower mosaic virus, Cd – Cadmium, dsRNA – Double-stranded RNA, GA – Gibberellin, GBSS – Granule-bound starch synthase, GST – Glutathione S-transferase, HDAC – Histone deacetylase, HD-RNAi – Host-delivered RNAi, hpRNA – Hairpin RNA, NNN – N′-Nitrosonornicotine, PCS – Phytochelatin synthase, PTGS – Post-transcriptional gene silencing, PSTVd – Potato spindle tuber viroid, PVY – Potato virus Y, miRNA – MicroRNA, piRNA – Piwi-interacting RNA, RKN – Root-knot nematode, RISC – RNA-induced silencing complex, RNAi – RNA interference, siRNA – Small interfering RNA, TRBP – TAR-RNA-binding protein, TSNA – Tobacco-specific nitrosamine, UTR – Untranslated region.

*Corresponding author: khanzeba02@gmail.com



DOI: 10.1079/9781800626638.0002

## 2.1 Introduction to RNA Interference (RNAi)

The term RNA interference, also known as RNAi, was first introduced by Fire and his colleagues (Fire *et al.*, 1998). Mello and his colleagues later suggested that rde-1 plays a role in RNAi. The term RNAi describes a conserved gene-silencing mechanism that short RNAs mediate. It has become clear that this is one of the key mechanisms controlling gene expression and suppressing transposable elements (TEs) in eukaryotes that are either transcriptionally or post-transcriptionally located.

RNAi is a gene-silencing process that occurs when double-stranded RNA (dsRNA) is introduced and regulates the expression of genes by silencing and suppressing the activity of specific genes. This phenomenon is found in a variety of creatures, including mammals and plants, and is conserved across species. Due to its notable specificity and effectiveness, it has been extensively employed as a proficient instrument for gene function analysis. Genome-wide studies have been conducted on worms and flies using RNAi techniques, utilizing entire genome sequences (Gönczy *et al.*, 2000). High-throughput analysis of cultured cells in studies has been facilitated by the development of an automated RNAi screening system that utilizes assay plates and a microarray-based RNAi screening system (Mousses *et al.*, 2003). Recent research has uncovered the inherent functions of RNA interference and related phenomena, including the inhibition of transposon activity, immunity against viral infections, control of gene expression at the post-transcriptional and post-translational levels, and regulation of chromatin structure through epigenetic mechanisms (Grewal and Moazed, 2003). The practical application of RNAi in enhancing the genetic traits of crop plants is anticipated. In this context, our primary focus is on RNAi as a technique for reducing gene expression, specifically in crop plants.

The molecular mechanisms and functions of RNA interference are not fully understood. Profound research has concentrated on creating RNAi technologies for treating human illnesses and enhancing plant characteristics. Nevertheless, the application of RNAi to improve the nutritional content of plants for both animal and human consumption, along with the progress in related RNAi technologies, is still in the early stages of development. In this chapter we will discuss the current comprehension of plant RNAi mechanisms, along with concepts and strategies for improving plant nutritional quality through the development of advanced plant RNAi technologies (Fig. 2.1).

## 2.2 Historical Development and Milestones in RNAi Research

In the last few decades, our understanding of the functions of RNAs in controlling gene expression has significantly expanded (Fig. 2.2). The RNA world received a significant boost when RNAi was discovered. The RNA interference phenomenon was first observed in the genetically modified plant *Petunia hybrida* L. (Napoli *et al.*, 1990) by introducing the chalcone synthase gene (CHSA), which encodes a crucial enzyme in the anthocyanin production pathway. This resulted in the enhancement of anthocyanin pigments in *Petunia*. Surprisingly, instead of dark purple blooms, transgenic plants were obtained that produced white or chimaera flowers. This was caused by the silencing of the endogenous homologous gene, and this occurrence was referred to as 'co-suppression'. Additionally referred to as post-transcriptional gene silencing (PTGS) in plants, quelling in fungi, and RNA interference in mammals, RNAi is a phenomenon that is observed in numerous organisms and is conserved across species (Romano and Macino, 1992; Fire *et al.*, 1998). In subsequent work, Fire *et al.* (1998) initially suggested the term 'RNA interference' and described the process of RNA interference in the nematode *Caenorhabditis elegans*.

RNAi encompasses a collection of mechanisms utilizing short RNAs (less than 30 bases in length) that control gene expression in various eukaryotic organisms. The rapid advancement in our comprehension of RNAi-based mechanisms has resulted in the use of this potent procedure in investigations of gene function and therapeutic interventions for disease treatment. RNAi-based therapeutics utilize two-dimensional

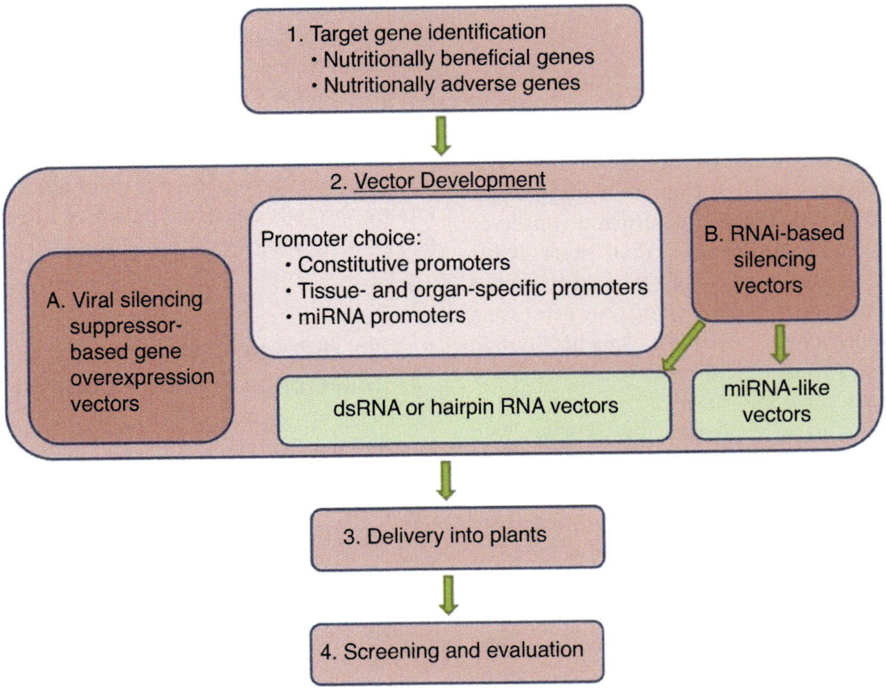


**Fig. 2.1.** RNA interference (RNAi)-based technologies for improving the nutritional qualities of plants. The first step is to identify and to clone the target genes, including nutritionally beneficial genes and adverse genes (1). The second step is to coordinatively overexpress nutritionally beneficial genes and to silence antinutritional and adverse genes in a spatially and temporally controlled manner by designing various RNAi suppressor or enhancer vectors (2). The third step is to produce transgenic plants by delivering into the plants the designed RNAi vectors specific for the designated target genes (3). Lastly, the desired transgenic plants are screened for improved nutritional value and for safety (4). mjRNA–microRNA.

pharmacological designs that rely solely on identifying strong Watson–Crick base pairing between the RNAi guide strand and the target. This approach allows for efficient design and testing of RNAi triggers. Currently, there are numerous clinical trials employing RNAi, and we may anticipate a remarkable expansion in the range of new applications (Kim and Rossi, 2008).

## 2.3 The Molecular Mechanism of RNAi

RNA interference mechanisms in *Drosophila* have been studied thoroughly utilizing *in vitro* systems, providing a comprehensive understanding of the underlying molecular processes (Waterhouse *et al.*, 2001; Hannon, 2002; Grewal and Moazed, 2003; Schwarz *et al.*, 2003) (Fig. 2.3). Initially, extended dsRNA known as Dicer is identified by a member of the RNase III family, and broken down into 21-nucleotide small interfering RNA (siRNA) duplexes. The duplex is unravelled, and typically one of the two strands is selectively integrated into the RNA-induced silencing complex (RISC). The antisense strand of the siRNA binds to mRNA as a guide, and the RISC enzyme cuts the mRNA at the middle of the siRNA. The siRNA duplex is composed of a 19-nucleotide double-stranded region with two nucleotide overhangs at the 3′ end. In *Drosophila*, the effectiveness of mRNA cleavage is significantly diminished when there are discrepancies between siRNA and the target

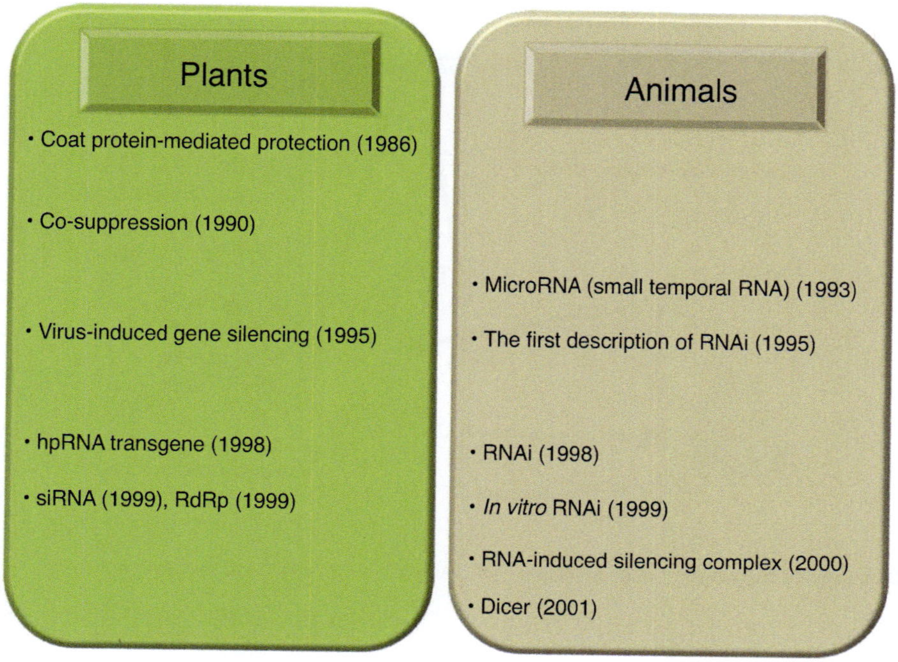


**Fig. 2.2.** A chronological history of the early work on RNAi and RNAi-related phenomena. hpRNA – hairpin RNA, siRNA – small interferRNA, RdRP – RNA-dependent RNA polymerase.

mRNA, especially if these discrepancies are in close proximity to the centre of the siRNA (Abdelgany *et al.*, 2003). It is important to acknowledge that plants can withstand a limited number of mismatches (Tang *et al.*, 2003).

## 2.4 Role of Small RNAs in Gene Regulation: siRNA, miRNA and piRNA

Small RNA molecules, including small interfering RNA (siRNA), microRNA (miRNA), and Piwi-interacting RNA (piRNA), are all part of the broader category of RNA interference (RNAi).

### 2.4.1 siRNA

The majority of eukaryotic cells employ RNAi as a regulatory mechanism. This method employs short molecules of dsRNA as triggers to direct control of gene activity based on homology (Almeida and Allshire, 2005) referred to as siRNA. These dsRNA molecules are around 21–22 base pairs long and possess distinctive 2-nucleotide 3′ overhangs. This enables them to be identified by the enzymatic machinery of RNAi, which ultimately results in the targeted mRNA being broken down through a process based on homology. In mammalian cells, siRNAs are created by cleaving larger dsRNA precursors using RNase III endonuclease Dicer-2 or through chemical or pharmacological techniques. Dicer forms complexes with RNA-binding proteins, namely the TAR-RNA-binding protein (TRBP), PACT, and Ago-2. These proteins play a role in transferring siRNAs to the RNA-induced silencing complex (RISC). The fundamental constituents of RISC consist of the members belonging to the Argonaute (Ago) family. Among humans, there are eight members within this family, although only Ago-2 exhibits an operational catalytic domain capable of cleavage activity (Liu *et al.*, 2004).

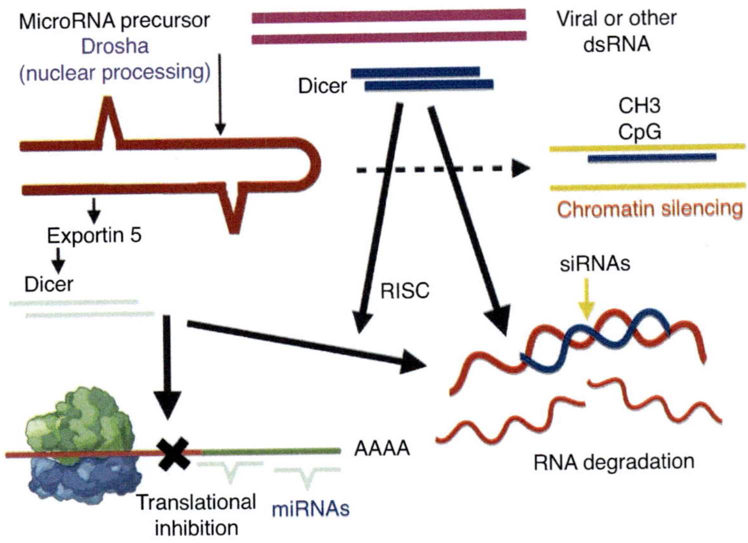


**Fig. 2.3.** Cellular pathways of gene silencing by RNAi. RISC – RNA-induced silencing complex.

### 2.4.2 miRNA

A crucial component of RNAi is the miRNAs. These are internal double-stranded molecules that control gene expression after transcription by forming complexes with RISC and attaching to the 3′-untranslated regions (UTRs) of target sequences using small segments of similarity called 'seed sequences' (Bartel, 2004; Bartel and Chen, 2004). The main mode of operation of miRNAs is to inhibit translation, which may be followed by the degradation of the mRNA. Since the antisense strand of miRNA duplexes cannot be selected by cleaving the passenger strand as siRNAs can, some other method must be used. The miRNA duplexes have defective Watson–Crick base pairing. RNA interference apparatus endogenous substrates are miRNAs. The miRNAs (pri-miRNAs) are initially expressed as lengthy primary transcripts (long RNAs). These miRNAs undergo nucleotide processing by the microprocessor complex, composed of Drosha and DGCR8, into 60–70 bp hairpins (Han *et al.*, 2004; Lee et al., 2006) or pre-miRNAs. Most significantly, this native gene silencing pathway can be utilized to regulate the desired gene or genes. When delivered to the cell, the siRNA effector will 'activate' RISC, leading to the designated mRNA being silencing in a potent and specific manner. As a result of its efficacy and specificity, RNAi has emerged as the preferred technique for inhibiting the expression of particular enzymes in mammalian cells.

### 2.4.3 piRNA

Piwi-interacting RNAs are a crucial type of short RNA molecule that are mainly found in mammals. They play a prime role in protecting the integrity of the genome. These molecules, which are usually 26–31 nucleotides long, function together with Piwi proteins, a subset of the Argonaute protein family. These piRNAs can be derived from certain genomic areas known as piRNA clusters. The clusters frequently contain fragments of transposon sequences, suggesting a possible evolutionary competition between transposons and the host genome. The generation of piRNAs includes an intricate mechanism. First, extended precursor transcripts are produced from piRNA clusters, which are subsequently converted into fully developed piRNAs. The specifics of this biogenesis route can differ between different species. The 'ping-pong' amplification cycle is a captivating process found in numerous organisms, including insects and vertebrates. In this process, piRNAs originating from several genomic strands direct Piwi proteins to produce secondary piRNAs in a mutually dependent manner. This method improves the efficiency of transposon silencing. Their main objective is to protect the genome against the potentially disruptive actions of transposons, which are mobile genetic components capable of moving within the genome. These piRNAs are mostly expressed in germline cells, specifically in sperm and egg cells. They have a vital function in maintaining the stability of genetic material that is inherited by the next generation. Their mode of operation is directing Piwi proteins towards transposon RNA's complimentary sequences, so promoting its breakdown and inhibiting translation into proteins. In addition to their involvement in post-transcriptional gene silencing, piRNAs have a function in epigenetic regulation by directing Piwi proteins to certain genomic sites, thus establishing long-lasting epigenetic modifications. Although piRNAs have been extensively studied in the field of animal biology, their occurrence and roles in plants have received less investigation. Disruption of the piRNA pathway has been associated with specific disorders, such as certain forms of cancer. The investigation of the function of piRNAs in both well-being and illness is a dynamic field of study that could have significant consequences for therapeutic treatments.

## 2.5 Techniques for Delivering Small RNAs to Plant Cells

In plants, RNA silencing occurs on multiple levels: cytoplasmic silencing through double-stranded RNA (dsRNA) causes mRNA breakage, endogenous mRNAs are silenced by micro-RNAs (miRNAs), and small RNA molecules that control gene expression by binding to specific messenger RNAs (mRNAs). This binding can lead to the cleavage of the RNA or the prevention of protein translation, a process known as PTGS. Also, RNA silencing is associated with the methylation of DNA in a sequence-specific manner, which results in the suppression of transcription, a process called transcriptional gene silencing (TGS) (Mansoor *et al.*, 2006).

The standard protocol for RNAi involves the cloning and insertion of a specific gene of interest into an appropriate plasmid, resulting in the formation of a recombinant plasmid. The recombinant plasmid is introduced into a suitable vector, such as *Agrobacterium*, which is well suited for plant transformation. Additional techniques for plant transformation include microprojectile bombardment, suction infiltration, syringing, and spraying infection approaches (Shao *et al.*, 2008). In plants, the RNAi process begins with the generation of siRNAs that are 21–24 nucleotides long. These siRNAs are produced within the cell by cleaving long double-stranded RNA molecules using an enzyme called Dicer, which belongs to the ribonuclease III-type family (Zamore *et al.*, 2000). These small interfering RNAs (21–24 nucleotides) are subsequently integrated into the RISC, which comprises many proteins such as Ago (Vaucheret, 2008) besides siRNAs. The ATP-activated RISC separates the double-stranded siRNA into single strands. The antisense strand of the siRNA molecule is integrated into a nuclease-containing RISC complex when the sense strand of the siRNA duplex is lost due to the activity of an RNA helicase (Kusaba, 2004). The RISC complex utilizes antisense siRNA sequences to specifically target and bind with the complementary transcript, resulting in the cleavage of mRNA or the prevention of translation, ultimately inhibiting protein synthesis (Bartel, 2004).

Efficient delivery systems are pivotal for the successful implementation of RNAi in plants, enabling precise and targeted gene silencing. Among the various strategies employed, *Agrobacterium*-mediated transformation stands out as a widely utilized method, leveraging the genetic engineering capabilities of *Agrobacterium tumefaciens* to introduce RNAi constructs into plant cells. Viral vectors, engineered to carry RNAi molecules, offer another potent approach, with modified plant viruses facilitating systemic gene silencing. Nanoparticle-based delivery systems, encompassing liposomes, polymeric nanoparticles and viral-like particles, shield RNA molecules from degradation and enhance uptake by plant cells. Physical methods, such as electroporation and the gene gun, provide means for transient expression studies by directly introducing RNAi constructs into plant cells. Root uptake strategies, involving soil drenching or hydroponic systems, focus on delivering RNAi molecules to plant roots, enabling systemic transport within the plant. *In planta* expression methods involve the integration of RNAi constructs into the plant genome, allowing the plant itself to produce and express the RNAi molecules. Additionally, RNAi spray technology, involving the direct application of RNAi sprays on to plant surfaces, offers a localized or systemic approach, particularly relevant for pest control. As research progresses, advancements in nanotechnology and synthetic biology continue to refine and expand the repertoire of RNAi delivery systems, enhancing their efficacy in plant studies and crop improvement endeavours.

## 2.6 Challenges and Advancements in RNAi Delivery

The transport of RNAi molecules encounters various obstacles that impact its effectiveness and selectivity. The occurrence of off-target effects, which involve the inadvertent suppression of genes, is a continuous problem that requires improved accuracy in the administration of RNAi. Furthermore, the durability of RNA molecules is at risk due to degradation caused by nucleases, which requires the use of inventive methods to preserve their integrity during transportation. Efficient cellular absorption, particularly the overcoming of cellular barriers, continues to be a challenge for the successful implementation of RNAi applications. Delivery of medicinal substances within living organisms faces challenges associated with biological barriers, immunological reactions, and concentrations at the intended spot. Accurate targeting of particular tissues or cells is essential for reducing potential adverse effects. Introducing external RNA molecules can activate immunological responses, necessitating the use of techniques to reduce immunogenicity. Notwithstanding these obstacles, notable progress has been achieved in the development of RNAi delivery systems. Nanoparticle-based methodologies, viral vectors, chemical alterations and inventive administration pathways have demonstrated potential in surmounting these obstacles. Intelligent delivery systems, which are sensitive to signals from cells, seek to improve the distribution of substances to specific targets while reducing unintended effects on other areas. Utilizing Clustered regularly interspaced short palindromic repeats/CRISPR-associated protein 9 (CRISPR-Cas9) technology allows for a dual gene-regulation strategy, enhancing specificity. The utilization of advanced bioinformatics techniques enhances the ability to create RNAi molecules with greater accuracy. As research advances, the continuous effort to find solutions to these issues and the integration of creative methodologies hold the potential to improve and enhance the effectiveness of RNAi distribution in various applications.

Introducing a gene into a plant cell by crossing its complex polymeric cell wall also sometimes presents difficulties for biotechnology researchers. In contrast, the limited effectiveness of genetic material transformation, cellular toxicity, and the incorporation of damaged DNA molecules resulting in the destruction of the host genome significantly hinder the efficiency of transgenic procedures (Gad *et al.*, 2020). In the past, a viral gene vector was commonly employed for the transportation of genetic elements into cells. This approach had numerous disadvantages, such as restricting the scope of hosts and the ability to transport and transit genetic elements within the nucleus, while also facing capacity limitations (Ghormade *et al.*, 2011). Nevertheless, the latest progress in nanotechnology has successfully addressed these limitations and drawbacks. Nanomaterials provide optimal potential as a substitute for gene delivery technologies in crop enhancement due to various compelling factors. A delivery system using nanoparticles inhibits the degradation of DNA by shielding it from nuclease enzymes, improves the ability to pass through the plant cell wall and transports the functional genetic components into the nucleus (Thomas and Klibanov, 2003). Prior studies have already reported the notable efficacy of mesoporous nanoparticles in delivering genetic elements to plant cells (Cebrián *et al.*, 2011).

## 2.7 Applications of RNAi in Crop Improvement (Fig. 2.4)

### 2.7.1 Modification of plant structure

The modification of plant structure, including height, stem elongation, leaf and inflorescence shape, and branching has been a subject of study for many years. These alterations have a significant impact on various agricultural parameters such as crop productivity, physiological and biochemical functions, lodging, and resilience to environmental challenges (Wang and Li, 2006). To mitigate the adverse effects of ongoing global climate change and the limited availability of water and nutrients on crop productivity, it is possible to manipulate the structure of the root system. This manipulation aims to achieve a consistent distribution of roots in the soil, which effectively maximizes the uptake of water and nutrients (de Dorlodot *et al.*, 2007). The recent research conducted on the model plants *Arabidopsis* and *Petunia*, as well as agricultural plants such as tomato, maize and rice, have provided valuable insights into the molecular mechanisms underlying plant design (Wang and Li, 2006, 2008). RNAi technology has been recently employed to enhance crop productivity through the alteration of plant architecture. The emergence and growth of adventitious roots were markedly suppressed in the OsPIN1 RNAi transgenic plants, resembling the phenotype observed in wild-type plants treated with N-1-naphthylphalamic acid (NPA), an inhibitor of auxin transport. Treatment with α-naphthylacetic acid (α-NAA) successfully reversed the mutant phenotypes observed in the RNAi plants. Manipulating the expression of OsPIN1 via genetic modification techniques led to alterations in the number of tillers and the ratio of shoots to roots. This indicates that OsPIN1 plays a significant role in the formation of adventitious roots and the process of tillering, which is dependent on the hormone auxin (Xu *et al.*, 2005).

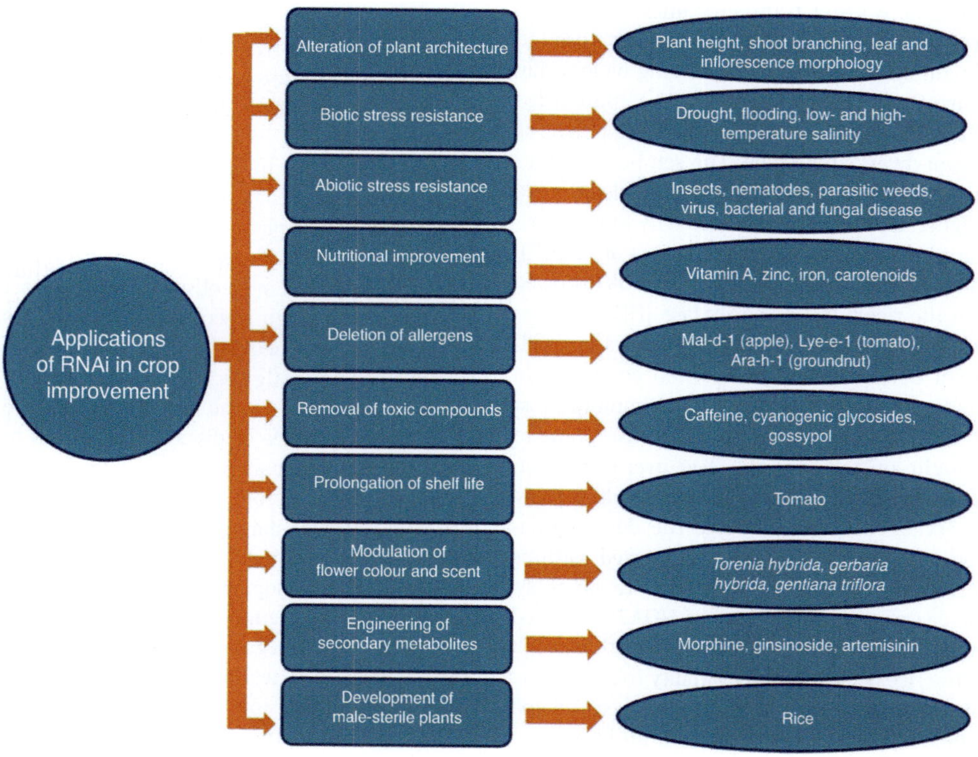


**Fig. 2.4.** Applications of RNAi in crop improvement.

### 2.7.2 Targeting specific genes for improved traits

Plant height is controlled by endogenous plant hormones, such as gibberellins (GAs) and brassinosteroides (BRs), which act as significant regulators in plants. Genetic modifications targeting genes involved in GA production are key objectives for modifying plant height (Hedden and Phillips, 2000). In their study, Zhou *et al.* (2006) successfully utilized interfering RNA to inhibit the expression of the OsGLU1 gene, which encodes a membrane-bound endo-1,4-β-D-glucanase in rice (*Oryza sativa* L.). This interference had a significant impact on the elongation of plant internodes, resulting in structural modifications to cell walls and changes in cell wall composition. Ultimately, these alterations led to the development of a dwarf phenotype. The technique of RNAi was employed to inhibit the expression of the OsGA20o×2 gene in rice. This gene is responsible for encoding the regulatory enzyme GA 20-oxidase, which is involved in the production of physiologically active GAs in plants. A set of transgenic lines with a phenotypic series were obtained, showing a drop in the amount of naturally occurring biologically active GA1. This reduction in GA1 content led to a fall in plant height and the development of a semi-dwarf phenotype. The semi-dwarf lines displayed reduced stem height, resistance to lodging, and improved productivity in comparison to the wild-type plants (Qiao *et al.*, 2007). In a separate research, Li *et al.* (2009) showed that the rice histone deacetylase (HDAC) genes exhibited distinct expression patterns and different developmental roles when compared to similar genes in *Arabidopsis*. Additionally, most of these genes were found to be susceptible to drought or salt stresses. The over-expression of many rice HDACs did not result in any observable phenotype, while the downregulation of select HDAC genes impacted various developmental characteristics. More precisely, the decrease in the activity of HDA703 caused by miRNA resulted in a decrease in the length of the rice peduncle and its ability to reproduce. Similarly, the suppression of a closely related gene, HDA710, using RNAi had an impact on the growth of the plant's non-reproductive parts.

The plant's height and the shape of its flag leaf were modified using RNAi. The suppression of HDT702 resulted in the development of narrower leaves and stems. The application of RNAi technology to manipulate plant architecture has broad potential in various sectors such as flowering, ornamental, plantation crops and forest trees. This technology can be used to facilitate convenient leaf plucking in tea or mulberry plants, enable the harvesting of fruits or seeds from tall trees, and to eliminate thorns in roses.

### 2.7.3 Enhancing stress tolerance in crops using RNAi

#### *2.7.3.1 Biotic stress resistance*

Biotic stress factors are induced by insects, nematodes and parasitic organisms. Weeds, as well as viral, bacterial and fungal diseases, impose various limitations on agricultural output, hence exerting an influence on the aggregate agriculture output.

**RNAI AS A NOVEL BIOTECHNOLOGICAL APPROACH TO MANAGE VIRAL INFECTIONS IN PLANTS.** Nevertheless, the primary reason for reduced plant productivity is viral infections. Viruses pose a challenge in terms of control due to their utilization of many mechanisms to proliferate and disseminate within and across plants. The primary causes responsible for viral illnesses are direct transfer of viruses from parent to offspring and indirect transmission through insect vectors, as well as the extensive production of monocultures. A crucial and frequently decisive feature of plant virus epidemiology is the assessment of the speed and magnitude of disease progression in economically significant plant populations, such as fruit crops, cash crops and food crops, resulting in substantial damages that rank second only to fungal infections. Hence, it is imperative to expedite the development of environmentally sustainable, non-chemical and socially acceptable methods for safeguarding crops. Pathogen-derived resistance techniques, which include obtaining resistance to a specific pathogen from its own genetic material, have been utilized to cultivate plants that are resistant to diseases. These procedures can be categorized into two types: (i) protein-dependent resistance, which involves the expression of viral proteins such as viral coat proteins and replication-associated proteins; and (ii) protein-independent resistance, which does not include protein expression and includes mechanisms such as gene silencing, PTGS using antisense methods, and hairpin RNA (hpRNA) processes (Shepherd *et al.*, 2009).

Several plant viruses possess RNA genomes and replicate through dsRNA intermediates, making them highly effective inducers of RNAi during early replication and targets for gene silencing in later stages of infection. Given that RNAi functions as an antiviral mechanism, it is not unexpected that numerous plant viruses possess genes that inhibit RNAi. Hence, the technique was first used by Waterhouse and Graham (1998) to create plants that are resistant to potato virus Y (PVY) by introducing vectors that can express both the sense and antisense transcripts of the viral helper-component proteinase (HCPro) gene. These plants will exhibit total immunity to PVY. The Spunta cultivar of potato was genetically modified using dsRNA obtained from the 3′-terminal region of the coat protein gene of PVY. The chosen transgenic lines exhibited significant resistance against three strains of PVY. Additionally, the infection of transgenic plants with potato virus X, either simultaneously or before being exposed to PVY, did not disrupt the resistance against PVY (Missiou *et al.*, 2004).

The papaya plant, which is renowned for its edible fruits and as a primary supplier of the papain enzyme, experiences significant reductions in yield due to viral infections (Figs 2.5 and 2.6). Transgenic plants expressing the coat protein were generated, and while the coat protein gene was found in all transgenic lines, only one line exhibited significant rearrangement of the inserted coat protein expression cassette. In the resistant lines, there was a notable reduction in the quantity of truncated coat protein mRNA. This confirms that the resistance to coat protein in papaya is most likely due to post-transcriptional gene silencing, which is mediated by RNA.

Haque *et al.* (2007) successfully grew *Nicotiana benthamiana* plants that expressed the coat protein gene of sweet potato feathery mottle virus. Two silenced and two non-silenced lines were chosen from these plants to examine the occurrence of transitive RNA silencing. This process involves the amplification of the RNAi signal throughout the plant by RNA-dependent RNA polymerase, resulting in the silencing of genes in other tissues and organs of the plant. Graft experiments were conducted to investigate this phenomenon. Scions that were not silenced and had the complete transgene were grafted on to rootstocks that induced silencing. When non-silenced scions were attached to 5′ silencing inducer rootstocks, it caused RNA silencing in the non-silenced scions and the silencing effect progressed towards the 3′ region of the transgenic mRNA. Likewise, when non-silenced scions were transplanted on to rootstocks that induce silencing, RNA silencing was triggered in the non-silenced scions. However, the silencing effect was limited to the 3′ region of the transgene and did not extend to the 5′ area. The findings were confirmed by the results obtained from crossing studies that involved non-silenced plants and plants induced with a 3′ silencing inducer (Haque *et al.* 2007).

In 2009, Schwind *et al.* created genetically modified tomato plants that were equipped with hairpin RNA. These plants demonstrated resistance to potato spindle tuber viroids. The overabundance of hairpin RNA is the primary component responsible for resistance. The authors' genome mapping revealed an uneven distribution of RNAi across the PSTVd genome, indicating the potential to genetically modify plants for resistance against viroids. The altered transgenic cassava plants demonstrated resistance to African cassava mosaic virus (ACMV) in a separate study, achieved by the expression of dsRNAs. Transgenic cassava lines containing elevated levels of AC1-homologous short RNAs exhibited ACMV replication linked to a protein-coding sequence that provides protection through the presence of a Rep/AC1-homologous hairpin double strand (Vanderschuren *et al.*, 2009).

**RNAI AS A NOVEL BIOTECHNOLOGICAL APPROACH TO MANAGE ROOT-KNOT NEMATODE INFECTIONS IN PLANTS.** Root-knot nematodes (RKNs), specifically *Meloidogyne* spp., are soil-dwelling plant parasites that inflict significant harm to crops and deplete plants of essential nutrients. Various parasitism proteins, which are produced by

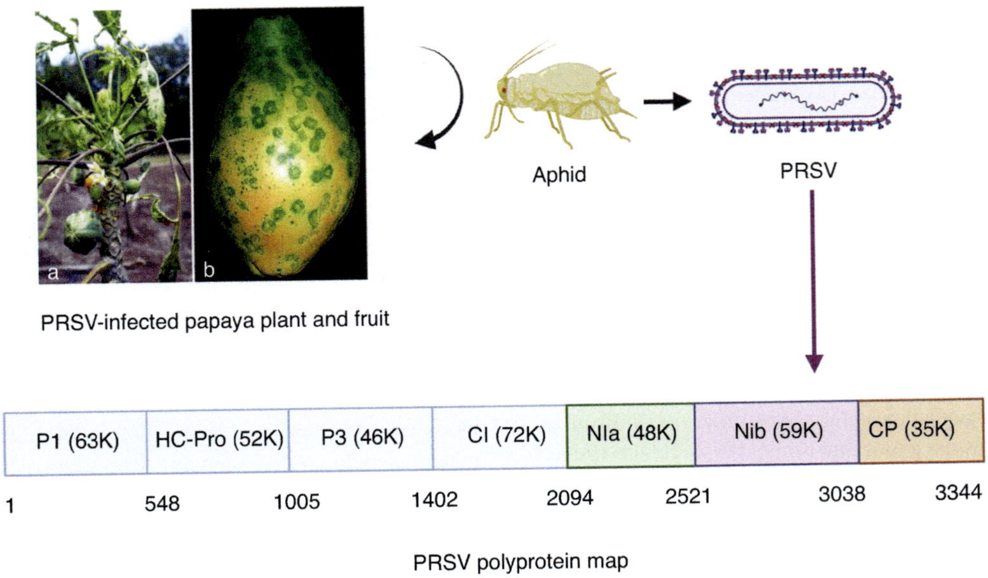


**Fig. 2.5.** Papaya ringspot virus (PRSV) infection symptoms and transformation and the PRSV polyprotein gene map. (From Mujtaba *et al*., 2021, with permission from the publisher.)

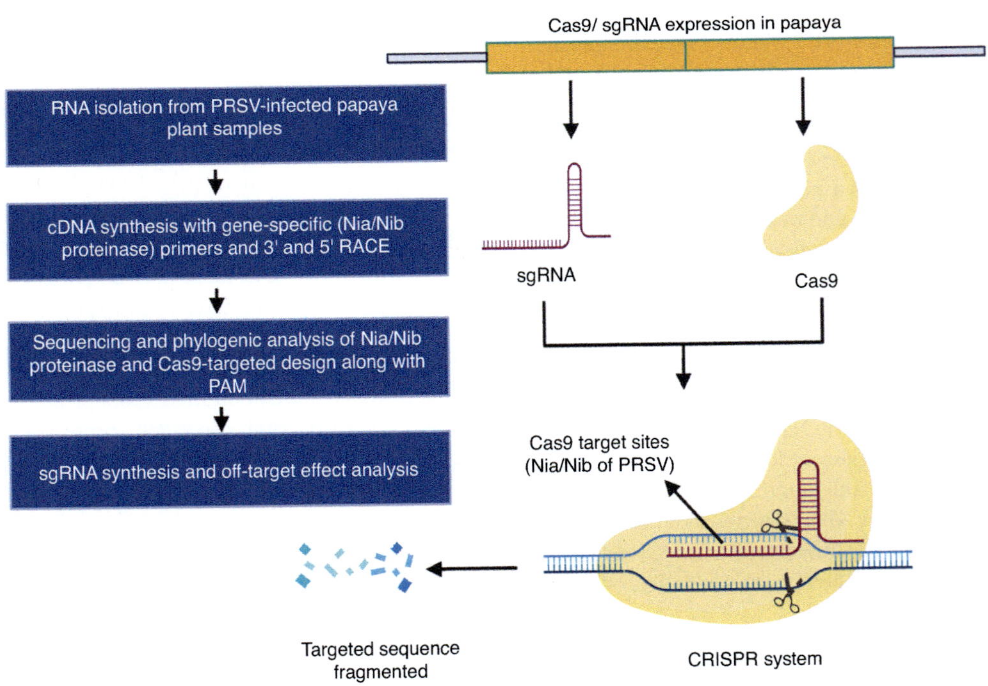


**Fig. 2.6.** Workflow for sgRNA Design and CRISPR-Cas9 Targeting of Nia/Nib Genes in Papaya Ring Spot Virus-Infected Plants.

parasitism genes and expressed in oesophageal gland cells, facilitate the infection and parasitism of plants by root knot nematodes (RKN). Bioengineered crops that produce dsRNA to specifically target genes involved in RKN parasitism offer a promising and adaptable approach for creating new, long-lasting RKN-resistant crops. This technology has the potential to deliver crops with an unparalleled level of wide resistance against RKNs (Huang *et al.*, 2006). Fairbairn *et al.* (2007) successfully obtained tobacco plants that produced dsRNA hairpin structures specifically designed to target a RKN. The authors determined that the use of host-delivered RNAi (HD-RNAi) to silence parasite genes is a new and promising strategy for disease resistance. This approach has significant potential for other biotechnological applications. The versatility of HD-RNAi is not limited to parasitic nematodes but can be applied to regulate various pests that feed on plants. Cyst nematodes employ parasitism proteins to effectively parasitize plants, enabling the formation of specialized feeding cells that are essential for the nematode's sustenance. Suppressing the expression of cyst nematode genes is expected to interrupt the parasitic life cycle and confer resistance to the host plant. RNAi data were used to investigate the expression of four parasitism genes in transgenic *Arabidopsis thaliana*, which serves as a host for the sugar beetroot cyst nematode *Heterodera schachtii*. The expression of RNAi constructs in plants specifically inhibited the activity of targeted nematode genes in nematodes that fed on these plants (Sindhu *et al.*, 2009). RNA interference targeting all four nematode parasitism genes resulted in a reduction in the population of mature adult females. Utilizing this method may prove to be a highly efficient approach for diagnosing cyst nematode issues (Sindhu *et al.*, 2009).

**RNAI AS A NOVEL BIOTECHNOLOGICAL APPROACH TO MANAGE FUNGAL INFECTIONS IN PLANTS.** Efforts to achieve RNAi-based resistance against fungal disease caused by *Phytophthora parasitica* var. *nicotianae* were successful by targeting the glutathione S-transferase (GST) enzyme. This approach led to a notable enhancement in the resistance of *Nicotiana tabacum* to infection after silencing the GST gene, as compared to control plants. Glutathione S-transferases are found in both eukaryotes and prokaryotes, where they facilitate a range of processes and process both natural and foreign substances. The suppression of GST was accomplished by inserting a GST gene in both the forward and reverse orientations into an RNAi vector. This was done with the intention of silencing the gene to halt the progression of black shank disease (Hernández *et al.*, 2009). It was shown that a small number of defence genes were activated at a higher level in plants with silenced GST during their interaction with the pathogen.

#### *2.7.3.2 Abiotic stress resistance*

Plant productivity is significantly reduced by environmental abiotic variables such as drought, flooding, temperature variations and salinity. The identification of new genes, analysis of their expression patterns under abiotic stress conditions, and enhanced comprehension of their functions through functional genomics will serve as the foundation for developing efficient engineering approaches that enhance stress tolerance (Pardo, 2010). Recent data suggest that RNAi plays a role in plants' responses to abiotic stress. Sunkar and Zhu (2004) studied the role of miRNAs in responding to abiotic stressors such as salinity, drought, cold and oxidative stress. In their study, *Arabidopsis* seedlings were exposed to different abiotic stimuli, revealing a notable increase in the expression of miR393 in response to cold, dehydration, high salinity, and abscisic acid (ABA) treatments. In addition, *Arabidopsis* plants responded to distinct abiotic stress treatments by regulating the expression of miR319c, miR389a, miR397b and miR402 to varying extents. Other investigations have demonstrated that miR395 and miR399 exhibit increased expression levels in response to sulfur and phosphate deprivation in *Arabidopsis* plants (Jones-Rhoades and Bartel, 2004). miR395 specifically targets the genes responsible for ATP sulfurylases, which are enzymes involved in the initial stage of inorganic sulfate assimilation. Additionally, it also targets the AST68 gene, which encodes a sulfate transporter (Jones-Rhoades and Bartel, 2004). This outcome was atypical for miRNA-mediated gene regulation, as miR395 was found to control two distinct sets of genes that operate inside the same metabolic pathway. Later, it was shown that miR399 decreases the activity of ubiquitin-conjugating enzyme (UBC24) by specifically targeting the 5′ UTR. This regulation was found to play a crucial role in how plants respond to phosphorus deficiency, as it affects the expression of miR399 or UBC24 in *Arabidopsis* (Chiou *et al.*, 2006). miR393 functions by inhibiting the expression of TIR1, which leads to the downregulation of auxin signalling and seedling growth in response to abiotic stress conditions. Additionally, miR393 also contributes to antibacterial resistance (Navarro *et al.*, 2006). Also, miR159 exhibited a response to hormone signalling and dehydration in *Arabidopsis* (Reyes and Chua, 2007). Furthermore, rice has been found to contain many miRNAs that are associated with stress. Specifically, only two miRNAs – miR393 and miR169g – have been identified as being linked to abiotic stress. It is worth noting that both of these miRNAs were found to be upregulated in response to dehydration (Zhao *et al.*, 2007). Jian *et al.* (2010) discovered new stress-related miRNAs in rice (*Oryza sativa* L. ssp. Japonica cv. 9522) seedlings that were subjected to cold, dehydration, salt and ABA. Recent research indicates that the receptor for activated C-kinase 1 (RACK1) is a widely preserved scaffold protein with multifaceted functions and significant involvement in the control of plant growth and development. The roles of RACK1 in response to drought stress in rice were studied by inhibiting the expression of the RACK1 gene using RNAi in transgenic rice plants. The transgenic rice plants exhibited a greater tolerance to drought stress in comparison to the non-transgenic rice plants. The many endogenous antioxidant systems found in plants, which consist of several enzyme scavengers such as superoxide dismutases, serve as the primary defence mechanism against the highly harmful superoxide radicals that build during periods of stress. Li *et al.* (2009) conducted a study on the activity of superoxide dismutase. They found that the activity of superoxide dismutase was significantly higher in transgenic rice plants compared to non-transgenic rice plants. The study proposed that RACK1 exerted a negative regulatory effect on the redox system-associated tolerance to drought stress in rice plants Li *et al.* (2009).

### 2.7.4 Contribution of RNAi to sustainable farming practices

RNA interference technology plays a crucial role in sustainable agriculture by providing precise methods for controlling pests and diseases, managing weeds and enhancing stress tolerance. Genetically modified crops that produce RNA molecules allow for accurate control of pests and diseases, hence decreasing the need for chemical pesticides. Also, RNAi improves agricultural productivity and quality by controlling nutrient usage and delaying the aging process, hence enhancing resistance to abiotic challenges such as drought, and decreases the ecological footprint of agriculture by minimizing the use of chemical inputs. This method enables precise customization of agricultural features and promotion of advantageous biotic interactions. Ensuring the safe and sustainable integration of RNAi in agriculture requires responsible deployment, which involves taking into account regulatory and ethical considerations. This highlights the importance of continuous research and collaboration among scientists, politicians and farmers.

### 2.7.5 Nutritional improvement

Advancement in the field of molecular biology, and contemporary methods in breeding, genetics, and biotechnology are employed to create crops with improved essential substances that provide nourishment to the body. Additional nutritional objectives are altering the fatty acid composition and increasing the quantities of antioxidants, specifically carotenoids like lycopene and flavonoids (Tucker, 2003).

In order to decrease the expression of two crucial fatty acid genes encoding desaturases in cottonseed oil, the technique of hpRNA-mediated gene silencing was utilized. The gene downregulation resulted in a significant rise in stearic acid levels, ranging from 2% to as high as 40%. Additionally, silencing another gene resulted in a huge enhancement of oleic acid content, reaching up to 77% compared to approximately 15% in seeds of untransformed plants (Liu *et al.*, 2002).

An RNAi technique was employed to augment the β-carotene concentration in potatoes by suppressing the activity of the β-carotene hydroxylase gene (BCH), responsible for the conversion of β-carotene to zeaxanthin. The process of *Agrobacterium tumefaciens*-mediated transformation was utilized to introduce two RNAi constructs into potato lines. One of the creations included the tuber-specific granule bound starch synthase (GBSS) promoter, while the other carried the strong constitutive cauliflower mosaic virus 35S (CaMV 35S) promoter. The transformants resulting from the GBSS construct exhibited a higher concentration of β-carotene compared to the CaMV 35S transformants. The findings indicated that suppressing BCH can enhance the levels of two beneficial carotenoids, β-carotene and lutein, in potatoes. This discovery offers a novel agricultural approach to reducing the occurrence of vitamin A deficiency in populations (Van Eck *et al.*, 2007).

### 2.7.6 Prolongation of fruit shelf life

Ethylene triggers several reactions throughout the ripening process through a signalling cascade, involving the participation of numerous genes. These genes not only initiated ripening but also contribute to its deterioration (Bapat *et al.*, 2010). Genetic engineering is perceived as the most viable and economical method to mitigate these losses, yet it lacks acceptance from the general public. Xiong *et al.* (2005) employed RNAi technology to enhance the longevity of tomatoes. The dsRNA of the tomato ACC oxidase expression unit was effectively inserted into tomato cultivar Hezuo 906 using *A. tumefaciens*-mediated transformation, with the cauliflower mosaic virus 35S promoter. The plants that were cultivated yielded fruits containing ethylene residues and exhibited an extended storage duration exceeding 120 days. These fruits had comparable quantities of total soluble sugar, titratable acid, amino acids and total soluble solids as the plants in the control group. The RNA and protein studies revealed the presence of non-RNA interference, semi-RNAi RNA interference, and full-RNA interference of ACC oxidase in the transgenic tomato plant population. Vrebalov *et al.* (2009) used RNAi repression to show that tomato AGAMOUS-LIKE1 (TAGL1) plays a crucial role in controlling the enlargement of fleshy fruits and the ripening process. These processes are essential for promoting the dispersal of seeds in fleshy fruits.

### 2.7.7 Elimination of toxic compounds

Both cultivated and wild plants possess diverse toxins that frequently impede the isolation of pure desired products. The elimination of undesirable substances from plants is a laborious and expensive procedure that necessitates the collaboration of engineering and chemistry. The utilization of RNAi would be a highly potent and effective approach in order to render plants devoid of toxins.

Cadmium (Cd), a very hazardous heavy metal, induces various health complications, even at minimal concentrations. In their research, Li *et al* (2007) conducted experiments aimed at decreasing the buildup of Cd in rice (*O. sativa* L. ssp. *japonica*) seeds. They achieved this by using RNAi to block the production of the phytochelatin synthase (PCS) gene OsPCS1. A hairpin construct consisting of a PCS fragment was created within the pRNAi-OsPCS1 vector, regulated by the ZMM1 promoter derived from maize, specifically active in seeds. *A. tumefaciens* was used to introduce the construct into rice. The RNAi rice plantlets were carefully chosen and grown in pots that were subjected to a concentration of 10 mg/kg of Cd. The expression of OsPCS1 at the transcriptional level decreased in the seeds of certain RNAi rice plants as compared to the wild type. The buildup of Cd in the seeds of RNAi rice was reduced by around 50%. No discernible disparity in growth was seen between RNAi and wild-type plants, as anticipated. This discovery is significant as it indicates that this novel method can be employed to regulate the buildup of heavy metals in crops.

Tobacco products, such as those derived from *Nicotiana tabacum* L., include several substances, including tobacco-specific nitrosamines (TSNAs), which have been found to cause negative health effects, including cancer.

TSNAs represent a group of compounds that are produced when pyridine alkaloids undergo nitrosation during the tobacco's curing and processing stages. Within this group, N′-nitrosonornicotine (NNN), recognized as a carcinogen, is formed through the nitrosation process of nornicotine. Nornicotine is a secondary alkaloid that is synthesized by the enzymatic N-demethylation of nicotine. Nornicotine is generated through the process of oxidative N-demethylation, which involves the conversion of nicotine by a nicotine N-demethylase enzyme during the stages of senescence and cure. Lewis *et al.* (2008) created genetically modified strains of burley tobacco (strain DH98-325-5) that contained an RNAi construct specifically engineered to suppress the activity of the nicotine demethylase gene. The transgenic lines displayed a significant reduction of 6-fold in nornicotine content compared to the untransformed controls. Examination of treated leaves indicated a proportional reduction in NNN and overall TSNAs. Suppressing the activity of nicotine demethylase is a highly effective method for drastically reducing the concentration of a specific carcinogenic compound found in tobacco products. The outcome showcased the efficacy of RNA interference (RNAi) technology in diminishing the concentration of an unwanted metabolite to a level that has not yet been attained by traditional approaches.

### 2.7.8 Development of seedless fruits

Fruits without seeds are preferred by consumers, both in fresh fruits (for example, grapes, citrus fruits and bananas) and in preserved or processed fruit products (for example, frozen aubergine and tomato sauce). This characteristic can enhance the quality of fruits, especially when the seeds are tough or have an unpleasant taste. RNAi technology is employed to inhibit the activity of ARF7 in tomato plants. The introduction of transgenic plants led to the generation of fruits that do not contain seeds (de Jong *et al.*, 2009). A novel gene family associated with the regulation of fruit set has been recently discovered. The AUCSIA family consists of two members that encode peptides consisting of 53 amino acids. These peptides are produced in the ovary and their expression is significantly reduced after pollination. The AUCSIA genes in tomato were effectively silenced by RNAi (Molesini *et al.*, 2009). Plants with suppressed AUCSIA gene expression exhibited facultative parthenocarpy, resulting in the development of fruit without fertilization, leading to the production of seedless fruits upon removal of male reproductive organs from the flowers. One AUCSIA-silenced line exhibited obligatory parthenocarpy, resulting in consistently seedless fruits. Seedless fruits have been achieved by suppressing the activity of chalcone synthase (CHS), which is the initial gene in the pathway responsible for the production of flavonoids (Schijlen *et al.*, 2007).

### 2.7.9 Reducing the need for chemical inputs through RNAi

To decrease the requirement for chemical inputs, RNAi is utilized to create genetically modified crops that produce certain RNA molecules aimed at pests, illnesses or weeds. These RNA molecules can disrupt the expression of crucial genes in the specific species they target, resulting in the suppression or eradication of these species without the need for conventional chemical pesticides or herbicides. This focused strategy avoids unintended harm to non-target species, fosters ecological sustainability, and diminishes the overall chemical burden in agricultural systems. Using RNAi to augment inherent defence mechanisms in crops, farmers can proficiently control pests and illnesses while reducing dependence on wide-ranging chemical interventions. The outcome is an agricultural practice that is both ecologically benign and sustainable, in accordance with the principles of integrated pest management and precision agriculture. RNAi has contributed to the elimination of allergens, development of male-sterile plants, engineering of secondary metabolites, and modulation of flower colour and scent.

## 2.8 Bioinformatic Approaches to siRNA Selection and Optimization

Bioinformatics is crucial for advancing RNAi research in plants by offering vital tools and procedures for designing, analysing and interpreting RNAi experiments. Bioinformatics tools play a crucial role in the design of siRNA and short hairpin RNA, by enhancing the specificity and effectiveness of sequences while reducing off-target effects. These methods are also essential for predicting and analysing possible off-target effects, thus guaranteeing the specificity of RNAi constructs. Bioinformatics facilitates the evaluation of the effectiveness of RNAi molecules by taking into account parameters such as target accessibility, thermodynamic stability and secondary structure. Computational methods are useful in genome-wide RNAi screens for analysing extensive datasets and identifying genes associated with particular biological processes or pathways (Table 2.1). By incorporating additional omics data, such as transcriptomics and proteomics, one can achieve a thorough comprehension of gene function and regulatory networks. Bioinformatics tools also aid in the examination of high-throughput RNAi data produced by advanced sequencing technologies, enabling the detection and measurement of short RNAs and overall alterations in gene expression. Bioinformatics is an essential part of plant RNAi research as it improves the effectiveness and accuracy of studies by including evolutionary conservation, structural analysis, and the use of publicly accessible databases (Table 2.2).

## 2.9 Challenges, Recommendations and Future Perspectives

The utilization of RNAi has great potential in improving crop characteristics; however, its extensive implementation encounters many obstacles. An important issue is the presence of off-target effects, which can lead to unwanted genes being unintentionally silenced. This raises doubts about the accuracy of RNAi in modifying agricultural genetics. Furthermore, attaining reliable and enduring transgenic expression in plants has posed difficulties, resulting in discrepancies in the effectiveness of gene silencing among various crops and conditions. In addition, the possible ecological consequences of RNAi-altered crops, combined with the ambiguous regulations governing genetically modified organisms, require meticulous evaluation and risk analysis. To tackle these issues, specific and inventive measures are necessary. Researchers should prioritize enhancing the accuracy of RNAi technologies by improving the design strategies for siRNAs and optimizing delivery methods to minimize off-target effects. Moreover, it is necessary to establish reliable and consistent techniques for transgenic expression, guaranteeing consistent RNA interference effects in various plant tissues and environmental circumstances. Nanotechnology-based techniques and upgraded viral vectors have the potential to greatly enhance the efficiency of RNAi in different crops, leading to major improvements in delivery systems. Thorough evaluations of environmental safety are essential for obtaining regulatory approval and gaining public support. These evaluations involve conducting extensive studies to analyse the potential ecological effects and inadvertent harm to non-target creatures.

The future prospects of RNAi in crop enhancement are highly promising. Scientists can investigate the concept of trait stacking, which involves mixing various RNAi traits to tackle intricate agricultural problems. This strategy may entail augmenting the ability of crops to withstand pests, diseases and environmental stressors, resulting in plants that possess increased resilience and yield. Furthermore, the use of RNAi for precise epigenetic alterations in crops presents novel opportunities for manipulating gene expression patterns without affecting the fundamental DNA sequence, providing added adaptability in crop enhancement approaches. With the ongoing progress of RNAi technologies, their utilization is expected to expand from model crops to non-model and orphan crops, thereby making a valuable contribution to global food security and sustainable agriculture. Public awareness and education activities are crucial for promoting clear and open communication regarding the advantages and safety of RNAi technology, hence promoting acceptance and comprehension among all parties involved.

**Table 2.1.** Comparative overview of miRNA identification approaches in plants.

| Approach | Principle | Advantages | Limitations |
|---|---|---|---|
| **Computational approaches** | | | |
| Homology based | Identify miRNAs by comparing to known miRNAs | Relies on evolutionary conservation; useful for well-studied species | May miss species-specific or novel miRNAs |
| *De novo* prediction | Predict miRNAs based on sequence and structure features | Suitable for identifying novel miRNAs in less-studied species | Relies heavily on accurate prediction algorithms; may result in false positives |
| Machine learning based | Utilize machine learning algorithms for classification | Can handle large datasets and learn complex patterns | Requires well-curated training datasets; may be computationally intensive |
| **Experimental approaches** | | | |
| Small RNA sequencing | High-throughput sequencing of small RNAs | Provides a comprehensive view; identifies novel miRNAs | Relies on sequencing technology; may miss low-abundance or tissue-specific miRNAs |
| Microarray analysis | Hybridization-based detection of miRNA expression | High-throughput, suitable for known miRNA profiling | Limited to known miRNAs; may have lower sensitivity compared to sequencing |
| Quantitative PCR | Experimental validation and quantification | Highly sensitive and specific; suitable for validation | Limited throughput; may not be ideal for large-scale studies |
| **Integrated approaches** | Combining computational and experimental methods | Reduces false positives/ negatives by combining strengths | Requires access to both computational tools and experimental techniques |
| **Functional approaches** | | | |
| Functional genomics | Study functional role through knockdown/ overexpression | Provides insights into miRNA function | Requires manipulation of gene expression; may not directly identify miRNAs |

**Table 2.2.** Comprehensive overview of publicly available databases and computational tools for plant miRNA and target identification, deposition and functional analysis[a].

| Category | Tool/database | Description | Website/access |
|---|---|---|---|
| **miRNA and target identification** | | | |
| Database | miRBase | Repository for annotated miRNA sequences | miRBase (www.mirbase.org) |
| Database | Plant small RNA Target Analysis Server (psRNATarget) | Predicts miRNA targets in plant genomes | psRNATarget (www.zhaolab.org/psRNATarget) |
| Tool | TargetFinder | Identifies miRNA targets using a scoring algorithm | TargetFinder (www.mirdb.org) |
| Tool | psRobot | Integrative tool for plant small RNA target prediction | psRobot (http://omicslab.genetics.ac.cn/psRobot/) |
| **miRNA deposition and retrieval** | | | |
| Database | Gene Expression Omnibus (GEO) | Repository for high-throughput functional genomics data | GEO (www.ncbi.nlm.nih.gov/geo) |
| Database | European Nucleotide Archive (ENA) | Archive for nucleotide sequence data | ENA (www.ebi.ac.uk/ena) |
| Database | Sequence Read Archive (SRA) | Archive for high-throughput sequencing data | SRA (www.ncbi.nlm.nih.gov/sra) |
| **Functional analysis** | | | |
| Tool | AgriGO | Analyses functional significance of gene lists | AgriGO (http://systemsbiology.cau.edu.cn/agriGOv2/) |
| Tool | PANTHER | Classifies genes and proteins based on function | PANTHER (https://pantherdb.org/) |
| Tool | PlantGSEA | Gene Set Enrichment Analysis tailored for plants | PlantGSEA (https://bioinformatics.cau.edu.cn/PlantGSEA/) |

[a]Please note that the availability and features of these tools and databases may change, and it is recommended to check the respective websites for up-to-date information.

## References


Abdelgany, A., Wood, M. and Beeson, D. (2003) Allele-specific silencing of a pathogenic mutant acetylcholine receptor subunit by RNA interference. *Human Molecular Genetics* 12(20), 2637–2644. DOI: 10.1093/hmg/ddg280.

Almeida, R. and Allshire, R.C. (2005) RNA silencing and genome regulation. *Trends in Cell Biology* 15(5), 251–258. DOI: 10.1016/j.tcb.2005.03.006.

Bapat, V.A., Trivedi, P.K., Ghosh, A., Sane, V.A., Ganapathi, T.R. and Nath, P. (2010) Ripening of fleshy fruit: Molecular insight and the role of ethylene. *Biotechnology Advances*, 28(1), 94–107. DOI: 10.1016/j.biotechadv.2009.10.002.

Bartel, D.P. (2004) MicroRNAs: Genomics, biogenesis, mechanism, and function. *Cell* 116(2), 281–297. DOI: 10.1016/s0092-8674(04)00045-5.

Bartel, D.P. and Chen, C.Z. (2004) Micromanagers of gene expression: The potentially widespread influence of metazoan microRNAs. *Nature Reviews Genetics* 5, 396–400. DOI: 10.1038/nrg1328.

Cebrián, V., Yagüe, C., Arruebo, M., Martín-Saavedra, F.M., Santamaría, J. *et al.* (2011) On the role of the colloidal stability of mesoporous silica nanoparticles as gene delivery vectors. *Journal of Nanoparticle Research* 13(9), 4097–4108. DOI: 10.1007/s11051-011-0353-8.

Chiou, T.-J., Aung, K., Lin, S.-I., Wu, C.-C., Chiang, S.-F. *et al.* (2006) Regulation of phosphate homeostasis by microRNA in *Arabidopsis*. *Plant Cell* 18(2), 412–421. DOI: 10.1105/tpc.105.038943.

de Dorlodot, S., Forster, B., Pages, L., Price, A., Tuberosa, R. *et al.* (2007) Root system architecture: Opportunities and constraints for genetic improvement of crops. *Trends in Plant Science* 12(10), 474–481. DOI: 10.1016/j.tplants.2007.08.012.

de Jong M., Wolters-Arts M., Feron R., Mariani C. and Vriezen W.H. (2009) The *Solanum lycopersicum* auxin response factor 7 (SlARF7) regulates auxin signaling during tomato fruit set and development. Plant Journal 57(1), 160–170. DOI: 10.1111/j.1365-313X.2008.03671.x.

Fairbairn, D.J., Cavallaro, A.S., Bernard, M., Mahalinga-Iyer, J., Graham, M.W. *et al.* (2007) Host-delivered RNAi: An effective strategy to silence genes in plant parasitic nematodes. *Planta* 226(6), 1525–1533. DOI: 10.1007/s00425-007-0588-x.

Fire, A., Xu, S., Montgomery, M.K., Kostas, S.A., Driver, S.E. *et al.* (1998) Potent and specific genetic interference by double-stranded RNA in *Caenorhabditis elegans*. *Nature* 391(6669), 806–811. DOI: 10.1038/35888.

Gad, M.A., Li, M.-J., Ahmed, F.K. and Almoammar, H. (2020) Nanomaterials for gene delivery and editing in plants: Challenges and future perspective. In: Abd-Elsalam, K.A. (ed.) *Multifunctional Hybrid Nanomaterials for Sustainable Agri-Food and Ecosystems*. Elsevier, pp. 135–153.

Ghormade, V., Deshpande, M.V. and Paknikar, K.M. (2011) Perspectives for nano-biotechnology enabled protection and nutrition of plants. *Biotechnology Advances* 29(6), 792–803. DOI: 10.1016/j.biotechadv.2011.06.007.

Gönczy, P., Echeverri, C., Oegema, K., Coulson, A., Jones, S.J. *et al.* (2000) Functional genomic analysis of cell division in *C. elegans* using RNAi of genes on chromosome III. *Nature* 408(6810), 331–336. DOI: 10.1038/35042526.

Grewal, S.I.S. and Moazed, D. (2003) Heterochromatin and epigenetic control of gene expression. *Science* 301(5634), 798–802. DOI: 10.1126/science.1086887.

Han, J., Lee, Y., Yeom, K.H., Kim, Y.K., Jin, H. *et al.* (2004) The Drosha – DGCR8 complex in primary microRNA processing. *Genes & Development* 18(24), 3016–3027. DOI: 10.1101/gad.1262504.

Hannon, G.J. (2002) RNA interference. *Nature* 418, 244–251. DOI: 10.1038/418244a.

Haque, A.K.M.N., Tanaka, Y., Sonoda, S. and Nishiguchi, M. (2007) Analysis of transitive RNA silencing after grafting in transgenic plants with the coat protein gene of sweet potato feathery mottle virus. *Plant Molecular Biology* 63(1), 35–47. DOI: 10.1007/s11103-006-9070-6.

Hedden, P. and Phillips, A.L. (2000) Gibberellin metabolism: New insights revealed by the genes. *Trends in Plant Science* 5(12), 523–530. DOI: 10.1016/s1360-1385(00)01790-8.

Hernández, I., Chacón, O., Rodriguez, R., Portieles, R., López, Y. *et al.* (2009) Black shank resistant tobacco by silencing of glutathione S-transferase. *Biochemical and Biophysical Research Communications* 387(2), 300–304. DOI: 10.1016/j.bbrc.2009.07.003.

Huang, G., Allen, R., Davis, E.L., Baum, T.J. and Hussey, R.S. (2006) Engineering broad root-knot resistance in transgenic plants by RNAi silencing of a conserved and essential root-knot nematode parasitism gene. *Proceedings of the National Academy of Sciences USA* 103:14302–14306. DOI: 10.1073/pnas.0604698103.

Jian, X., Zhang, L., Li, G., Zhang, L., Wang, X. *et al.* (2010) Identification of novel stress-regulated micro RNAs from *Oryza sativa* L. *Genomics* 95, 47–55. DOI: 10.1016/j.ygeno.2009.08.017.

Jones-Rhoades, M.W. and Bartel, D.P. (2004) Computational identification of plant microRNAs and their targets, including a stress-induced miRNA. *Molecular Cell* 14, 787–799. DOI: 10.1016/j.molcel.2004.05.027.

Kim, D. and Rossi, J. (2008) RNAi mechanisms and applications. *Biotechniques* 44(5), 613–616. DOI: 10.2144/000112792.

Kusaba, M. (2004) RNA interference in crop plants. *Current Opinion in Biotechnology* 15, 139–143. DOI: 10.1016/j.copbio.2004.02.004.

Lee, Y., Hur, I., Park, S.Y., Kim, Y.K., Suh, M.R. *et al.* (2006) The role of PACT in the RNA silencing pathway. *EMBO Journal* 25, 522–532. DOI: 10.1038/sj.emboj.7600942.

Lewis, R.S., Jack, A.M., Morris, J.W., Robert, V.J., Gavilano, L.B. et al. (2008). RNA interference (RNAi)-induced suppression of nicotine demethylase activity reduces levels of a key carcinogen in cured tobacco leaves. *Plant Biotechnology Journal* 6(4), 346–354. DOI: 10.1111/j.1467-7652.2008.00324.x.

Li, D.-H., Liu, H., Yang, Y.-L., Zhen, P.-P. and Liang, J.-S. (2009) Down-regulated expression of RACK1 gene by RNA interference enhances drought tolerance in rice. *Rice Science* 16(1), 14–20. DOI: 10.1016/S1672-6308(08)60051-7.

Liu, J., Carmell, M.A., Rivas, F.V., Marsden, C.G., Thomson, J.M. *et al.* (2004) Argonaute2 is the catalytic engine of mammalian RNAi. *Science* 305, 1437–1441. DOI: 10.1126/science.1102513.

Liu, Q., Singh, S. P. and Green, A.G. (2002) High-stearic and high-oleic cottonseed oils produced by hairpin RNA-mediated post-transcriptional gene silencing. *Plant Physiology* 129(4), 1732–1743. DOI: 10.1104/pp.001933.

Mansoor, S., Amin, I., Hussain, M., Zafar, Y. and Briddon, R.W. (2006) Engineering novel traits in plants through RNA interference. *Trends in Plant Science Journal* 11, 559–565. DOI: 10.1016/j.tplants.2006.09.010.

Missiou, A., Kalantidis, K., Boutla, A., Tzortzakaki, S., Tabler, *et al.* (2004) Generation of transgenic potato plants highly resistant to potato virus Y (PVY) through RNA silencing. *Molecular Breeding* 14, 185–197. DOI: 10.1023/B:MOLB.0000038006.32812.52.

Molesini, B., Pandolfini, T., Rotino, G.L., Dani, V. and Spena, A. (2009) *Aucsia* gene silencing causes parthenocarpic fruit development in tomato. *Plant Physiology* 149(1), 534–548. DOI: 10.1104/pp.108.131367.

Mousses, S., Caplen, N.J., Cornelison, R., Weaver, D., Basik, M. *et al.* (2003) RNAi microarray analysis in cultured mammalian cells. *Genome Research* 13(10), 2341–2347. DOI: 10.1101/gr.1478703.

Mujtaba, M., Wang, D., Carvalho, L.B., Oliveira, J.L., Espirito Santo Pereira, A. do *et al.* (2021) Nanocarrier-mediated delivery of miRNA, RNAi, and CRISPR-Cas for plant protection: current trends and future directions. *ACS Agricultural Science & Technology* 1(5), 417–435. DOI: 10.1021/acsagscitech.1c00146.

Napoli, C., Lemieux, C. and Jorgensen, R. (1990) Introduction of a chimeric chalcone synthase gene into petunia results in reversible cosuppression of homologous genes in trans. *Plant Cell* 2, 279–289. DOI: 10.1105/tpc.2.4.279.

Navarro, L., Dunoyer, P., Jay, F., Arnold, B., Dharmasiri, N. *et al.* (2006) A plant miRNA contributes to antibacterial resistance by repressing auxin signaling. *Science* 312, 436–439. DOI: 10.1126/science.1126088.

Pardo, J.M. (2010) Biotechnology of water and salinity stress tolerance. *Current Opinion in Biotechnology* 21, 185–196. DOI: 10.1016/j.copbio.2010.02.005.

Qiao, F., Yang, Q., Wang, C.L., Fan, Y.L., Wu, X.F. *et al.* (2007) Modification of plant height via RNAi suppression of *OsGA20ox2* gene in rice. *Euphytica* 158, 35–45. DOI: 10.1007/s10681-007-9422-6.

Reyes, J.L. and Chua, N.H. (2007) ABA induction of miR159 controls transcript levels of two MYB factors during Arabidopsis seed germination. *Plant Journal* 49, 592–606. DOI: 10.1111/j.1365-313X.2006.02980.x.

Romano, N. and Macino, G. (1992) Quelling: Transient inactivation of gene expression in *Neurospora crassa* by transformation with homologous sequences. *Molecular Microbiology* 22, 3343–3353. DOI: 10.1111/j.1365-2958.1992.tb02202.x.

Schijlen, E.G., de Vos, C.R., Martens, S., Jonker, H.H., Rosin, F.M., Molthoff, J.W. *et al.* (2007) RNA interference silencing of chalcone synthase, the first step in the flavonoid biosynthesis pathway, leads to parthenocarpic tomato fruits. *Plant Physiology* 144(3), 1520–1530. DOI: 10.1104/pp.107.100305.

Schwarz, D.S., Hutvágner, G., Du, T., Xu, Z., Aronin, N. *et al.* (2003) Asymmetry in the assembly of the RNAi enzyme complex. *Cell* 115, 199–208. DOI: 10.1016/s0092-8674(03)00759-1.

Schwind, N., Zwiebel, M., Itaya, A., Ding, B., Wang, M.-B. *et al.* (2009) RNAi-mediated resistance to potato spindle tuber viroid in transgenic tomato expressing a viroid hairpin RNA construct. *Molecular Plant Pathology* 10(4), 459–469. DOI: 10.1111/j.1364-3703.2009.00546.x.

Shao, Y., Zhu, H.L., Tian, H.Q., Wang, X.G., Lin, X.J. *et al.* (2008) Virus-induced gene silencing in plant species. *Russian Journal of Plant Physiology* 55(2), 168–174. DOI: 10.1134/S1021443708020027.

Shepherd, D.N., Martin, D.P. and Thomson, J.A. (2009) Transgenic strategies for developing crops resistant to geminivirus. *Plant Sci* 176, 1–11. DOI: 10.1016/j.plantsci.2008.08.011.

Sindhu, A.S., Maier, T.R., Mitchum, M.G., Hussey, R.S., Davis, E.L. *et al.* (2009) Effective and specific in planta RNAi in cyst nematodes: Expression interference of four parasitism genes reduces parasitic success. *Journal of Experimental Botany* 60(1), 315–324. DOI: 10.1093/jxb/ern289.

Sunkar, R. and Zhu, J.K. (2004) Novel and stress-regulated microRNAs and other small RNAs from Arabidopsis. *Plant Cell* 16, 2001–2019. DOI: 10.1105/tpc.104.022830.

Tang, G., Reinhart, B.J., Bartel, D.P. and Zamore, P. D (2003) A biochemical framework for RNA silencing in plants. *Genes & Development* 17(1), 49–63. DOI: 10.1101/gad.1048103.

Thomas, M. and Klibanov, A. (2003) Non-viral gene therapy: Polycation-mediated DNA delivery. *Applied Microbiology and Biotechnology* 62(1), 27–34. DOI: 10.1007/s00253-003-1321-8.

Tucker, G. (2003) Nutritional enhancement of plants. *Current Opinion in Biotechnology* 14(2), 221–225. DOI: 10.1016/S0958-1669(03)00031-4.

Vanderschuren, H., Alder, A., Zhang, P. and Gruissem, W. (2009) Dose-dependent RNAi-mediated geminivirus resistance in the tropical root crop cassava. *Plant Molecular Biology* 64, 549–557. DOI: 10.1007/s11103-009-9472-3.

Van Eck, J., Conlin, B., Garvin, D.F., Mason, H., Navarre, D.A. *et al.* (2007) Enhancing beta-carotene content in potato by RNAi-mediated silencing of the beta-carotene hydroxylase gene. *American Journal of Potato Research* 84: 331–342. DOI: 10.1007/BF02986245.

Vaucheret, H. (2008) Plant ARGONAUTES. *Trends in Plant Science* 13, 350–358. DOI: 10.1016/j.tplants.2008.04.007.

Vrebalov, J., Pan, I.L., Arroyo, A.J., McQuinn, R., Chung, M. *et al.* (2009) Fleshy fruit expansion and ripening are regulated by the Tomato *SHATTERPROOF* gene *TAGL1*. *Plant Cell* 21(10), 3041–3062. DOI: 10.1105/tpc.109.066936.

Wang, Y. and Li, J. (2006) Genes controlling plant architecture. *Current Opinion in Biotechnology* 17(2), 123–129. DOI: 10.1016/j.copbio.2006.02.004.

Wang, Y. and Li, J. (2008) Molecular basis of plant architecture. *Annual Review of Plant Biology* 59(1), 253–279. DOI: 10.1146/annurev.arplant.59.032607.092902.

Waterhouse, P.M. and Graham, M.W. (1998) Virus resistance and gene silencing in plants can be induced by simultaneous expression of sense and antisense RNA. *Proceedings of the National Academy of Sciences USA* 95, 13959–13964. DOI: 10.1073/pnas.95.23.13959.

Waterhouse, P.M., Ming-Bo, W. and Lough, T. (2001) Gene silencing as an adaptive defence against viruses. *Nature* 411, 834–842. DOI: 10.1038/35081168.

Xiong, A.-S., Yao, Q.-H., Peng, R.-H., Li, X., Han, P.-L. *et al.* (2005) Different effects on ACC oxidase gene silencing triggered by RNA interference in transgenic tomato. *Plant Cell Reports* 23, 639–646. DOI: 10.1007/s00299-004-0887-7.

Xu, M., Zhu, L., Shou, H. and Wu, P. (2005) *A PIN1* family gene, *OsPIN1*, involved in auxin-dependent adventitious root emergence and tillering in rice. *Plant and Cell Physiology* 46(10), 1674–1681. DOI: 10.1093/pcp/pci183.

Zamore, P.D., Tuschl, T., Sharp, P.A. and Bartel, D.P. (2000) RNAi: Double-stranded RNA directs the ATP-dependent cleavage of mRNA at 21 to 23 nucleotide intervals. *Cell* 101, 25–33. DOI: 10.1016/S0092-8674(00)80620-0.

Zhao, B.T., Liang, R.Q., Li, W., Xiao, H.S., Lin, H.X. *et al.* (2007) Identification of drought-induced microRNAs in rice. *Biochemical and Biophysical Research Communications* 354(2), 585–590. DOI: 10.1016/j.bbrc.2007.01.022.

Zhou, H.L., He, S.J., Cao, Y.R., Chen, T., Du, B.X. *et al.* (2006) OsGLU1, a putative membrane-bound endo-1, 4-b-d-glucanase from rice, affects plant internode elongation. *Plant Molecular Biology* 60(1), 137–151. DOI: 10.1007/s11103-005-2972-x.

# 3 Gene Pyramiding to Increase the Sustainability of Crops Under Biotic and Abiotic Stresses

**Vishal B. Mhetre[1]*, Vishal Dinkar[2]*, Hatkari Vittal[3], Bhargav Kiran Sirasapalli[4], Shikha Jain[3], Chaithra T.S.[5] and Manjunath S. Patil[6]**

*[1]Defence Institute of High Altitude Research, Leh, UT of Ladakh – 194101, India; [2]ICAR-Central Institute of Temperate Horticulture, Srinagar – 191132, J&K, India; [3]ICAR-Indian Agricultural Research Institute, New Delhi – 110012, Delhi, India; [4]Centurion University of Technology and Management, R. Sitapur – 761211, Odisha, India; [5]ICAR-Indian Institute of Soil and Water Conservation, Research Centre, Kota – 324002, Rajasthan, India; [6]Zonal Agriculture Research Station (MPKV, Rahuri), Solapur – 413002, Maharashtra, India*

**Abstract**

Sustainable agricultural production faces numerous ecological threats, like high temperatures, drought, excess salinity, diseases and insect pests. The agricultural productivity in many countries faces significant threats due to such environmental challenges. The increasing global food demand cannot be met sustainably using only conventional crop-breeding methods. However, the evolution of molecular genetics and their related technologies presents promising avenues for identifying and enhancing new crop species. Conventional crop-breeding methods are not enough to address the complexities of modern agriculture. Developing effective and smart crop-breeding techniques is essential to enhance several agronomic and nutritional characteristics of crops at the same time. These advanced techniques will help enhance the quality and quantity of crops, making them more nutritious and better suited to meet the demands of an ever-expanding global population. Advancements in molecular genetics have opened new possibilities in crop improvement. Techniques such as gene pyramiding through marker-assisted selection (MAS), marker-assisted backcrossing (MABC), marker-assisted recurrent selection (MARS), and various omics approaches are being utilized to produce improved crop varieties. These game-changing approaches are helping us address food security and sustainability challenges. These methodologies have expedited the creation of resilient and tolerant crop lines with remarkable precision, contributing to agricultural sustainability. However, gene stacking is an underutilized strategy with significant potential for strengthening crops against biotic and abiotic stresses. By improving crop quality and tolerance to these stresses, gene stacking can play a crucial role in achieving sustainable advancements in crop production.

**Keywords:** sustainability, drought, gene stacking, marker, MAS, diseases, pests, salinity, climate change, yield

*Corresponding authors: vishaldinkar1991@gmail.com and vishalmhetre8@gmail.com

 *Plant Breeding Technology: Future Trends and Challenges* (ed. Zeba Khan)
DOI: 10.1079/9781800626638.0003

**Abbreviations**

APR – Adult-plant resistance, BLB – Bacterial leaf blight, CRISPR – clustered regularly interspaced short palindromic repeats, CTF – Cold tolerance at the fertilization stage, DP – Donor parent, GM – Genetically modified, GWAS – Genome-wide association analysis, HR – Hypersensitive response, MABB – Marker-assisted backcross breeding, MABC – Marker-assisted backcrossing, MARS – Marker-assisted recurrent selection, MAS – Marker-assisted selection, MTA – Marker-trait association, NIL – Near isogenic line, PPNs – Plant-parasitic nematodes, PR – Pathogenesis-related, QTL – Quantitative trait locus, R-gene – Resistance gene, RIL – Recombinant inbred line, ROS – Reactive oxygen species, RP – Recurrent parent, SMV – Soybean mosaic virus, TF – Transcription factor, TOSPO – Tomato spotted wilt virus, TYLCV – Tomato yellow leaf curl virus.

## 3.1 Introduction

Defining crop sustainability, as with every other sustainability concept, is challenging. Nevertheless, it is a common agreement that crop sustainability should at least address the three fundamental pillars of sustainable development, by simultaneously appraising environmental, economic and social issues related to agricultural practices (Pham and Smith, 2014). However, the sustainability assessment of agricultural practices generally can be challenging since it involves many case-specific variables. In a sustainable crop production system, it is essential to consider, not just the crops themselves, but also the agricultural ecosystem in which they are grown, and the other ecosystems that may be impacted by agricultural production practices, whether directly or indirectly (Lewandowski *et al.*, 1999). This means taking a holistic approach to agriculture that considers the interconnections between various ecosystem components, such as soil health, water quality, biodiversity and climate. By doing so, we can ensure that our agricultural practices are productive, profitable, environmentally responsible and socially beneficial in the long run.

Stress is the after-effect of environmental and external conditions that affect crop plants' growth, development and yielding ability (Verma *et al.*, 2013). When subjected to various environmental stresses, crop plants exhibit altered physiology and metabolism, ultimately compromising crop yield and productivity. Both abiotic and biotic stresses, either directly or indirectly, impact different developmental stages: seed germination, seedling establishment, tillering, photosynthesis, photosynthate mobilization, flowering, fertility, pollination and fruiting/grain, in grain as well as in horticultural crops (Kushwaha *et al.*, 2021; Adhikari *et al.*, 2023). This perturbs agricultural production and impacts ecosystems and human health, often requiring management strategies to mitigate their effects. However, if a particular crop plant is tolerant to a specific stress condition, it acclimates to the subjected stress conditions in a time-dependent manner (Verma *et al.*, 2013). This acclimatization allows the crop plant to maintain its growth and productivity even under stressful conditions. Therefore, it is the prime objective to engineer agricultural crops with smart genetic make-up to readily adapt or be tolerant to various stressors to ensure sustainable crop production (Dinkar *et al.*, 2020; Sun *et al.*, 2022).

Abiotic stressors are inflicted by non-living environmental factors such as extreme temperatures, moisture deficit, floods, salinity, nutrient deficit, heavy metal contamination, pollution or physical damage, which can directly impact the physiological processes of an organism, affecting its growth, development and overall health (Dresselhaus and Hückelhoven, 2018; Mhetre *et al.*, 2022; Dinkar *et al.*, 2024a). For example, prolonged exposure to high temperatures can lead to heat stress in plants, causing wilting, reduced photosynthesis and the deterioration of quality characteristics (Mhetre *et al.*, 2022), as well as limiting the crop production by altering developmental growth, metabolism and productivity (Khan and Shahwar, 2020). As we move towards a future of global sustainability, it is important to recognize the potential impact of abiotic threats and their intensification due to climate change and population growth (Kopecká *et al.*, 2023). Predicting the exact ramifications of these alterations on agricultural systems is exceedingly challenging, as it hinges on numerous parameters often overlooked in predictive models. While certain forecasts indicate a possible balance between positive and negative effects on crop production in the medium term, numerous studies assert that negative consequences are expected to prevail in the long run (Bulgari *et al.*, 2019).

Biotic stress, on the other hand, is caused by living organisms or their products, such as pests, pathogens, parasites and competing plants, which can directly damage an organism's tissues, disrupt its metabolism or compete for essential resources, such as nutrients or sunlight (Dresselhaus and Hückelhoven, 2018). For instance, plant pathogens like fungi or bacteria can infect crops, leading to diseases that reduce yield and quality. Cereals, pulses and a variety of horticultural crops, such as fruits and vegetables, face a diverse array of potential pests and are susceptible to their threats (Nicoletto *et al.*, 2019). Fungal, bacterial and viral phytopathogens stand out as primary contributors to the decline in both the quality and quantity of horticultural produce. The existence of these biotic stresses poses a significant challenge for plants, yet understanding their precise impact on plant response remains elusive (Nicoletto *et al.*, 2019). Gaining a thorough understanding of plants, pests and their interactions presents a significant challenge. Furthermore, accurately quantifying damage is a complex process, especially when considering factors such as sucking pests, pathogens and weeds (Peterson and Higley, 2001).

Gene pyramiding is a crop-breeding technique where multiple desirable quantitative trait loci (QTLs)/candidate genes from multiple sources/parents are combined into a single genotype (Ye and Smith, 2009). Gene pyramiding finds its roots in the traditional practice of employing insecticide mixtures to widen the range of insects targeted during a single spraying session (Muthurajan and Balasubramanian, 2010). Gene pyramiding is an extension of backcross breeding, which has been traditionally used for breeding disease-resistant cultivars. Mostly, the pyramiding technique is employed for introgression resistance against different races of a single pathogen or multiple pathogens simultaneously, conferred by a single major gene against each race/pathogen, thereby making the resistance viable for a very short period (Kottapalli *et al.*, 2010; Steiner *et al.*, 2019). However, the same concept can be employed to achieve multiple stress-tolerance crops, as evident from numerous studies on the pyramiding of desirable genes/QTLs conferring multiple stress tolerance into a single genotype (Muthurajan and Balasubramanian, 2010; Suresh and Malathi, 2013; Ruengphayak *et al.*, 2015; Bai *et al.*, 2018; Rana *et al.*, 2019; Pancaldi and Trindade, 2020; Kushwaha *et al.*, 2021). Nevertheless, the utilization of this gene pyramiding technique is not up to the mark for breeding against biotic stresses and quality improvement in many commercial crops (Balconi *et al.*, 2012; Zhu *et al.*, 2012; Piquerez *et al.*, 2014; Rana *et al.*, 2019). As described by Ye and Smith (2009), in broad terms, gene pyramiding aims to achieve four major goals: (i) to enhance the performance of traits by combining compatible genes; (ii) to address deficiencies by introducing genes from multiple sources; (iii) to enhance the varietal longevity against diseases; and (iv) to expand the genetic diversity of cultivated varieties. While there is ongoing debate about the feasibility of the third objective, evidence strongly supports the attainment of the others (Ye and Smith, 2009).

## 3.2 Gene Pyramiding: Concept and Techniques

### 3.2.1 Concept and technical overview

Gene pyramiding is a valuable crop-breeding technique that involves stacking multiple genes into a single plant, thereby creating novel genetic combinations that can confer desirable traits such as disease resistance, improved yield, and tolerance to environmental stress. This technique can be applied in both traditional and advanced molecular breeding methods, making it a versatile tool for crop improvement (Dormatey *et al.*, 2020). By introducing new genetic lines through gene pyramiding, plant breeders can develop crops that are better suited to the needs of farmers and consumers alike. Adding a single resistant gene in newly bred varieties forms the basis of many crop-breeding techniques for tolerance against stressors (Kottapalli *et al.*, 2010; Steiner *et al.*, 2019). As a result, there is now a focus on creating genotypes resistant to multiple stresses by stacking multiple genes from different sources (Suresh and Malathi, 2013; Rana *et al.*, 2019; Pancaldi and Trindade, 2020). The idea of gene pyramiding was introduced by Nelson (1978) with the aim of creating varieties with enduring disease resistance by incorporating multiple distinct major genes. Since it is an introgression of genes coding for more than two traits, the same process can be termed as 'multi-trait introgression' (Rana *et al.*, 2019). The concept of integrating multiple QTLs/ genes and studying its consequences has been put forward in several agriculturally important plants, including rice, barley, wheat, mustard, pearl millet, soybean, maize, groundnut, sorghum and many more.

The whole gene pyramiding process has two distinct phases: (i) 'pedigree' development, where all targeted loci are accumulated in a 'root genotype'; and (ii) 'genetic fixation', where targeted loci are fixed into the homozygous state to develop an 'ideal genotype'. Gene pyramiding can be achieved by three backcrossing schemes: (i) 'stepwise transfer' where pedigree and fixation for each gene are completed in a stepwise manner in a single backcrossing programme. This takes the longest period to complete; (ii) 'simultaneous transfer' where root genotype is obtained by including each donor in a stepwise manner and then followed by fixation of each gene simultaneously; and (iii) 'simultaneous and stepwise transfer' where separate backcrossing programmes are run simultaneously for each gene and fixation of the genes is obtained by one or two backcrossings. These fixed lines are then used once more, for root genotype construction and final fixation to obtain an ideal genotype with all the targeted loci. This is the most commonly used method of gene pyramiding and has the least chance of missing target genes during backcrossing. Fig. 3.1 illustrates a hypothetical model for enhancing a crop variety's resistance to biotic and abiotic stress factors through gene pyramiding.

### 3.2.2 Conventional breeding techniques

Traditional breeding methods often use the well-established technique of backcross breeding. Through this method, desirable traits from a parent plant are gradually introduced into the genetic makeup of a new plant by repeatedly crossing it with one of its parents (Kumar *et al.*, 2023). This process creates new plant varieties that inherit the best traits of their parents while maintaining genetic consistency. Recurrent selection, pedigree breeding and backcrossing are essential methods for transferring inherited traits and required genes from donor parents to recipient lines (Dormatey *et al.*, 2020). This meticulous approach not only facilitates the infusion of desired traits but also underscores the crucial role of selection in pinpointing traits of interest. The backcross breeding method is also employed in the process of pyramiding resistance genes (R-genes) (Ramalingam *et al.*, 2020; Bharadwaj *et al.*, 2022; Pradhan *et al.*, 2023). Pedigree breeding is a systematic approach to breeding plants that involves tracking and analysing the ancestry of individual plants to select desirable parent plants for subsequent generations (Koutsika-Sotiriou *et al.*, 2010; Pikunova *et al.*, 2022). By selectively breeding plants with desirable pedigrees, breeders aim to develop new varieties with improved performance and specific characteristics, particularly in self-pollinated plants. Recurrent selection focuses on continuously improving populations of plants over multiple generations and subjecting a population to selection cycles, where plants with desirable traits are chosen as parents for the next generation's breeding pool (Hallauer and Darrah, 1985; Yusop *et al.*, 2020; Chen *et al.*, 2022). This process maintains genetic diversity while gradually enhancing desired traits such as yield, disease resistance or adaptation to specific environmental conditions. Recurrent selection allows breeders to continually refine plant populations, ensuring they remain adaptable and productive in evolving agricultural landscapes.

### 3.2.3 Molecular breeding techniques

Although the phenotypic recurrent selection is widely recognized as a valuable strategy for enhancing polygenic traits by enriching the frequency of favourable alleles associated with diverse economic attributes and preserving genetic diversity within heterozygous populations, its efficacy is often limited (Kushwaha *et al.*, 2021). This limitation stems from the susceptibility of phenotypic selection to environmental influences and the time-consuming nature of genotypic selection, which typically requires at least two to three crop seasons to complete a single selection cycle. Conversely, molecular

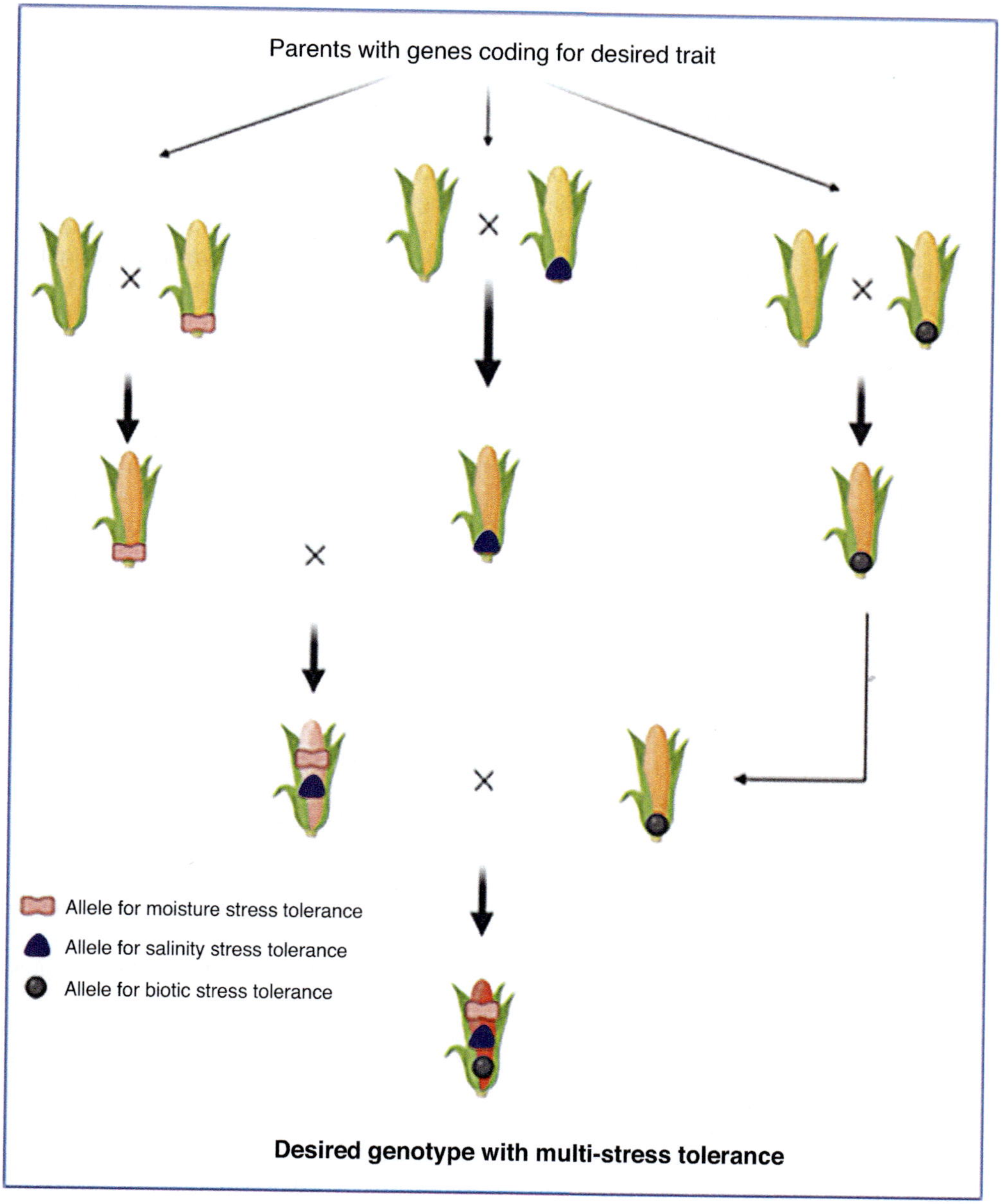


**Fig. 3.1.** Diagrammatic representation of gene pyramiding for sustainable crop development.

marker technology presents an opportunity to expedite this process and enhance selection efficiency, as it is independent of developmental stages, inheritance patterns and environmental factors, and is equally effective for above and below-ground plant parts. Marker-assisted selection (MAS) techniques have reduced the time required to develop new resistant crop varieties. MAS is an indirect selection strategy where the inheritance of a target gene is traced by genomic regions termed molecular markers, which are closely linked/co-segregated with it (Kushwaha *et al.*, 2021). Thus, precise dissection of underlying polymorphic sites in the genome influencing the trait variance is a prerequisite before leveraging DNA-based markers in the selection process. Identifying a genotype, selecting parents based on thorough diversity and purity assessments, and conducting a comprehensive evaluation of hybrid progeny for their vigour, constitute key stages for marker-assisted selection (Collard and Mackill, 2008). Typically, a mapping population/panel is screened with different types of DNA-based markers (genotyping) and phenotypic measurements taken (phenotyping). With the help of sophisticated software, significantly linked/associated markers are identified, which are supposed to boost breeding efficiency. For successful MAS, robust marker – trait associations (MTAs) are needed, which has led to the evolution of several techniques, such as marker-assisted backcrossing (MABC), marker-assisted recurrent selection (MARS), genomic selection (GS) and genome-wide association analysis (GWAS), which has proven successful in enhancing stress resistance and improving quality traits in numerous agriculturally significant crops, such as wheat (Sun *et al.*, 2020; Pandurangan *et al.*, 2021; Mallick *et al.*, 2022a,b; Song *et al.*, 2023), rice (Gouda *et al.*, 2020; Haque *et al.*, 2021; Zheng *et al.*, 2021) and cotton (Lopes *et al.*, 2020; Kushanov *et al.*, 2021). Marker-assisted selection, thus facilitates systematic improvements in breeding cycles, minimizes reliance on traditional phenotypic assessments and guarantees accurate genetic alterations (Shahwar *et al.*, 2024).

#### 3.2.3.1 *Marker-assisted backcrossing (MABC)*

MABC stands out as a precise and highly efficient approach for transferring a single locus governing a trait of interest while preserving the recurrent parent's fundamental attributes (Collard and Mackill, 2008). MABC has demonstrated effectiveness, particularly with genes/QTLs explaining substantially high phenotypic variations. The primary goal of marker-assisted backcross breeding (MABB) is to introduce a specific gene of interest from an agronomically inferior source (donor parent (DP)) into a selected breeding line (recurrent parent (RP)) (Hasan *et al.*, 2015). The desired outcome is to develop an improved line that predominantly inherits the targeted gene from the DP while retaining the genetic background of the RP throughout the rest of the genome. MABC entails the use of molecular markers for the transfer and selection of the specific loci (foreground selection), accelerated recovery of the RP genome during backcrossing (background selection), and reducing the length of the genomic regions from the DP flanking the target loci and introgressed across the genome (recombinant selection) (Hospital and Charcosset, 1997; Hospital, 2011; Kumar *et al.*, 2023). With adherence to selection criteria, marker-assisted gene pyramiding holds the potential to yield resilient and environmentally sustainable enhancements in crop varieties (Dormatey *et al.*, 2020; Kushwaha *et al.*, 2021). This comprehensive approach not only ensures durability in crop improvement but also fosters long-term agricultural sustainability by reducing reliance on agri-inputs such as fertilizers and pesticides (Panigrahi *et al.*, 2013; Jiang *et al.*, 2016; Rana *et al.*, 2019; Mallick *et al.*, 2021).

#### 3.2.3.2 *Marker-assisted recurrent selection (MARS)*

This term describes a molecular breeding technique that uses molecular markers to identify and select multiple loci associated with complex traits simultaneously. The approach aims to assemble the most desirable genotype by utilizing genomic information within a population or among genetically related populations (Ribaut *et al.*, 2010). MARS facilitates the seamless execution of genotypic selection and controlled intercrossing among carefully chosen individuals within the confines of a single crop season, all within the framework of a singular selection cycle (Jiang *et al.*, 2007). Hence, MARS can potentially augment the effectiveness of recurrent selection processes, expediting the overall procedure, particularly by incorporating numerous beneficial genes or QTLs from diverse origins (Kushwaha *et al.*, 2021). This is achieved through recurrent selection strategies implemented within multi-parent populations. In the context of complex traits like yield, resistance against various stresses, nutrition, etc., MARS has been suggested as a method for the 'forward breeding' of indigenous genes and the simultaneous stacking of multiple QTLs (Ribaut *et al.*, 2010; Gantait *et al.*, 2019; Dixit *et al.*, 2020).

## 3.3 Gene Pyramiding for Biotic Stress Management

### 3.3.1 Menace of biotic stresses

Biotic stresses, such as diseases caused by pathogens, such as fungi, bacteria and viruses; or damage from pests, such as insects or nematodes, pose significant threats to crop yield and quality. Traditional methods of controlling these stresses, such as chemical pesticides, can have adverse environmental and health effects. Moreover, pests and pathogens can evolve to overcome single-gene resistance in crops, rendering them ineffective over time. Numerous ecological challenges threaten sustainable agricultural production, including drought, extreme temperatures, soil salinity, diseases and pest infestations. Traditional breeding methods alone are insufficient to feed the burgeoning population. However, advancements in molecular genetics offer promising avenues for developing resilient crop varieties. MAS facilitates the rapid and accurate development of breeding lines with enhanced tolerance to various stresses and high yields. Despite its potential, gene pyramiding/stacking remains underutilized in several important crops, for improving stress resistance with enhanced durability and effectiveness of resistance by stacking multiple R-genes, better yield and quality traits.

### 3.3.2 Impact of gene pyramiding in disease control

The success of multiple gene introgression relies heavily on the effect size of the gene under introgression, the strength of MTAs (i.e. marker–gene linkage strength), and the selection intensity imposed. Major QTLs and candidate genes with qualitative variation are relatively easy to score phenotypically and, thus, closely linked or genic markers can be developed easily. For example, R-genes conferring vertical resistance are the most commonly pyramided genes in major field crops (Mallick *et al.*, 2021, 2022a,b). The genes encoding phytopathogen molecular pattern receptors (i.e. R-genes), pathogenesis-related (PR) proteins or transcription factors (TFs) involved in pathogen recognition, activation of downstream signal transduction and triggering defence responses such as antimicrobial compound secretion, outbursts of reactive oxygen species (ROS) or the hypersensitive response (HR), can be targeted for creating crop plants with specific to broad-spectrum resistance (Van Loon and Strien, 1999). Genes with race specificity generally involve localized host cell death (i.e. HR) to restrict pathogen spread. Transgenic overexpression of PR-proteins has shown a high resistance level against various pathogens. Transgenics expressing several insect-resistance genes simultaneously can be a quick and economical way to tackle insect pests, e.g. Bt-crops, but their environmental and ecological impacts are always questionable (Kushwaha *et al.*, 2021). Stacking multiple genes offers broad-spectrum resistance against pathogens by incorporating genes with various modes of action. This broadens the range of resistance, protecting against a wider array of biotic stresses. Because pests and pathogens must overcome multiple resistance mechanisms at once, the risk of resistance breakdown is lower than with single-gene approaches.

### 3.3.3 Gene pyramiding in horticultural crops for the management of biotic stresses

Horticultural crops are less hardy, and their economic part is vulnerable to different diseases and pests, which mostly lead to poor pricing in the market. Thus, plant protection is the most important factor with respect to horticultural crops. The following examples demonstrate the profound impact of gene pyramiding on enhancing disease resistance in various horticultural and agronomic crops, thereby contributing to global food security and sustainable agriculture (Fig. 3.2; Table 3.1).

#### *3.3.3.1 Tomato*

In tomatoes, gene pyramiding has been employed to combat diseases such as late blight (*Phytophthora infestans*) and tomato yellow leaf curl virus (TYLCV). By combining multiple R-genes, such as those from wild relatives like *Solanum habrochaites* and *Solanum pimpinellifolium*, breeders have developed tomato varieties with durable resistance to late blight and TYLCV. This approach has contributed to reduced reliance on chemical fungicides and pesticides, promoting sustainable disease management practices in tomato production (Thomson and Pfeiffer, 2010). At the World Vegetable Centre (WorldVeg, previously AVRDC), Taiwan tomato lines with multiple disease tolerance and enhanced nutritional value have been created using MABB (Hanson *et al.*, 2016). Expectedly, these genetic stocks can be valuable breeding/donor material worldwide.

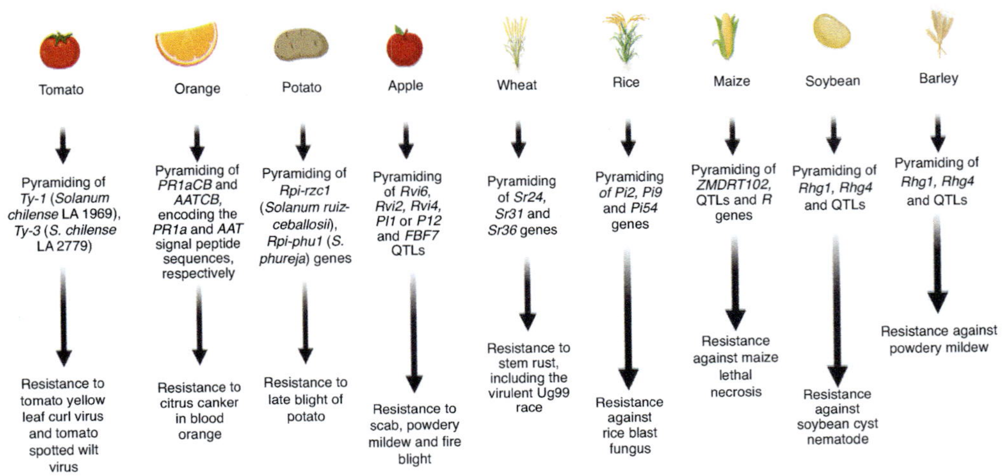


**Fig. 3.2.** Applications of gene pyramiding for sustainable management of crops under various biotic stresses. (Created with Biorender.com. Based on Singh *et al.*, 2011, 2016; Cook *et al.*, 2012, 2014; Consuegra *et al.*, 2015b; Baumgartner *et al.*, 2015; Wang *et al.*, 2016; Acevedo-Garcia *et al.*, 2017; Stefańczyk *et al.*, 2020; Olweny *et al.*, 2020; Peng *et al.*, 2021.)

**Table 3.1.** Gene pyramiding in vegetable crops for biotic stress resistance.

| Pathogen–Host System | Genes/QTL | Donor parent | References |
|---|---|---|---|
| *Meloidogyne hapla* in potato | *RMh-tarand*, *RMh-chca* | *S. tarijense*, *S. chacoense* | Tan *et al.*, 2008 |
| Late blight (*P. infestans*) in potato | *Rpi-mcd1*, *Rpi-ber* | *S. microdontum*, *S. berthaultii* | Tan *et al.*, 2010 |
| TYLCV (Mono- and bipartite) in tomato | *Ty-2*, *Ty-3* | LA3473, CLN2585D, CA4, GC171, TY-172 | Prasanna *et al.*, 2015 |
| TYLCV and TOSPO in tomato | *Ty-1*, *Ty-3*, *Sw-5* | LA2779, LA1969 | Consuegra *et al.*, 2015a |
| Late blight (*P. infestans*), TYLCV, bacterial wilt (*R. solanacearum*), wilt (*F. oxysporum* f. sp. *lycopersici* race 2), grey leaf spot (*Stemphylium* spp.) and TMV in tomato | *Bwr-12*, *Ty-2*, *Ty-3*, *Tm2*, *Ph-2*, *Ph-3*, *Sm* | CLN2777G, G2-6-20-15B, LBR-11 | Thomson and Pfeiffer, 2010; Hanson *et al.*, 2016 |

TYLCV, Tomato Yellow leaf curl virus; TOSPO, Tomato Spotted Wilt Virus; TMV, Tobacco Mosaic Virus

#### *3.3.3.2 Potato*

Gene pyramiding has been crucial in developing potato varieties resistant to late blight (*P. infestans*). Incorporating multiple R-genes, such as those from wild potato species like *Solanum demissum* and *Solanum bulbocastanum*, has resulted in varieties with enhanced and durable resistance to late blight. Genes such as *Rpi-vnt1.1*, *Rpi-blb1* and *Rpi-blb2* have been pyramided to confer durable resistance (Vleeshouwers *et al.*, 2008). These resistant varieties offer farmers an effective and environmentally friendly means of controlling late blight, reducing yield losses and chemical inputs (Gebhardt and Valkonen, 2001).

#### *3.3.3.3 Apple*

The *Venturia inaequalis* cause of apple scab is a fungus that is a major problem for apple production worldwide. Through gene pyramiding, apple breeders have developed cultivars with enhanced resistance to apple scab by incorporating multiple R-genes from wild apple relatives (Gessler *et al.*, 2006).

#### *3.3.3.4 Citrus*

Citrus canker, caused by the bacterium *Xanthomonas citri* subsp. *citri*, poses a significant threat to citrus production worldwide. Gene pyramiding has been a better strategy to enhance citrus resistance against canker by incorporating multiple R-genes from citrus and related species. While progress in developing resistant citrus cultivars through gene pyramiding is ongoing, it holds promise for sustainable disease management in citrus production (Gottwald *et al.*, 2002). By harnessing the power of gene pyramiding, horticultural crops can achieve enhanced and durable resistance to devastating diseases, leading to improved yield stability, reduced reliance on chemical inputs and more sustainable production practices.

### 3.3.4 Gene pyramiding in field crops for the management of biotic stresses

#### *3.3.4.1 Wheat*

Gene pyramiding involving multiple R-genes such as *Sr24*, *Sr31* and *Sr36* has significantly enhanced resistance against the devastating stem rust pathogen (*Puccinia graminis* f. sp. *tritici*), including the virulent *Ug99* race (Singh *et al.*, 2011, 2016) (Fig. 3.2).

#### *3.3.4.2 Rice*

Introgression of multiple R-genes (*Pi2*, *Pi9*, and *Pi54*) has enhanced resistance against the rice blast fungus (*Magnaporthe oryzae*), which is a global threat to rice production (Fukuoka *et al.*, 2009; Wang *et al.*, 2016).

#### *3.3.4.3 Maize*

Pyramiding of multiple genes such as *ZmDRT102* QTLs, and R-genes has conferred resistance against Maize Lethal Necrosis (MLN) caused by the combination of Maize Chlorotic Mottle Virus (MCMV) and Sugarcane Mosaic Virus (SCMV) (Olweny *et al.*, 2020).

#### *3.3.4.4 Soybean*

Pyramiding of R-genes such as *Rhg1*, *Rhg4* and QTLs has enhanced resistance against the soybean cyst nematode, a major yield-limiting factor in soybean production (Cook *et al.*, 2012, 2014).

#### *3.3.4.5 Barley*

Pyramiding of multiple R-genes such as *Mla*, *Mlo* and QTLs has enhanced resistance against powdery mildew (*Blumeria graminis* f. sp. *hordei*), a common fungal disease in barley (Acevedo-Garcia *et al.*, 2017).

### 3.3.5 Gene pyramiding for insect and nematode pest management

Phytophagous insects and plant-parasitic nematodes (PPNs) are specialized pests that directly damage crop plants by feeding and serving as virus carriers. Gene pyramiding can be a sustainable breeding strategy against insect and nematode pests' tolerance. This strategy is particularly important in horticultural crops where pest pressure can be high and where chemical control methods may be undesirable due to environmental concerns or consumer preferences. Cry- proteins are toxins encoded by *cry* genes, originally derived from the bacterium *Bacillus thuringiensis* (Crickmore *et al.*, 1998; Bravo *et al.*, 2007). They are effective against a wide range of insect pests. Some key *cry* genes include:

- *cry1Ab* – effective against lepidopteran pests such as caterpillars.
- *cry2Ab* – also effective against lepidopteran pests, often used in combination with *cry1Ab* for enhanced efficacy.
- *cry3Aa* – effective against coleopteran pests such as rootworms.
- *vip3Aa* – produces a protein toxic to lepidopteran pests and has been used in combination with *cry* toxins to broaden the spectrum of resistance (Estruch *et al.*, 1996).

Likewise, the cowpea trypsin inhibitor (*CpTI*) gene confers resistance against certain lepidopteran pests by inhibiting their ability to digest plant proteins, thus reducing damage. However, these *cry* toxin genes are only available with transgenics and have little practical value in countries like India due to policy and legal issues.

Nematodes are specialized polyphagous pests in crop fields. Several R-genes have been reported against agriculturally important PPNs, but their spectrum of control and effectiveness is always questionable. For example, the *Mi-1.2* gene derived from the wild tomato species *Solanum peruvianum* provides resistance against root-knot nematodes (*Meloidogyne* spp.), which are significant pests in many horticultural crops (Williamson and Kumar, 2006). The *Hs1pro-1* gene provides resistance against cyst nematodes (*Heterodera schachtii*) by inhibiting their ability to establish feeding sites in plant roots (Goverse *et al.*, 2000). Unfortunately, chemical control alone, or along with integrated pest management practices, is widely adopted for PPN management. Alternatively, *in planta*, RNAi-based knockdown of the known susceptibility genes and pathways known as advantageous hubs for PPNs has shown potential in engineering resistance against the nematodes, but it involves a transgenic approach, and their robustness under field conditions is yet to be fully established. Therefore, engineering host resistance through backcross breeding and gene pyramiding can be a cost-effective, eco-friendly, durable and sustainable strategy against PPNs (Fig. 3.2).

### 3.3.6 Gene pyramiding to achieve exemplary qualitative resistance against major diseases

#### *3.3.6.1 Gene pyramiding in rice against blast and BLB disease*

Bacterial leaf blight (BLB) caused by the bacterium *Xanthomonas oryzae* pv. *oryzae* and blast caused by the fungus *Magnaporthe oryzae* are major biotic constraints in rice (Zeigler *et al.*, 1994). Host resistance is the best way to manage such obnoxious diseases. More than 140 R-genes for blast and BLB resistance have been characterized in cultivated and wild rice accessions, and several genes have been fine-mapped and cloned, which serve as sources of breeding host resistance against these pathogens (Hittalmani *et al.*, 2000; Ellur *et al.*, 2016). Marker-assisted gene pyramiding using four BLB R-genes (*Xa21*, *xa5*, *xa4*, and *xa13*) at the International Rice Research Institute (IRRI), Philippines, resulted in a broader spectrum and a better level of resistance against BLB in elite but susceptible cultivars than breeding lines with monogenic resistance (Huang *et al.*, 1997). Similarly, a high level of BLB resistance was realized by Sanchez *et al.* (2000) after introgressing three BLB R-genes into susceptible rice lines using a marker-aided backcrossing procedure. For blast resistance, Hittalmani *et al.* (2000) pyramided three R-genes (*Pi1*, *Piz5* and *Pitai*) into susceptible rice cv. Co39 through MAS. In India, the Indian Agricultural Research Institute (IARI), New Delhi, and the National Rice Research Institute (NRRI), Cuttack, have made significant strides in rice disease resistance breeding through MABB (Singh *et al.*, 2012; Ellur *et al.*, 2016). Two lines bred via gene pyramiding, 'Pusa 1601' (PRR78 + *Xa13* + *Xa21*) and 'Pusa 1609' (PRR78 + *Piz5* + *Pi54*), are improved versions of a popular fertility restorer for basmati type rice cv. PRR78 (Singh *et al.*, 2012). Later, lines 'Pusa 1601' and 'Pusa 1609' were used for combining all four R-genes, and MAS led to the development of line 'Pusa 1790', which carried *Xa13*, *Xa21*, *Piz5* and *Pi54* in PRR78 background. Likewise, pyramiding of *xa13*, *Xa21*, *Pi2* and *Pi54 was used* for conferring BLB and blast resistance in highly demanded basmati rice cv. PB1121 (Ellur *et al.*, 2016).

#### 3.3.6.2 Gene pyramiding in rice against gall midge

The Asian rice gall midge (*Orseolia oryzae*) presents another significant challenge to rice production, particularly in the rice-growing hotspots of Asia. Since no effective chemical control measures are available, resistance breeding can be considered an economically and environmentally viable strategy. Several genetic resources and 11 R-genes have been known to confer resistance against gall midge (Kumar *et al.*, 2005; Leelagud *et al.*, 2020). Different R-gene combinations have been utilized for rice gall midge resistance breeding, e.g. *Gm2* + *Gm6*t (Katiyar *et al.*, 2001), dominant R-genes (*Gm1* + *Gm4*) into popular landraces in Tamil Nadu (Kumaravadivel *et al.*, 2006), and *Gm1* + *Gm4* into popular cultivars such as Tapaswini and Swarna at NRRI, Cuttack. Recently, the Indian Council of Agricultural Research, under the Ministry of Agriculture and Farmers' Welfare, Government of India, has launched an expansive network project aimed at gene pyramiding to cultivate multiple biotic stress-resistant rice varieties. Multiple R-gene introgression through MABB in rice breeders aims to enhance the durability and stability of disease control, ensuring sustainable production. Kumar *et al.* (2017) successfully introgressed BLB (*Xa21*) and gall midge (*Gm4* and *Gm8*) R-genes into restorer line 'RPHR-1005'. The promising near isogenic lines (NILs) with triple gene introgression and >92% RP genome recovery exhibited complete fertility restoration as cv. RPHR-1005. The collaborative effort of IRRI, Philippines, and Indian research institutions underscores the importance of leveraging genetic resources and modern breeding techniques to enhance the resilience of rice crops against a range of biotic stresses, including blast, BLB and gall midge in rice.

#### 3.3.6.3 Gene pyramiding in common wheat against wheat rusts

Three wheat rust diseases, leaf rust (*Puccinia triticina*), stripe rust (*P. striiformis* f. sp. *tritici*) and stem rust (*P. graminis* f. sp. *tritici*) are critical for global wheat production, including India. Genetic resistance has been acknowledged as a fundamental long-term strategy for managing this disease. Various diploid, tetraploid and hexaploid wheat landraces, cultivars, wild species and related species from the Triticeae tribe are documented to harbour rust resistance traits (McIntosh *et al.*, 2013, 2020). Collectively, more than 200 genes/QTLs have been characterized, which serve as a broad genetic base for the management of all three rust pathogens. Since wheat rust pathogens can undergo a sexual cycle, the chances of the emergence of virulent races are always high (Mundt, 2018). Selection and combining one or more *all-stage resistant* (ASR; also, *seedling resistance*) genes with *adult-plant resistance* (APR; also, *partial rusting*) gene(s) is the key to attaining durable rust resistance in wheat. For example, against new *Ug99* stem rust races in Asia/Africa, major R-genes viz., (*Sr22*, *Sr25*, *Sr26*, *Sr33*, *Sr35*, *Sr45* and *Sr50*) are effective when deployed in combination (Mundt, 2018). However, the *Ug99* race-resistant varieties possessing 4 to 5 *APR* genes are known to deliver high resistance levels in the field (Singh *et al.*, 2011). Genomics has revolutionized the marker development and rust resistance breeding in wheat. In recent years, several popular Indian wheat varieties such as HD2967 (*Lr37/Sr38/Yr17* as 2NS/2AS translocation + *LrTrk/YrTrk*), HD2733 (*Lr19/Sr25* + *Lr24/Sr24* + *Yr10*), HD2932 (*Lr19/Sr25* + *Sr26* + *Yr10*), HD2687 (*Lr19/Sr25* + *Lr28*), and VLGehun 738 and VL Gehun 907 (*Yr5* + *Yr10*) have been improved for their rust resistance using MABB (Bhawar *et al.*, 2011; Mallick *et al.*, 2021, 2022a,b). These wheat varieties were improved for two or more rust pathogens by incorporating a combination of seedling and/or APR genes, which has increased the spectrum and durability of the resistance under field conditions. Since the rust pathogen evolves rapidly into new, more virulent races, the chances of epidemics always threaten wheat production. Therefore, the durability of resistance in wheat varieties is of prime importance. For example, once the widely cultivated bread wheat cv. PBW343 became uncultivable due to its susceptibility to yellow and leaf rust races. The cv. Unnat PBW343 is the improved NIL of cv. PBW343, where two leaf rust resistance genes, *Lr37/Yr17/Sr38* and *Lr76/Yr70*, are incorporated by marker-assisted gene pyramiding (Suryavanshi and Singh, 2020). Complex loci such as *Lr34/Yr18/Sr57/Pm38/Sb1/Bdv1*, *Lr46/Yr29/Sr58/Pm39*, *Lr67/Yr46/Sr55/Pm46*, *Sr2/Yr30/Lr27/Pm* and *Sr31/Lr26/Yr9/Pm8* are known APR genes, and confer quantitative and multi-pathogen resistance, which is one of the most common strategies to increase the durability of resistance against wheat rust diseases (McIntosh *et al.*, 2013, 2020). The *Sr2* gene is the most extensively used R-gene against *Ug99* races in the wheat breeding programme at CIMMYT, Mexico (Mundt, 2018). Therefore, new sources of rust resistance are continuously searched, and linked markers for underlying R-genes are developed and introgressed into cultivated varieties in the modern breeding programme (Dinkar *et al.*, 2020; McIntosh *et al.*, 2020; Rani *et al.*, 2020; Nymagoud *et al.*, 2022). These findings emphasize the significance of leveraging genetic diversity from wild relatives to expand the genetic pool of resistance and formulate enduring solutions for addressing wheat rusts on a global scale.

##### *3.3.6.4 Gene pyramiding in soybean against mosaic disease*

Soybean mosaic disease is attributed to infection by the soybean mosaic virus (SMV) (Wang *et al.*, 2022). SMV is a *Potyvirus* that has a limited host range but is the most prevalent and devastating soybean-infecting virus across the world. SMV can infect *Glycine max* (cultivated) and *G. soja* (wild) soybeans (Widyasari *et al.*, 2020). The genetic basis of soybean – SMV interaction is well studied, several resistance sources and R-genes are mapped, and linked molecular makers are available for MAS (Shi *et al.*, 2009; Jin *et al.*, 2022). Currently, four R-genes for G-strains and >10 R-genes for SC-strains of SMV are known (Jin *et al.*, 2022; Wang *et al.*, 2022). Gene pyramiding holds promise in achieving resilient and wide-ranging resistance against multiple strains of SMV. Shi *et al.* (2009) combined three R-genes *Rsv1* and *Rsv3* (from cv. J05) and *Rsv4* (from cv. V94-5152) using linked microsatellite markers, which conferred resistance against seven SMV strains (G1 to G7). However, *Rsv1*, *Rsv3*, *Rsv4*, *GmRmv*, etc. are complex loci and may harbour several R-genes conferring resistance to SMV and other related viruses. Thus, these loci needed to be fine-mapped and cloned to identify the candidate gene and functional marker development and proper deployment based on SMV strain prevalence (Wang *et al.*, 2022).

## 3.4 Gene Pyramiding for Abiotic Stress Management

### 3.4.1 Menace of abiotic stresses

Abiotic stress is any adverse situation or substance from non-living agents, such as atmospheric and edaphic factors, that can hinder a plant's metabolism, growth or development (Álvarez and Acosta-Motos, 2022). Atmospheric factors such as temperature, light, rainfall, wind and edaphic factors such as soil pH, mineral nutrients availability, soil moisture, etc. are essential for better crop growth. These abiotic stressors reduce plant productivity, affecting their morphological, physiological, phenological, biochemical and molecular reactions and leading to abiotic stresses estimated to cause 50–70% losses in crop productivity (Boyer, 1982; Francini, 2019). Furthermore, it is anticipated that these stresses will be exacerbated due to climate change and land degradation. The application of gene pyramiding in abiotic stress management is schematically represented in Fig. 3.3.

#### *3.4.1.1 Moisture stress*

Moisture stress includes water deficit (drought) and waterlogging/flood (submergence), and both are disastrous for global agricultural production and food security. Drought driven by irregularities in hydrological patterns, expanding drought-prone areas due to uneven and uncertain rainfall, water source shortages and other environmental changes is chronic abiotic stress affecting plant growth and development

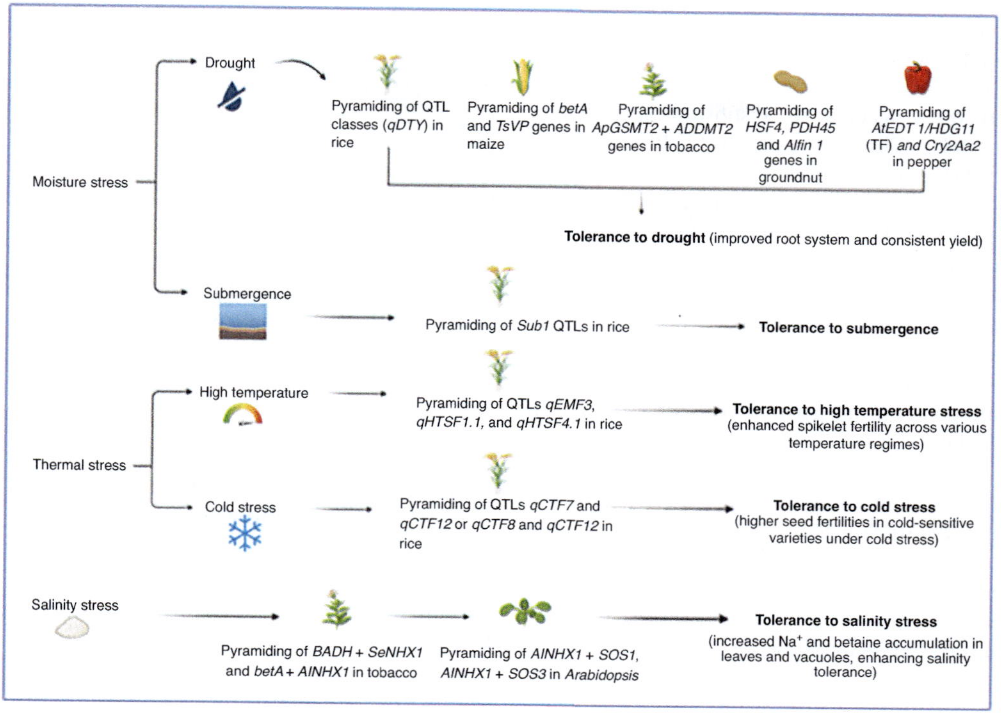


**Fig. 3.3.** Applications of gene pyramiding for sustainable management of abiotic stresses. (Created with Biorender.com. Based on Zhou *et al*., 2008; Duan *et al*., 2009; Yang *et al*., 2009; He *et al*., 2011, 2013; Wei *et al*., 2011; Ma *et al*., 2014; Shinada *et al*., 2014; Zhu *et al*., 2015; Pehlivan *et al*., 2016; Endo *et al*., 2016; Ramu et al., 2016; Vivitha *et al*., 2017; Biradar *et al*., 2018; Kumar *et al*., 2018; Fan *et al*., 2019; Haque *et al*., 2021; Venkatesh *et al*., 2022; Ye *et al*., 2022).

(Tamanna *et al.*, 2023). It accounts for about 70% of potential global crop yield losses (Bashir *et al.*, 2021). Water deficit hampers growth and physiological processes, decreases net photosynthesis, transpiration and leaf water potential, and substantially reduces crop biomass and yield. Drought conditions change cell water potential, decrease the activity of anabolic enzymes associated with photosynthesis and nitrate assimilation, and stimulate the generation of reactive oxygen species. This can lead to membrane damage, lipid peroxidation, DNA impairment and protein degradation (Tamanna *et al.*, 2023).

Likewise, waterlogging also poses significant economic losses by reducing crop yields, diminishing produce quality and increasing production costs due to the need for additional inputs and management practices. Extended flooded conditions lead to oxygen deficiency in the soil, restricting root and other living organism respiration, altering soil chemical properties, and resulting in chlorophyll loss in leaves in the lower canopy, arrested crop growth, and surface root proliferation with the retreat of water levels (An *et al.*, 2022). Under waterlogged/anaerobic conditions, plant transpiration is perturbed, and prolonged depletion in root oxygen supply can cause root death; this also impairs nutrient uptake. Waterlogging during sowing or germination stages generally kills seeds and seedlings.

#### *3.4.1.2 Thermal stress*

Thermal stress, encompassing both heat and cold stress, arises when temperatures deviate from the optimal range for plant growth, resulting in irreversible harm. The detrimental effects of high temperatures on cereals vary depending on the severity, timing and duration of the heat stress. Each degree increase in temperature causes an ~6% reduction in global wheat production globally. It affects plant vegetative stages, resource allocation and reproductive processes, leading to a substantial reduction in yields (Chandra *et al.*, 2023; Bharati *et al.*, 2024). High temperatures can cause reduced seed germination potential, inhibited plant growth, leaf and twig scorching, leaf and fruit discoloration, premature leaf senescence and sunburn, and particularly affect the number of fertile florets, spikes and seed set per plant. Pollen and anthers are more vulnerable to heat than ovules; this results in poor anther dehiscence, low pollen grain germination and poor pollen shedding, especially in tropical climates with intense radiation and high temperatures, which are major limitations to plant development and growth (Mitra *et al.*, 2021). Additionally, Higher temperatures result in rapid cellular injury and cell death, impairing membrane systems, denaturing proteins and inactivating enzymes in the plant body (Kopecká *et al.*, 2023; Bharati *et al.*, 2024). High-temperature stress results in a reduction in yield and quality characteristics of the field as well as horticultural crops (Dresselhaus and Hückelhoven, 2018; Mhetre *et al.*, 2022).

Like heat stress, cold stress also negatively impacts the cellular and developmental processes of plants. This is a serious problem in crop production in temperate and high-altitude regions, which encompasses approximately 15 million hectares constrained annually by low-temperature stress globally. Cold stress includes damage by chilling (0–20°C) and freezing (<0°C), hampering critical biological processes such as impairing chloroplast function, reducing photosynthesis and causing cell death. Freezing temperatures form micro-crystals of ice in the cells, leading to freezing-induced dehydration and electrolyte leakage. Increased leakage of electrolytes due to raised membrane permeability and decreased cell tolerance to temperature changes potentially cause plant death (Jha *et al.*, 2017).

##### *3.4.1.3 Salinity stress*

Salinity stress is severely impacting global crop productivity and food security. Approximately 50% of the world's agricultural land is projected to be affected by 2050 due to rising salinity levels (Jiménez-Mejía *et al.*, 2022). Accumulation of salts in the soil limits the root's ability to absorb water and nutrients from the soil, causing ion cytotoxicity and osmotic stress, negatively affecting microbial diversity in the plant rhizosphere. It affects enzymatic activities, stomatal function and photosynthetic rates, increases the synthesis of reactive oxygen species (ROS), damages cell membranes, lipids, proteins, DNA and RNA, and induces programmed cell death, promoting hypertonic stress due to the accumulation of $Na^+$ and $Cl^-$ ions (Zhao *et al.*, 2020; Jiménez-Mejía *et al.*, 2022).

### 3.4.2 Gene pyramiding for moisture stress tolerance

Moisture stress tolerance is a complex trait, and for better water deficit tolerance, plants resort to less water loss via evapotranspiration, better stomatal and vacuolar activity, accumulation of solutes, maintenance of turgor pressure, better root characteristics, better photosynthetic activity, etc. Therefore, the characterization of genes influencing these component traits is of prime importance prior to breeding for enhanced tolerance. Several gene pyramiding studies have shown significantly enhanced drought tolerance in various model and crop plant species. For instance, enhanced drought tolerance in tobacco and maize genotypes was realized through the pyramiding of glycine betaine synthesis pathway-related genes (*ApGSMT2* + *ApDMT2*), which resulted in lower electrolyte leakage, reduced malondialdehyde content, and increased levels of glycine betaine and chlorophyll (He *et al.*, 2011, 2013). Likewise, Zhu *et al.* (2015) reported enhanced drought tolerance in pepper (*Capsicum annuum* L.) by constitutive co-expression of *Arabidopsis thaliana* homodomain-leucine zipper class TF *Enhanced Drought Tolerance/HOMEODOMAIN GLABROUS11* (*AtEDT1/HDG11*) and *Bacillus thuringiensis* derived *Cry2Aa2*, which led to improved root systems, reduced stomatal density, and increased chlorophyll content, fruit yield and resistance to insects. Introgression of several stress tolerance related genes may lead to yield penalty, but this is not always true (Kumar *et al.*, 2018; Mallick *et al.*, 2022a,b).

Successful introgression of several major *qDTY* class QTLs along with the *Sub1* locus in various rice varieties (Swarna-Sub1, IR64-Sub1, Samba Mahsuri, TDK1-Sub, and MR219) led to the consistent and superior performance of the varieties under drought stress conditions (Kumar *et al.*, 2018; Sandhu *et al.*, 2019a). Co-expression of *eIF4A1* with *NHX1* (a $Na^+/H^+$ antiporter) in sweet potato (Zhang *et al.*, 2019c) and pyramiding of the rice QTL *qDTY* in rice genotypes MR219 and MRQ74 (Shamsudin *et al.*, 2016a,b) also underscored the effectiveness of pyramiding of synergistically acting genes in improved drought tolerance. Under the transgenic approach, Ramu *et al.* (2016) simultaneously expressed regulatory genes such as *Heat Shock TF (HSF4)*, *Pea DNA helicase (PDH45)* and *Alfalfa zinc finger 1 (Alfin 1)* in groundnut, and Venkatesh *et al.* (2022) stacked multiple TFs *viz.*, *MuMYB96*, *MuWRKY3*, and *MuNAC4*, from horse gram (*Macrotyloma uniflorum*) in groundnut and reported enhanced drought tolerance without yield compensation. Likewise, Wei *et al.* (2011) showcased the superiority of transgenic maize lines expressing choline dehydrogenase (*betA* from *Escherichia coli*) and vacuolar $H^+$-pyrophosphatase (*TsVP* from *Thellungiella halophila*), which were better in drought tolerance and showed reduced electrolyte leakage and higher grain yield compared to single gene-expressing lines and/or wild-type plants.

Farmers with small landholdings in South and South-east Asia face a hard challenge, where ~15 million hectares of rice-growing land are vulnerable to excess water conditions like waterlogging and submergence (Luo *et al.*, 2016; Haque *et al.*, 2021). Many rely on local rice landraces, which, while moderately tolerant to submergence, often yield poorly. Conversely, some farmers opt for high-yielding varieties vulnerable to submergence stress, which suffer losses from heavy monsoon rains and periodic flash floods. Thus, the best way to address such issues is to develop a stress-tolerant version of the cultivated/popular varieties acceptable to farmers in the region through MAS (Haque *et al.*, 2021). Primarily, the rice landrace FR13A is considered as a good submergence tolerant donor in the rice breeding programme. The *Sub1* locus from the FR13A-derived line may carry three key genes encoding ethylene-responsive factors (*Sub1A*, *Sub1B*, and *Sub1C*), identified as significant submergence tolerance indicators. Rice varieties *viz.*, IR64-Sub1, Samba Mahsuri-Sub1, Thadokkamq-Sub1, BR11-Sub1, and Swarna-Sub1 have been developed through MABB using the *Sub1* locus (Kumar *et al.*, 2018; Haque *et al.*, 2021). The *Sub1* locus has also been introgressed into various rice varieties such as Hom Mali 821, Pink3, MH725, and Improved Lalat through MABC along with additional genes conferring blast, BLB, and brown plant hopper (BPH) resistance in rice (Korinsak *et al.*, 2016; Luo *et al.*, 2016). These varieties are the saviour of rice production in flooding-prone areas of the world, including India.

### 3.4.3 Gene pyramiding for thermal stress tolerance

Population growth, industrial revolution and huge carbon emissions have led to global warming, which has elevated the earth's average temperature by ~1.13°C in the past 110 years and is estimated to increase by 1.4–5.8°C by the end of this century (Chandra *et al.*, 2023). Thermal stress tolerance is a complex trait defined by several plant strategies and traits to minimize cellular injury and yield stability. Exploring strategies to bolster crop resilience against heat and cold stress through the strategic combination of specific genes/QTLs offers promising avenues for improving crop productivity in challenging environmental conditions. Ye *et al.* (2022) tested the efficacy of pyramiding two QTLs for early morning flowering (*qEMF3*) and heat tolerance (*qHTSF4.1*) into rice cv. IR64. NILs exhibited enhanced spikelet fertility across various temperature regimes in controlled indoor growth chambers, highlighting the potential of both QTLs in maintaining high spikelet fertility during the critical flowering stage under elevated temperatures. Likewise, Vivitha *et al.* (2017) utilized MAS to introgress heat-tolerance QTLs *qHTSF1.1* and *qHTSF4.1* from rice cv. N22 into cv. Improved White Ponni, which improved the spikelet fertility under heat stress. These experiments highlight the potential of QTL pyramiding in mitigating heat-induced yield losses in rice cultivation.

Effective management of cold stress during fertilization is crucial for maintaining seed fertility in rice production under cold temperature conditions. Cold tolerance at the fertilization stage (CTF) is a quantitatively inherited trait that makes MAS and MABB of QTLs a promising approach in breeding programmes. Shinada *et al.* (2014) demonstrated the effectiveness of QTL pairs *qCTF7* + *qCTF12* or *qCTF8* + *qCTF12* in significantly higher seed fertilities compared to plants lacking these QTLs and the cold-sensitive rice cv. Suisei. Endo *et al.* (2016) successfully pyramided QTLs *qCT-4* from the Bhutanese rice cv. Kuchum and *qLTB3* from the Chinese cv. Lijiangxintuanheigu into the widely grown Japanese cv. Hitomebore to improve cold tolerance. Dormatey *et al.* (2020) highlighted the importance of introgressing the cold-tolerance genes *viz.*, *9PssT-3*, *9PssT-7*, *9PssT9*, *9SCT1a*, *9SCT2*, *qPSST-3*, *qPSST-7*, *qPSST-9*, *qSCT1a* and *TSF4-1* for development of cold-tolerant rice genotypes. Likewise, different combinations of the QTLs viz., *qCTBB-5*, *qCTBB-6*, *qCTS-6*, *qCTF7*, *qCTF8*, and *qCTS12* were transferred into different rice cultivars in Japan and China and reported to enhance the cold stress tolerance (Shinada *et al.*, 2014; Yang *et al.*, 2016). These studies collectively underscore the availability of variation and genes for cold stress tolerance in rice, which can potentially be used in pyramiding strategies to enhance cold tolerance across diverse environments.

### 3.4.4 Gene pyramiding for salinity stress tolerance

The looming threat of climate change is set to exacerbate existing challenges posed by salinity alongside other abiotic stresses like submergence, high temperatures and drought, significantly impacting food sustainability and crop production. Various genetic approaches have been employed to enhance salinity tolerance across different plant species in response to this pressing issue. In *A. thaliana* and *Festuca arundinacea*, the stacking of genes from the salt overly sensitive pathway, including combinations such as *AtNHX1* + *SOS1*, *AtNHX1* + *SOS3*, *SOS2* + *SOS3*, and *SOS1* + *SOS2* + *SOS3*, has shown promising results. These stacked genes have led to reduced $Na^+$ accumulation, minimized growth retardation, fewer yellow leaves and decreased death rates in resultant plants (Yang *et al.*, 2009; Ma *et al.*, 2014; Pehlivan *et al.*, 2016; Fan *et al.*, 2019). Similarly, in tobacco, stacking genes from the betaine synthesis pathway, such as *BADH* + *SeNHX1* and *betA* + *AtNHX1*, has resulted in increased $Na^+$ and betaine accumulation in leaves and vacuoles, enhancing salinity tolerance (Zhou *et al.*, 2008; Duan *et al.*, 2009). Furthermore, in *Oryza sativa*, the pyramiding of genes from the ion sequestration pathway, specifically *SaSRP3-1* + *SaVHAc1*, has led to improvements in $K^+/Na^+$ ratio, chlorophyll content, relative water content and shoot root growth (Biradar *et al.*, 2018). In fact, numerous genes encoding receptors, TFs, ion transporters ($H^+$, $Ca^{2+}$, $Na^+$, $K^+$, $NO^{3-}$, etc.), channel proteins, ROS scavengers, and osmoticum (proline, trehalose, glycine betaine, myoinositol), salinity-responsive miRNA and players of SOS pathways are known and can be selectively included in introgression breeding schemes for engineering salinity stress (reviewed in Reddy *et al.*, 2017).

### 3.4.5 Gene pyramiding for enhanced tolerance to multiple stresses

The continuous cultivation of one or few varieties with a single or few related R-genes has caused a notable change in the frequency of pathogens or insect races, leading to the eventual breakdown of resistance (boom and bust). The susceptibility of single-gene host resistance in crops is higher due to pathogen or insect race frequency shifts. Race non-specific/quantitative R-genes are the best strategy to control several races of a pathogen, whereas several plant traits can provide multi-pathogen tolerance. For example, maize chitinases, phytoalexin, reactive oxygen species (ROS) induction, and known R-genes such as *Lr34* and *Yr36* against wheat leaf and yellow rusts, respectively, can confer broad-spectrum quantitative resistance (Pilet-Nayel *et al.*, 2017). Evidence suggests that combining multiple race-specific genes can provide broad-spectrum resistance. Therefore, it is currently prioritized to pyramid several R-genes from different sources into a single plant to develop genotypes with resistance against multiple pathogens. However, gene pyramiding has not yet been completely used for superior crop quality, along with increased resistance to multiple biotic factors in most of the major crops. To maintain agricultural productivity under various climatic circumstances, crops should have more than one strategy/trait for the most important stress factors. Since most of the tolerance-related plant features are regulated by the tandem action of a few major and several genes, it is essential to modify numerous genes to check biological (phytopathogens, insects, PPNs, weeds and parasitic plants) and non-biological factors (chemicals, toxic metals, pollutants, etc.) below an economic threshold level. Compared to traditional breeding, MABC and gene pyramiding for combining resistant genes into a single genetic background take the shortest time with high precision and serve the purpose. Some successful studies on gene pyramiding for tolerance against biotic and abiotic stressors are summarized in Table 3.2.

Multiple abiotic stress factors can occur simultaneously at different intensities, which collectively inflict heavy damages due to reduced photosynthesis, impaired nutrient uptake, cellular damage and even plant death, ultimately compromising the yield and quality of crops. Marker-assisted gene pyramiding has enabled the transfer of multiple stress tolerance genes into a single plant, thus creating a more robust and durable crop variety effective against specific to diverse stress environments. Notably, this approach has been successfully applied to incorporate genes conferring higher resistance to biotic and abiotic stresses. For instance, a pyramiding of *AVP1*, *OsSIZ1* and *Fld* in creeping bentgrass showed synergistic action and significant enhancement in plant growth with better resilience against drought, salinity and heat, and under nitrogen and phosphate starvation, leading to better overall performance under both normal and stress environments (Zhao *et al.*, 2024). Similarly, in various crop species such as rice, tobacco, maize and poplar, the stacking of genes involved in fructan biosynthesis, mannitol biosynthesis, ion sequestration, the glyoxalase pathway, glutamine synthesis and ascorbate biosynthesis has led to remarkable improvements in salinity and drought tolerance (Su *et al.*, 2011; Nguyen *et al.*, 2013; Biradar *et al.*, 2018; Gupta *et al.*, 2018; James *et al.*, 2018; Li *et al.*, 2019). Furthermore, the co-expression of abiotic stress-responsive TFs and the stacking of multiple stress-tolerance genes have demonstrated enhanced tolerance to salinity, drought and other stresses in crops like groundnut (Pruthvi *et al.*, 2014) and potatoes (Ahmad *et al.*, 2010). Most agricultural plants are sensitive to salts at all stages of their lives, but this is particularly true when they are seedlings (Thitisaksakul *et al.*, 2015). Additionally, some crop varieties that are resistant to excessive salts express genes that are sensitive to salt; these genes/QTLs can be mapped using molecular markers to facilitate the breeding of salt-tolerant lines (Tiwari *et al.*, 2016). The $Na^+/K^+$ symporter-encoding major QTL *Saltol* mapped in cv. Pokkali and *SKC1* (also *OsHKT1;5*) characterized in cv. Nona Bokra is a major source of salinity tolerance in rice, maintaining $K^+$ homeostasis in salinity (reviewed in Reddy *et al.*, 2017). Genes from the same class may act differently under different abiotic stress environments, e.g. Ras-related nuclear protein (*RAN*) GTPases in rice. Rice *OsRAN1* positively regulates cold tolerance, whereas *OsRAN2* negatively regulates the salt stress response in rice (Xu and Cai, 2014). These advancements highlight the potential of marker-assisted gene pyramiding as a powerful tool for developing multiple stress-tolerant crop varieties, thereby contributing to sustainable agriculture in the face of changing environmental conditions. However, limited success has been achieved in developing crop varieties with persistent resistance against multiple stresses through gene pyramiding. Thus, through the application of genomics-assisted breeding, selective introgression of two or more genes that encode signalling molecules acting via different pathways under different abiotic stress acting synergistically can be a better strategy for engineering varieties with multiple stress tolerance.

### 3.4.6 Gene pyramiding for higher crop productivity and yield stability

Climate change and global warming inflict unprecedented droughts, heat waves and flash floods followed by dry spells, and cause evolution of novel pathogen variants and parasite diseases, and huge insect pest multiplication

**Table 3.2.** Successful gene pyramiding experiments for sustainable crop production.

| Crops | Traits | Pyramided genes | References |
|---|---|---|---|
| **Biotic stress resistance** | | | |
| Potato | Late blight | *Rpi-phu1*, *Rpi-rzc* | Sliwka *et al.*, 2012 |
| Cotton | Bacterial blight/sheath<br>Bollworm<br>Weeds and pathogens<br>Insect- pests | *Chi11*, *t1p*, *Xa21*<br>*Cry1Ac*, *Cry2AbptxD/*<br>*Phi*<br>*Cry1Ac*, *Cry2Ac* | Jackson *et al.*, 2004; Maruthasalam *et al.*, 2007; Puspito *et al.*, 2015; Pandey *et al.*, 2017 |
| Wheat | Leaf and stem rust<br>Cereal cyst nematode<br>Aphids | *SrCad*, *Sr33*, *Lr34*,<br>*FhbCreX*, *CreY*<br>*Gn2*, *Gn4* | Ali *et al.*, 2019; Zhang *et al.*, 2019a; Liu *et al.*, 2000 |
| Rice | Gall midge<br>Blast<br>Brown plant hopper<br>Blight<br>Bacterial sheath blight<br>Stemborer | *Gm1*, *Gm2*, *Gm4*<br>*Pi(2)t*, *Pi25*, *Pi(t)a*, *Xa4*,<br>*Xa5*, *Xa13*, *Xa21*<br>*Bph1*, *Bph2*, *Xa5*,<br>*Xa13*, *Xa21*<br>*Xa12*, *Rc7*, *Cry1AB1*,<br>*Cry14c* | Das and Rao, 2015; Joseph *et al.*, 2004; Singh *et al.*, 2001; Datta *et al.*, 2002; Narayanan *et al.*, 2002; Liu *et al.*, 2016 |
| Soybean | Mosaic virus | *Rsv1, Rsv3, Rsv4* | Shi *et al.*, 2006 |
| Tomato | Leaf curl/spotted virus | *Ty-1, Ty-3, Sw-5* | Verlaan *et al.*, 2013 |
| Barley | Mosaic virus<br>Stripe rust | *rym4, rym5, rym9,*<br>*rym113QTL* | Castro *et al.*, 2003; Werner *et al.*, 2005 |
| Maize | Corn borer | *Cry1le, Cry1Ac* | Jiang *et al.*, 2016 |
| Chickpea | Lepidopteron | *Cry1Ac, Cry1Ab* | Mehrotra *et al.*, 2011 |
| Pea | Nodulation ability | *Sym9, Sym10* | Schneider, 2002 |
| Broccoli | Diamond back-moth | *Cry1Ac, Cry1c* | Cao *et al.*, 2002 |
| Pepper | Root-knot nematode | *Me1, Me2* | Djian-Caporalino *et al.*, 2014 |
| **Abiotic stress resistance** | | | |
| Rice | Cold | *9PssT-3, 9PssT-7,*<br>*9PssT9, 9SCT1a,*<br>*9SCT2, qPSST-3,*<br>*qPSST-7, qPSST-9,*<br>*qSCT1a, TSF4-1* | Yang *et al.*, 2013; Kim *et al.*, 2014; Fujino *et al.*, 2019 |
| | Drought | *Saltol* | Lin *et al.*, 2007; Das and Rao, 2015 |
| | Heat, drought, salt and cold | *OsHSP18.6* | Wang *et al.*, 2015 |
| | Waterlogging | *Sub1, qDTY* QTLs | Kumar *et al.*, 2018 |

and attacks. Biotic and abiotic stressors have been reported to produce annual losses of roughly 25% and 6–25%, respectively (Waddington *et al.*, 2010; Pandey *et al.*, 2017). For instance, only a single polyphagous insect pest, fall armyworm (*Spodoptera frugiperda* L.), has caused significant damage to over 80 plant species in sub-Saharan Africa, China, India, Sri Lanka, Bangladesh and Thailand in recent years. These include fruits, vegetables, cash crops like cotton and sugarcane, and staple foods like millet, rice and maize. The outbreak has also endangered food security and resulted in significant financial losses (Kansiime *et al.*,

2019; Womack *et al.*, 2020). The Green Revolution varieties significantly increased the yield and productivity of major food grains in starving countries like India. The Food and Agriculture Organization states that by 2050, global food production needs to rise by 70% to fulfil the increasing need for food from the growing population (Mwobobia *et al.*, 2020). Therefore, high yield potential along with stress tolerance is the primary breeding objective in most of the crop. In order to promote agricultural sustainability, MAS and gene pyramiding have accelerated the creation of need-based, disease-and pest-resistant, durable cultivars (Table 3.3). A difficulty for sustainable farming practices is the wide range of different insect pests and pathogens. Therefore, adequate and effective control techniques are needed to combat these biological risks to prevent food shortages and financial losses worldwide. Using traditional breeding strategies to improve crops is challenging because some unwanted genes persist even after numerous backcross generations, and some desired ones also do. This makes it difficult to screen for undesirable genes conveniently.

Numerous QTL analyses and GWAS have discovered and characterized several QTLs for yield and yield-related traits in major crops (Table 3.3). For example, information on more than 600 genes and 2000 QTLs related to the yield traits is present in the Gramene database (Pulindala *et al.*, 2022). Both cultivated and wild species have been the treasure of yield-enhancing genetic factors. For example, an *Oryza longistaminata*-derived *O. sativa* NIL carrying QTL *qPB8.1* conferred ~50% more spikelets and >27% higher grain yield compared to the RP (Long *et al.*, 2023). Likewise, introgressions from soybean progenitor *Glycine soja* (Li *et al.*, 2008) and wheat D-genome donor *Aegilops tauschii* (Ma *et al.*, 2023) have yield-increasing genes. However, before their application in varietal development, candidate gene analysis, analysis of effect size and interactions, and robust markers need to be developed for easy and fast introgression. Major yield effect *qDTY* series QTLs in rice work additively (synergistically) in the pyramided lines with a 10–30% yield advantage (Shamsudin *et al.*, 2016a). Likewise, pyramiding of *qDTYs* along with the *Sub1* locus in rice varieties e.g. CR dhan 801 (*qDTY1.1* + *qDTY2.1* + *qDTY3.1* + *Sub1*) in India, and Bahuguni dhan-1 (*qDTY1.1* + *qDTY3.1* + *Sub 1*) and Bahuguni dhan-2 (*qDTY3.1* + *Sub1*) in Nepal, have conferred multiple stress tolerance with stable yield under drought and flood conditions (Sandhu *et al.*, 2019b). However, background effects can change the effect size of these QTLs, and this area is largely unexplored (Shamsudin *et al.*, 2016a,b).

### 3.4.7 Reduction in pesticide and fertilizer usage

Although chemical pesticides have long been employed to manage biotic stress factors on crops, their use has recently been discouraged due to several disadvantages, including high costs and the requirement for multiple applications, which are often unaffordable for small-scale farmers. Moreover, chemical pesticides have been found to have potential negative impacts on human and animal health and on non-target organisms in the ecosystem through environmental contamination. Pest resistance to chemical pesticides is another problem. Herbicide resistance is reported in 262 weed species worldwide. Therefore, the idea of developing host plant resistance is a long-lasting, affordable and environmentally safer way, and gene pyramiding offers a more effective way to broaden this scope by reducing reliance on chemical control methods, safeguarding valuable microfauna, maintaining ecological balance and restricting the evolution of super races, ensuring sustainable food security (Mundt, 2014; Hanson *et al.*, 2016; Dinkar *et al.*, 2020).

Higher production is impossible without proper plant nutrition. Several well-characterized nutrient transporters are known, which are being targeted for enhancing the nutrient use efficiency of crop plants, and this can potentially decrease the fertilizer input cost. For example, tomato *SlAMT1* (an ammonium transporter) can directly uptake $NH_4^+$ from soil and *NRT2.1* root-specific transporters are involved in nitrate ($NO_3^-$) uptake, characterized by cabbage and cucumber (Li *et al.*, 2018; Dinkar *et al.*, 2024a). Phosphate transporters and their regulator phosphate starvation response (*PHR*) gene are involved in inorganic

**Table 3.3.** Studies highlighting the yield improvement by QTL analysis and introgression.

| Crop | Trait | QTL/Gene | Donor | Recipient | Change | Reference |
|---|---|---|---|---|---|---|
| Rice | Grain size/weight | *GS3, GS5, qsw5* | MTU3626 | BPT5204 | 22–50% increase in yield | Pulindala *et al.*, 2022 |
| | Grain number | *LP1* | NLR33892 | | | |
| | Yield under drought | $qDTY_{2.2}$ | IR 84984-83-15-18-B | MRQ74 | Yield increased by >1000 kg/ha | Shamsudin *et al.*, 2016a,b |
| | | $qDTY_{3.1}$ | IR 77298-14-1-2-10 | | | |
| | | $qDTY_{12.1}$ | IR 81896-B-B-195 | | | |
| | Yield under drought | $qDTY_{1.1}$ | IR 91659-54-35, from N22/Swarna population | Swarna-Sub1 | Nine promising NILs with different QTL combinations showed an average grain yield advantage of 0.2–1.7 t/ha under reproductive-stage drought and 0.1–1.0 t/ha under submergence with no yield penalty under non-stress conditions | Sandhu *et al.*, 2019a |
| | | $qDTY_{2.1}$ and $qDTY_{3.1}$ | IR 81896-B-B-195, from Apo/Swarna population | | | |
| | Submergence tolerance | *Sub1* | *Swarna-Sub1* | | | |
| | Lodging tolerance and yield | *SCM2* (a gain-of-function mutant of *APO1* gene) | Habataki | Koshihikari | NILs showed enhanced culm strength, and increased spikelet number and grain yield | Ookawa *et al.*, 2010 |
| Wheat | Grain width and 1000-kernel weight | *TaGW2* | Rialto | Spark ×-Rialto derived RILs | Yield increased by 5.5% | Simmonds *et al.*, 2014 |
| | Yield under severe drought and hot climate | *QYld.aww-1B.2* | Excalibur | Excalibur × Kukri-derived NIL | 54.5% increase in grain yield, 43.5% in biomass and 32% in fertile tillers | Tura *et al.*, 2020 |
| | Kernel length | *AetT093_2Dv1G100900.1* gene at *QKl-2D.5 locus* | *Aegilops tauschii* acc. T015 | Zhoumai 18-derived IL | Explained 5.91–19.83% of phenotypic variations at seven environments | Ma *et al.*, 2023 |
| | Yield under drought | *Qyld.csdh.7AL* | SQ1 (breeding line) | HUW234 | ~25% higher yield than RP HUW234 when rainfed | Gautam *et al.*, 2021 |

| | | | | | | |
|---|---|---|---|---|---|---|
| Soybean | Seed yield | *Glyma.09G048900* (an oxygen-binding protein on chr. 17) | GWAS panel | - | Consistently high seed yield across all locations tested | Ayalew *et al.*, 2022 |
| | Seed yield | QTL on LG A1 | *Glycine soja* PI 245331 | *G. max* acc. 7499 | Seed yield increased by 6.3% | Li *et al.*, 2008 |
| Maize | Yield | *NAC7* (NAC-domain TF) | Plants from Illinois High Protein (IHP) and Illinois Low Protein (ILP) populations | IHP1 × ILP1 hybrid | Hybrids showed prolonged stay-green and increased grain yield by an average of 4.6 bushels per acre | Zhang *et al.*, 2019b |
| Peanut | Yield (pod weight and seed weight) | *Arahy.3R9A5K* gene *at qHYF_B06* locus | JH6 and KX01-6 | JH6 × KX01-6-derived RIL | Explained 8.77–31.40% variation in pod number per plant and % shelling | Yang *et al.*, 2023 |
| Chickpea | Yield under drought | *QTL-hotspot* region | Acc. ICC 4958 | Pusa 372 | BC-IL BGM 10216 had 16% more yield than Pusa 372; released as cv. Pusa Chickpea 10 216 | Bharadwaj *et al.*, 2021 |

NIL, Near isogeneic line; RIL, Recombinant inbred line; Acc., Accession; *SCM2*, *Strong Culm2; APO1*, *Aberrant Panicle Organization 1*.

phosphorus uptake homeostasis maintenance and *PHR* expression (Zhang *et al.*, 2021). Several $K^+$ transporter gene families are well known in crop plants, such as *LKT1* (tomato), *SKT1* (potato), *HAK1* (barley), etc. (Ragel *et al.*, 2019). Likewise, sulfate transporters in roots are involved in sulfate uptake from soil (Dinkar *et al.*, 2024a). The homologues of these nutrient transporters are supposed to be present in all crop plants. However, their precise characterization is lacking. Moreover, several genotypes have been characterized as having the inherent capacity to tolerate the low nutrient availability in the soil. However, their genetic characterization and regulation are still a matter of research (He *et al.*, 2022). Thus, robust marker development and its usage for gene pyramiding in the agricultural industry is crucial to take crop production to the next level.

## 3.5 Advantages of Gene Pyramiding for Sustainable Crop Production

Gene pyramiding is one of the breeding strategies, combining several genes, mostly with major genes for preferred attributes, into one plant variety. It offers an array of benefits in relation to sustainable agricultural production, such as:

**1. Broad spectrum resistance:**

By stacking multiple R-genes against different pests and pathogens, gene pyramiding can provide broad-spectrum resistance, reducing the reliance on chemical pesticides. This can lead to more sustainable pest and disease management practices, minimizing environmental damage and reducing health risks for farmers and consumers. For example, pyramiding multiple R-genes against rice blast fungus (*Magnaporthe oryzae*) has enhanced durable resistance in rice varieties (Joseph *et al.*, 2004).

**2. Enhanced stress tolerance:**

Pyramiding multiple genes conferring tolerance to abiotic stresses such as drought, salinity or extreme temperatures can improve crop performance under adverse environmental conditions. This enhances the resilience of crops to climate variability, ensuring more stable yields and reducing vulnerability to climate change impacts (Wang *et al.*, 2015).

**3. Delayed resistance development:**

Stacking multiple R-genes in a crop variety can delay the development of resistance in target pests and pathogens. The presence of multiple mechanisms of resistance makes it more challenging for pests and pathogens to overcome plant defences, thereby prolonging the effectiveness of crop protection measures (Poland *et al.*, 2012; Das and Rao, 2015).

**4. Reduced input requirements:**

Resilient crop varieties resulting from gene pyramiding often require fewer inputs such as water, fertilizers and pesticides. This can lead to cost savings for farmers, as well as reduced environmental pollution and resource depletion associated with intensive agricultural practices.

**5. Sustainable intensification:**

Gene pyramiding reduces reliance on chemical pesticides and promotes environmentally friendly pest management practices and, thus, supports the concept of sustainable intensification by enabling the production of higher yields on existing agricultural land while minimizing negative environmental impacts. By reducing yield variability and enhancing resource use efficiency, pyramided varieties contribute to more sustainable and resilient farming systems.

**6. Extended shelf life and post-harvest quality:**

Incorporating genes associated with enhanced shelf life, post-harvest storage and quality traits can reduce food waste and improve marketability. For instance, pyramiding genes for delayed fruit ripening and improved firmness in tomatoes have resulted in varieties with extended shelf life and reduced losses during transportation and storage.

## 3.6 Challenges and Limitations Faced in Implementing Gene Pyramiding

While gene pyramiding offers significant advantages in crop improvement, it also comes with its own set of challenges and limitations:

**1. Trait compatibility:**

One of the primary challenges is ensuring that the traits stacked through pyramiding are compatible. Some traits may interact negatively, leading to reduced performance or even detrimental effects on the plant's growth and development.

**2. Linkage drag:**

Genes controlling different traits may be genetically linked, making it difficult to select each trait independently during the breeding process. This linkage can slow down the progress of pyramiding as it requires multiple rounds of crossing and selection to break the linkage.

**3. Labour-intensive and time-consuming:**

Pyramiding multiple genes typically requires extensive breeding efforts over several generations. It involves crossing multiple parent lines, selecting the desired traits in each generation, and genotyping to confirm the presence of the target genes. This process can be labour-intensive and time-consuming.

**4. Availability of genetic resources:**

The availability of genetic resources with the desired traits can be a limitation. Suppose the traits of interest are not present in the germplasm pool. In that case, sourcing genetic material from wild relatives or other breeding programmes may be necessitated, which can introduce additional challenges due to differences in genetic backgrounds.

**5. Regulatory and social acceptance:**

Genetically modified (GM) crops obtained through gene pyramiding may face regulatory hurdles and public acceptance issues in some regions. Strict regulations, labelling requirements and consumer perceptions regarding GM crops can impact the commercialization and adoption of pyramided varieties.

**6. Trait stability:**

It is crucial to ensure the stability of stacked traits across different environments and over multiple generations. Environmental factors and genetic interactions may influence the expression of the desired traits, leading to variability in performance.

**7. Off-target effects:**

Introducing multiple transgenes into a plant genome through gene pyramiding can potentially result in unintended effects, such as gene silencing, disruption of endogenous gene expression or activation of unintended pathways. Thorough molecular characterization and safety assessments are necessary to mitigate these risks.

**8. Cost considerations:**

The cost of developing pyramided varieties, including research, development, regulatory compliance and intellectual property rights, can be significant. This cost may limit the accessibility of these improved varieties to small-scale farmers or resource-constrained regions.

Despite these challenges, gene pyramiding remains a powerful tool for crop improvement against different stressors (Tables 3.2 and 3.3). It offers the potential to address complex traits and enhance resilience against biotic and abiotic stresses in agricultural systems. Advances in new molecular marker technology and genomics-assisted breeding are continuously improving the selection efficiency, precision of pyramiding approaches, and cost cutting, and overcoming most of these limitations.

## 3.7 Conclusion

Gene pyramiding has become a pivotal strategy in the modern plant breeding industry for constructing genotypes with better agronomic features and enhanced resilience against stresses. Converging the genes affecting different metabolic pathways broadens the protection spectrum, enhancing the effectiveness and durability of resistance/tolerance against stress factors, thereby contributing to sustainable agricultural practices and global food security. Modern high-throughput phenomics facilities combined with cheaper genome sequencing platforms have revolutionized genomics and have made available saturated linkage maps, QTL and candidate gene characterization, allele mining, easier and cheaper SNP discovery and genotyping (genotype-by-sequencing, SNP chips, KASP assays), development and establishment of robust MTAs for MAS and MABB, which expedite the selection process, identification of best haplotypes and linkage drag minimization, and, thus, enhance the overall breeding efficiency. Although gene pyramiding is practically viable for a limited number of genes/QTLs with high to moderate phenotypic effect size and with genic and functional markers, combining more loci with minor effect size and under epistatic interactions is practically impossible. In such a situation, other MAS methods, like MARS, are better than MABC (Varshney *et al.*, 2013). However, to counter the problem of the evolution of virulence against R-genes, strategic spatial and temporal deployment of different gene-pyramided varieties can impede the pathogen evolution and extend their durability. Likewise, multiline varieties or varietal mixtures can be a solution. However, these methods have their own limitations. Other approaches of gene pyramiding/stacking include cisgenesis, which involves the transfer of a natural gene from a sexually crossable plant species, and transgenesis, – involving any sexually incompatible and non-plant species. However, both these approaches have serious limitations of national and international regulatory policies and high research investment. Modern genome editing technology, such as CRISPR-based systems, can edit and modify native plant genes to bring desirable phenotypic changes. The CRISPR systems have shown their potential in engineering biotic and abiotic stress factors via precise activation (CRISPRa) or interference (CRISPRi) of genetic and epigenetic factors/pathways in a target-specific manner and offer multiplexing to target several genes/traits simultaneously (Najera *et al.*, 2019; Dinkar *et al.*, 2024a,b). Success has been achieved in several grain and vegetable crops such as rice, wheat, sorghum, tomato, potato, etc. (reviewed in Najera *et al.*, 2019). Thus, guide RNA constructs targeting several genes regulating plant response to stress can be an alternative to the gene pyramiding strategy. However, the first genome-edited commercial variety is awaited, and even a standard transformation and regeneration protocol for successful genome editing is lacking in most crops. Hence, the development of robust MTAs, marker-assisted gene(s) introgression, accelerated RP genome recovery and selection of best-performing haplotype is the best available strategy to improve upon the existing production potential of the cultivated crop varieties under diverse and fluctuating environments.

## References

Acevedo-Garcia, J., Spencer, D., Thieron, H., Reinstädler, A., Hammond-Kosack. K. *et al.* (2017) Mlo-based powdery mildew resistance in hexaploid bread wheat generated by a non-transgenic TILLING approach. *Plant Biotechnology Journal* 15(3), 367–378. DOI: 10.1111/pbi.12638.

Adhikari, S., Joshi, A., Chandra, A.K., Bharati, A., Sarkar, S. *et al.* (2023) SMART plant breeding from pre-genomic to post-genomic era for developing climate-resilient cereals. In: Sharma, D., Singh, S., Sharma, S.K. and Singh, R. (eds) *Smart Plant Breeding for Field Crops in Post-Genomics Era*. Springer, Singapore, pp. 41–97. DOI: 10.1007/978-981-19-8218-7_2.

Ahmad, R., Kim, Y., Kim, M., Kwon, S., Cho, K. *et al.* (2010) Simultaneous expression of choline oxidase, superoxide dismutase and ascorbate peroxidase in potato plant chloroplasts provides synergistically enhanced protection against various abiotic stresses. *Physiologia Plantarum* 138, 520–533. DOI: 10.1111/j.1399-3054.2010.01348.x.

Ali, A., Cao, J., Jiang, H., Chang, C., Zhang, H.P. *et al.* (2019) Unravelling molecular and genetic studies of wheat (*Triticum aestivum* L.) resistance against factors causing pre-harvest sprouting. *Agronomy* 9, 117. DOI: 10.3390/agronomy9030117.

Álvarez, S. and Acosta-Motos, J.R. (2022) Miscellaneous sets of abiotic stresses and plant strategies to cope with them. *Agronomy* 12(11), 2727. DOI: 10.3390/agronomy12112727.

An, Y., Jiao, X., Gu, Z., Shi, C. and Liu, K. (2022) Effects of straw return and aeration on oxygen status and redox environment in flooded soil. *Soil and Water Research* 17, 29–35. DOI: 10.17221/87/2020-SWR.

Ayalew, H., Schapaugh, W., Vuong, T. and Nguyen, H.T. (2022) Genome-wide association analysis identified consistent QTL for seed yield in a soybean diversity panel tested across multiple environments. *Plant Genome* 15(4), e20268. DOI: 10.1002/tpg2.20268.

Bai, Y., Kissoudis, C., Yan, Z., Visser, R.G. and Linden, G. (2018) Plant behaviour under combined stress: Tomato responses to combined salinity and pathogen stress. *Plant Journal* 93(4), 781–793. DOI: 10.1111/tpj.13800.

Balconi, C., Stevanato, P., Motto, M. and Biancardi, E. (2012) Breeding for biotic stress resistance/tolerance in plants. In: Ashraf, M., Öztürk, M., Ahmad, M. and Aksoy, A. (eds) *Crop Production for Agricultural Improvement*. Springer, Dordrecht, The Netherlands, pp. 57–114. DOI: 10.1007/978-94-007-4116-4_4.

Bashir, S.S., Hussain, A., Hussain, S.J., Wani, O.A., Zahid Nabi, S. *et al.* (2021) Plant drought stress tolerance: understanding its physiological, biochemical and molecular mechanisms. *Biotechnology and Biotechnological Equipment* 35, 1912–1925. DOI: 10.1080/13102818.2021.2020161.

Baumgartner, I.O., Patocchi, A., Frey, J.E., Peil, A. and Kellerhals, M. (2015) Breeding elite lines of apple carrying pyramided homozygous resistance genes against apple scab and resistance against powdery mildew and fire blight. *Plant Molecular Biology Reporter* 33, 1573–1583. DOI: 10.1007/s11105-015-0858-x.

Bharadwaj, C., Tripathi, S., Soren, K.R., Thudi, M., Singh, R.K. *et al.* (2021) Introgression of "QTL-hotspot" region enhances drought tolerance and grain yield in three elite chickpea cultivars. *Plant Genome* 14(1), e20076. DOI: 10.1002/tpg2.20076.

Bharadwaj, C., Jorben, J., Rao, A., Roorkiwal, M., Patil, B.S. *et al.* (2022) Development of high yielding fusarium wilt resistant cultivar by pyramiding of "genes" through marker-assisted backcrossing in chickpea (*Cicer arietinum* L.). *Frontiers in Genetics* 13, e924287. DOI: 10.3389/fgene.2022.924287.

Bharati, A., Kumar, A., Chandra, A.K., Phogat, S., Poudel, A. *et al.* (2024) Breeding and molecular approaches for heat-resilient crops. In: Singh, A. and Pandey, S. (eds) *Climate-Resilient Agriculture*. Apple Academic Press, New York, pp. 193–226. DOI: 10.1201/9781003455271.

Bhawar, K.B., Sharma, J.B., Singh, A.K., Sivasamy, M., Singh, M. *et al.* (2011) Molecular marker assisted pyramiding of leaf rust resistance genes *Lr19* and *Lr28* in bread wheat (*Triticum aestivum* L.) variety HD2687. *Indian Journal of Genetics and Plant Breeding* 71(4), 304–311. Available at: https://www.isgpb.org/journal/index.php/IJGPB/article/view/840.

Biradar, H., Karan, R. and Subudhi, P.K. (2018) Transgene pyramiding of *Salt Responsive Protein 3-1* (*SaSRP3-1*) and *SaVHAc1* from *Spartina alterniflora* L. enhances salt tolerance in rice. *Frontiers in Plant Science* 9, 1304. DOI: 10.3389/fpls.2018.01304.

Boyer, J.S. (1982) Plant productivity and environment. *Science* 218, 443–448. DOI: 10.1126/science.218.4571.443.

Bravo, A., Gill, S.S. and Soberón, M. (2007) Mode of action of *Bacillus thuringiensis* Cry and Cyt toxins and their potential for insect control. *Toxicon* 49(4), 423–435. Available at: DOI: 10.1016/j.toxicon.2006.11.022.

Bulgari, R., Franzoni, G. and Ferrante, A. (2019) Biostimulants application in horticultural crops under abiotic stress conditions. *Agronomy* 9(6), 306. DOI: 10.3390/agronomy9060306.

Cao, L.Y., Zhuang, J.Y., Zhan, X.D., Zheng, K.L. and Cheng, S.H. (2002) Hybrid rice resistant to bacterial blight developed by marker assisted selection (Zhongguo Shuidao Kexue). *Chinese Journal of Rice Science* 17, 184–186. Available at: https://www.cabidigitallibrary.org/doi/full/10.5555/20033079217

Castro, A.J., Capettini, F., Corey, A.E., Filichkina, T., Hayes, P.M. *et al.* (2003) Mapping and pyramiding of qualitative and quantitative resistance to stripe rust in barley. *Theoretical and Applied Genetics* 107, 922–930. DOI: 10.1007/s00122-003-1329-6.

Chandra, A.K., Joshi, A., Tripathi, A., Kumar, A., Pandey, S. *et al.* (2023) Climate-resilience maize: heat stress, signaling, and molecular interventions. *Journal of Plant Growth Regulation* 42, 6349–6366. DOI: 10.1007/s00344-022-10844-6.

Chen, L., Wang, P., Cheng, X., Yan, Z., Wu, F. *et al.* (2022) Recurrent selection of new breeding lines and yield potential, nutrient profile and *in vitro* rumen characteristics of *Melilotus officinalis*. *Field Crops Research* 287, 108657. DOI: 10.1016/j.fcr.2022.108657.

Collard, B.C.Y. and Mackill, D.J. (2008) Marker-assisted selection: an approach for precision plant breeding in the twenty-first century. *Philosophical Transactions of the Royal Society B: Biological Sciences* 363, 557–572. DOI: 10.1098/rstb.2007.2170.

Consuegra, O.G., Gómez, M.P. and Zubiaur, Y.M. (2015a) Pyramiding *TYLCV* and *TSWV* resistance genes in tomato genotypes. *Revista de Protección Vegetal* 30(2), 161–161. Available at: http://scielo.sld.cu/scielo.php?script=sci_arttext&pid=S1010-27522015000200011&lng=es&nrm=iso&tlng=en.

Consuegra, S., John, E., Verspoor, E. and deLeaniz, C.G. (2015b) Patterns of natural selection acting on the mitochondrial genome of a locally adapted fish species. *Genetics Selection Evolution* 47, 58. DOI: 10.1186/s12711-015-0138-0.

Cook, D.E., Lee, T.G., Guo, X., Melito, S., Wang, K. *et al.* (2012) Copy number variation of multiple genes at *Rhg1* mediates nematode resistance in soybean. *Science* 338(6111), 1206–1209. DOI: 10.1126/science.1228746.

Cook, D.E., Bayless, A.M., Wang, K., Guo, X., Song, Q. *et al.* (2014) Distinct copy number, coding sequence, and locus methylation patterns underlie *Rhg1*-mediated soybean resistance to soybean cyst nematode. *Journal of Plant Physiology* 171(13), 1265–1273. DOI: 10.1016/j.jplph.2014.04.014.

Crickmore, N., Zeigler, D.R., Feitelson, J., Schnepf, E., Van Rie, J. *et al.* (1998) Revision of the nomenclature for the *Bacillus thuringiensis* pesticidal crystal proteins. *Microbiology and Molecular Biology Reviews* 62(3), 807–813. DOI: 10.1128/mmbr.62.3.807-813.1998.

Das, G. and Rao, G.J.N. (2015) Molecular marker assisted gene stacking for biotic and abiotic stress resistance genes in an elite rice cultivar. *Frontiers in Plant Science* 6, 698. DOI: 10.3389/fpls.2015.00698.

Datta, K., Baisakh, N., Thet, K.M., Tu, J. and Datta, S.K. (2002) Pyramiding transgenes for multiple resistance in rice against bacterial blight, yellow stem borer and sheath blight. *Theoretical and Applied Genetics* 106, 1–8. DOI: 10.1007/s00122-002-1014-1.

Dinkar, V., Jha, S.K., Mallick, N., Niranjana, M., Agarwal, P. *et al.* (2020) Molecular mapping of a new recessive wheat leaf rust resistance gene originating from *Triticum spelta*. *Scientific Reports* 10(1): 22113. DOI: 10.1038/s41598-020-78679-3.

Dinkar, V., Kushwaha, A.K., Singh, A.K., Sahu, K.P., Sarkar, S. *et al.* (2024a) Recent advances in CRISPR/Cas9 for climate-resilient agriculture in vegetable crops. In: Singh, A. and Pandey, S. (eds) *Climate-Resilient Agriculture*. Apple Academic Press, New York, pp. 373–414. DOI: 10.1201/9781003455271.

Dinkar, V., Pandey, S., Kumar, A., Shiv, A., Lal, D. *et al.* (2024b) Epigenetic regulations under plant stress: a cereals perspective. *Environmental and Experimental Botany* 220, e105688. DOI: 10.1016/j.envexpbot.2024.105688.

Dixit, S., Singh, U.M., Singh, A.K., Alam, S., Venkateshwarlu, C. *et al.* (2020) Marker assisted forward breeding to combine multiple biotic-abiotic stress resistance/tolerance in rice. *Rice* 13, 29. DOI: 10.1186/s12284-020-00391-7.

Djian-Caporalino, C., Palloix, A., Fazari, A., Marteu, N., Barbary, A. *et al.* (2014) Pyramiding, alternating or mixing: comparative performances of deployment strategies of nematode resistance genes to promote plant resistance efficiency and durability. *BMC Plant Biology* 14, 53. DOI: 10.1186/1471-2229-14-53.

Dormatey, R., Sun, C., Ali, K., Coulter, J.A., Bi, Z. *et al.* (2020) Gene pyramiding for sustainable crop improvement against biotic and abiotic stresses. *Agronomy* 10, 1255. DOI: 10.3390/agronomy10091255.

Dresselhaus, T. and Hückelhoven, R. (2018) Biotic and abiotic stress responses in crop plants. *Agronomy* 8(11), 267. DOI: 10.3390/agronomy8110267.

Duan, X., Song, Y., Yang, A. and Zhang, J. (2009) The transgene pyramiding tobacco with betaine synthesis and heterologous expression of *AtNHX1* is more tolerant to salt stress than either of the tobacco lines with betaine synthesis or *AtNHX1*. *Physiologia Plantarum* 135, 281–295. DOI: 10.1111/j.1399-3054.2008.01194.x.

Ellur, R.K., Khanna, A., Yadav, A., Pathania, S., Rajashekara, H. *et al.* (2016) Improvement of basmati rice varieties for resistance to blast and bacterial blight diseases using marker assisted backcross breeding. *Plant Science* 242, 330–341. DOI: 10.1016/j.plantsci.2015.08.020.

Endo, T., Chiba, B., Wagatsuma, K., Saeki, K., Ando, T. *et al.* (2016) Detection of QTLs for cold tolerance of rice cultivar 'Kuchum' and effect of QTL pyramiding. *Theoretical and Applied Genetics* 129, 631–640. DOI: 10.1007/s00122-015-2654-2.

Estruch, J.J., Warren, G.W., Mullins, M.A., Nye, G.J., Craig, J.A. *et al.* (1996) Vip3A, a novel *Bacillus thuringiensis* vegetative insecticidal protein with a wide spectrum of activities against lepidopteran insects. *Proceedings of the National Academy of Sciences USA* 93(11), 5389–5394. DOI: 10.1073%2Fpnas.93.11.5389.

Fan, Y., Yin, X., Xie, Q., Xia, Y., Wang, Z. *et al.* (2019) Co-expression of *SpSOS1* and *SpAHA1* in transgenic *Arabidopsis* plants improves salinity tolerance. *BMC Plant Biology* 19, 74. DOI: 10.1186/s12870-019-1680-7.

Francini, S. (2019) Abiotic stress effects on performance of horticultural crops. *Horticulturae* 5, 67. DOI: 10.3390/horticulturae5040067.

Fujino, K., Hirayama, Y. and Kaji, R. (2019) Marker-assisted selection in rice breeding programs in Hokkaido. *Breeding Science* 69, 383–392. DOI: 10.1270/jsbbs.19062.

Fukuoka, S., Saka, N., Koga, H., Ono, K., Shimizu, T. *et al.* (2009) Loss of function of a proline-containing protein confers durable disease resistance in rice. *Science* 325(5943), 998–1001. DOI: 10.1126/science.1175550.

Gantait, S., Sarkar, S. and Verma, S.K. (2019) Marker-assisted selection for abiotic stress tolerance in crop plants. In: Roychoudhury, A. and Tripathi, D. (eds) *Molecular Plant Abiotic Stress: Biology and Biotechnology*. John Wiley & Sons Ltd, USA, pp. 335–368. DOI: 10.1002/9781119463665.ch18.

Gautam, T., Amardeep, S., Rakhi, G., Kumar, A. Gahlaut, V. *et al*. (2021) Introgression of a drought insensitive grain yield QTL for improvement of four Indian bread wheat cultivars using marker assisted breeding without background selection. *Journal of Plant Biochemistry and Biotechnology* 30, 172–183. DOI: 10.1007/s13562-020-00553-0.

Gebhardt, C. and Valkonen, J.P.T. (2001) Organization of genes controlling disease resistance in the potato genome. *Annual Review of Phytopathology* 39(1), 79–102. DOI: 10.1146/annurev.phyto.39.1.79.

Gessler, C., Pertot, I. and Perazzolli, M. (2006) Plasmid stability in the biocontrol agent *Pseudomonas fluorescens* strain CHA0 and its derivatives. *Phytopathology* 96(4), 401–406. DOI: 10.1128/aem.62.3.1076-1080.1996.

Gottwald, T.R., Graham, J.H. and Schubert, T.S. (2002) Citrus canker: the pathogen and its impact. *Plant Health Progress* 3(1), 15. DOI: 10.1094/PHP-2002-0812-01-RV.

Gouda, G., Gupta, M.K., Donde, R., Mohapatra, T., Vadde, R. and Behera, L. (2020) Marker-assisted selection for grain number and yield-related traits of rice (Oryza sativa L.). *Physiology and Molecular Biology of Plants* 26, 885–898. DOI: 10.1007/s12298-020-00773-7.

Goverse, A., Overmars, H., Engelbertink, J., Schots, A., Bakker, J. and Helder, J. (2000) Both induction and morphogenesis of cyst nematode feeding cells are mediated by auxin. *Molecular Plant – Microbe Interactions* 13(10), 1121–1129. DOI: 10.1094/MPMI.2000.13.10.1121.

Gupta, B.K., Sahoo, K.K., Ghosh, A., Tripathi, A.K., Anwar, K. *et al*. (2018) Manipulation of glyoxalase pathway confers tolerance to multiple stresses in rice. *Plant, Cell and Environment* 41, 1186–1200. DOI: 10.1111/pce.12968.

Hallauer, A.R. and Darrah, L.L. (1985) Compendium of recurrent selection methods and their application. *Critical Reviews in Plant Sciences* 3(1), 1–33. DOI: 10.1080/07352688509382202.

Hanson, P., Lu, S.F., Wang, J.F., Chen, W., Kenyon, L. *et al*. (2016) Conventional and molecular marker-assisted selection and pyramiding of genes for multiple disease resistance in tomato. *Scientia Horticulturae* 201, 346–354. DOI: 10.1016/j.scienta.2016.02.020.

Haque, M.A., Rafii, M.Y., Yusoff, M.M., Ali, N.S., Yusuff, O. *et al*. (2021) Recent advances in rice varietal development for durable resistance to biotic and abiotic stresses through marker-assisted gene pyramiding. *Sustainability* 13, 10806. DOI: 10.3390/su131910806.

Hasan, M.M., Rafii, M.Y., Ismail, M.R., Mahmood, M., Rahim, H.A. *et al*. (2015) Marker-assisted backcrossing: a useful method for rice improvement. *Biotechnology and Biotechnological Equipment* 29(2), 237–254. DOI: 10.1080/13102818.2014.995920.

He, B., Hu, F., Du, H., Cheng, J., Pang, B. *et al*. (2022) Omics-driven crop potassium use efficiency breeding. *Frontiers in Plant Science* 13, 1076193. DOI: 10.3389/fpls.2022.1076193.

He, C., He, Y., Liu, Q., Liu, T., Liu, C. *et al*. (2013) Co-expression of genes *ApGSMT2* and *ApDMT2* for glycine betaine synthesis in maize enhances the drought tolerance of plants. *Molecular Breeding* 31, 559–573. DOI: 10.1007/s11032-012-9815-7.

He, Y., He, C., Li, L., Liu, Z., Yang, A. *et al*. (2011) Heterologous expression of *ApGSMT2* and *ApDMT2* genes from *Aphanothece halophytica* enhanced drought tolerance in transgenic tobacco. *Molecular Biology Reports* 38, 657–666. DOI: 10.1007/s11033-010-0152-9.

Hittalmani, S., Parco, A., Mew, T.V., Zeigler, R.S. and Huang, N. (2000) Fine mapping and DNA marker-assisted pyramiding of the three major genes for blast resistance in rice. *Theoretical and Applied Genetics* 100, 1121–1128. DOI: 10.1007/s001220051395.

Hospital, F. (2011) Size of donor chromosome segments around introgressed loci and reduction of linkage drag in marker assisted backcross programs. *Genetics* (3), 158–1379. DOI: 10.1093/genetics/158.3.1363.

Hospital, F. and Charcosset, A. (1997) Marker-assisted introgression of quantitative trait loci. *Genetics* 147, 1469–1485. DOI: 10.1093/genetics/147.3.1469.

Huang, N., Angeles, E.R., Domingo, J., Magpantay, G., Singh, S. *et al*. (1997) Pyramiding of bacterial blight resistance genes in rice: marker-assisted selection using RFLP and PCR. *Theoretical and Applied Genetics* 95, 313–320. DOI: 10.1007/s001220050565.

Jackson, R.E., Bradley, J.R., Van Duyn, J.W. and Gould, F. (2004) Comparative production of *Helicoverpa zea* (Lepidoptera: Noctuidae) from transgenic cotton expressing either one or two *Bacillus thuringiensis* proteins with and without insecticide oversprays. *Journal of Economic Entomology* 97, 1719–1725. DOI: 10.1603/0022-0493-97.5.1719.

James, D., Borphukan, B., Fartyal, D., Ram, B., Singh, J. *et al.* (2018) Concurrent overexpression of *OsGS1;1* and *OsGS2* genes in transgenic rice (*Oryza sativa* L.): impact on tolerance to abiotic stresses. *Frontiers in Plant Science* 9, 786. DOI: 10.3389/fpls.2018.00786.

Jha, U.C., Bohra, A. and Jha, R. (2017) Breeding approaches and genomics technologies to increase crop yield under low-temperature stress. *Plant Cell Reports* 36, 1–35. DOI: 10.1007/s00299-016-2073-0.

Jiang, F., Zhang, T., Bai, S., Wang, Z. and He, K. (2016) Evaluation of *Bt* corn with pyramided genes on efficacy and insect resistance management for the Asian corn borer in China. *PLOS ONE* 11, 0168442. DOI: 10.1371/journal.pone.0168442.

Jiang, G.L., Dong, Y., Shi, J. and Ward, R.W. (2007) QTL analysis of resistance to Fusarium head blight in the novel wheat germplasm CJ 9306. II. Resistance to deoxynivalenol accumulation and grain yield loss. *Theoretical and Applied Genetics* 115, 1043–1052. DOI: 10.1007/s00122-007-0630-1.

Jiménez-Mejía, R., Medina-Estrada, R.I., Carballar-Hernández, S., Orozco-Mosqueda, M., Santoyo, G. *et al.* (2022) Teamwork to survive in hostile soils: use of plant growth-promoting bacteria to ameliorate soil salinity stress in crops. *Microorganisms* 10, 150. DOI: 10.3390/microorganisms10010150.

Jin, T., Yin, J., Xue, S., Li, B., Zong, T. *et al.* (2022) *Rsc3K*, a SMV resistance gene localized on chromosome 2 in soybean cv. Kefeng No.1 interacts with virulence determinant P3. *bioRxiv*, preprint. DOI: 10.1101/2021.12.30.474612.

Joseph, M., Gopalakrishnan, S., Sharma, R.K. and Singh, A.K. (2004) Combining bacterial blight resistance and basmati quality characteristics by phenotypic and molecular marker assisted selection in rice. *Molecular Breeding* 13, 377–387. DOI: 10.1023/B:MOLB.0000034093.63593.4c.

Kansiime, M.K., Mugambi, I., Rwomushana, I., Nunda, W., Lamontagne-Godwin, J. *et al.* (2019) Farmer perception of fall armyworm (*Spodoptera frugiderda* JE Smith) and farm-level management practices in Zambia. *Pest Management Science* 75(10), 2840–2850. DOI: 10.1002/ps.5504.

Katiyar, S., Verulkar, S., Chandel, G., Zhang, Y., Huang, B. *et al.* (2001) Genetic analysis and pyramiding of two gall midge resistance genes (*Gm 2* and *Gm 6t*) in rice (*Oryzae sativa* L.). *Euphytica* 122, 327–334. DOI: 10.1023/A:1012965915812.

Khan, Z. and Shahwar, D. (2020) Role of heat shock proteins (HSPs) and heat stress tolerance in crop plants. In: Roychowdhury, R., Choudhury, S., Hasanuzzaman, M. and Srivastava, S. (eds) *Sustainable Agriculture in the Era of Climate Change*. Springer, pp. 211–234.

Kim, Y.S., Kim, I.S., Shin, S.Y., Park, T.H., Park, H.M. *et al.* (2014) Overexpression of dehydroascorbate reductase confers enhanced tolerance to salt stress in rice plants (*Oryza sativa* L. *japonica*). *Journal of Agronomy and Crop Science* 200, 444–456. DOI: 10.1111/jac.12078.

Kopecká, R., Kameniarová, M., Černý, M., Brzobohatý, B. and Novák, J. (2023) Abiotic stress in crop production. *International Journal of Molecular Sciences* 24, 6603. DOI: 10.3390/ijms24076603.

Korinsak, S., Siangliw, M., Kotcharerk, J., Jairin, J., Siangliw, J.L. *et al.* (2016) Improvement of the submergence tolerance and the brown planthopper resistance of the Thai jasmine rice cultivar KDML105 by pyramiding *Sub1* and *Qbph12*. *Field Crops Research* 188, 105–112. DOI: 10.1016/j.fcr.2015.10.025.

Kottapalli, K.R., Lakshmi Narasu, M. and Jena, K.K. (2010) An effective strategy for pyramiding three bacterial blight resistance genes into fine grain rice cultivar, Samba Mahsuri, using sequence tagged site markers. *Biotechnology Letters* 32, 989–996. DOI: 10.1007/s10529-010-0249-1.

Koutsika-Sotiriou, M., Mylonas, I.G., Ninou, E. and Traka-Mavrona, E. (2010) The cultivation revival of a landrace: pedigree and analytical breeding. *Euphytica* 176(1), 15–24. DOI: 10.1007/s10681-010-0206-z.

Kumar, A., Jain, A., Sahu, R.K., Srivastava, M.N., Nair, S. *et al.* (2005) Genetic analysis of resistance genes for the rice gall midge in two rice genotypes. *Crop Science* 5, 1631–1635. DOI: 10.2135/cropsci2004.0406.

Kumar, A., Balachiranjeevi, C.H., Bhaskar Naik, S., Rekha, G., Rambabu, R. *et al.* (2017) Marker-assisted pyramiding of bacterial blight and gall midge resistance genes into RPHR-1005, the restorer line of the popular rice hybrid DRRH-3. *Molecular Breeding* 37: 86. DOI: 10.1007/s11032-017-0687-8.

Kumar, A., Sandhu, N., Dixit, S., Yadav, S., Swamy, B.P.M. *et al.* (2018) Marker-assisted selection strategy to pyramid two or more QTLs for quantitative trait-grain yield under drought. *Rice* 11, 35. DOI: 10.1186/s12284-018-0227-0.

Kumar, P.K., Bellundagi, A., Krishna, H., Mallikarjuna, M.G., Thimmappa, R.K. *et al.* (2023) Development of bread wheat (*Triticum aestivum* L) variety HD3411 following marker-assisted backcross breeding for drought tolerance. *Frontiers in Genetics* 14, 1046624. DOI: 10.3389/fgene.2023.1046624.

Kumaravadivel, N., Uma, M.D., Saravanan, P.A. and Suresh, H. (2006) Molecular marker-assisted selection and pyramiding genes for gall midge resistance in rice suitable for Tamil Nadu Region. In: *Abstracts of the 2nd International Rice Congress, New Delhi,* p. 257.

Kushanov, F.N., Turaev, O.S., Ernazarova, D.K., Gapparov, B.M., Oripova, B.B. *et al.* (2021) Genetic diversity, QTL mapping, and marker-assisted selection technology in cotton (*Gossypium* spp. *Frontiers in Plant Science* 12, 779386. DOI: 10.3389/fpls.2021.779386.

Kushwaha, A.K., Dinkar, V., Thribhuvan, R., Mohan, N., Malpuri, S. *et al.* (2021) Emerging tools and techniques in crop improvement for higher productivity and multiple stress tolerance. In: Rao, C.S., Balakrishnan, M., Krishnan, P. and Kumar, V.V.S. (eds) *Agricultural Research, Technology and Policy: Innovations and Advances. ICAR-National Academy of Agricultural Research Management (NAARM)*. Hyderabad, Telangana, India, pp. 41–78. Available at: http://krishi.icar.gov.in/jspui/handle/123456789/74980

Leelagud, P., Kongsila, S., Vejchasarn, P., Darwell, K., Phansenee, Y. *et al.* (2020) Genetic diversity of Asian rice gall midge based on *mtCOI* gene sequences and identification of a novel resistance locus *gm12* in rice cultivar MN62M. *Molecular Biology Reports* 47(6), 4273–4283. DOI: 10.1007/s11033-020-05546-9.

Lewandowski, I., Härdtlein, M. and Kaltschmitt, M. (1999) Sustainable crop production: definition and methodological approach for assessing and implementing sustainability. *Crop Science* 39(1), 184–193. DOI: 10.2135/cropsci1999.0011183X003900010029x.

Li, D., Pfeiffer, T.W. and Cornelius, P.L. (2008) Soybean QTL for yield and yield components associated with Glycine soja alleles. *Crop Science* 48(2), 571–581. DOI: 10.2135/cropsci2007.06.0361.

Li, X., Ye, J., Munir, S., Yang, T. and Chen, W. (2019) Biosynthetic gene pyramiding leads to ascorbate accumulation with enhanced oxidative stress tolerance in tomato. *International Journal of Molecular Sciences* 20, 1558. DOI: 10.3390/ijms20071558.

Li, Y., Li, J., Yan, Y., Liu, W., Zhang, W. *et al.* (2018) Knock-down of *CsNRT2.1*, a cucumber nitrate transporter, reduces nitrate uptake, root length, and lateral root number at low external nitrate concentration. *Frontiers in Plant Science* 9, 722. DOI: 10.3389/fpls.2018.00722.

Lin, R., Zhao, W., Meng, X., Wang, M. and Peng, Y. (2007) Rice gene *OsNAC19* encodes a novel NAC-domain transcription factor and responds to infection by *Magnaporthe grisea Plant Science* 172, 120–130. DOI: 10.1016/j.plantsci.2006.07.019.

Liu, J., Liu, D., Tao, W., Li, W., Wang, S. *et al.* (2000) Molecular marker-facilitated pyramiding of different genes for powdery mildew resistance in wheat. *Plant Breeding* 119, 21–24. DOI: 10.1046/j.1439-0523.2000.00431.x.

Liu, Y., Chen, L., Liu, Y., Dai, H. and He, J.K. (2016) Marker assisted pyramiding of two brown plant hopper resistance genes, *Bph3 and Bph27 (t)*, into elite rice cultivars. *Rice* 9, 1–7. DOI: 10.1186/s12284-016-0096-3.

Long, W., Li, N., Jin, J., Wang, J. and Dan, D. (2023) Resequencing-based QTL mapping for yield and resistance traits reveals great potential of *Oryza longistaminata* in rice breeding. *Crop Journal* 11(5), 541–1549. DOI: 10.1016/j.cj.2023.03.017.

Lopes, C.M.L., Suassuna, N.D., Cares, J.E., Gomes, A.C.M.M., Perina, F.J. *et al.* (2020) Marker-assisted selection in *Gossypium* spp. for *Meloidogyne incognita* resistance and histopathological characterization of a near immune line. *Euphytica* 216, 19. DOI: 10.1007/s10681-020-2554-7.

Luo, Y., Ma, T., Zhang, A., Ong, K.H., Li, Z. *et al.* (2016) Marker-assisted breeding of the rice restorer line wanhui 6725 for disease resistance, submergence tolerance and aromatic fragrance. *Rice* 9, 66. DOI: 10.1186/s12284-016-0139-9.

Ma, D.M., Xu, W.R., Li, H.W., Jin, F.X., Guo, L.N. *et al.* (2014) Co-expression of the Arabidopsis *SOS* genes enhances salt tolerance in transgenic tall fescue (*Festuca arundinacea schreb*). *Protoplasma* 251, 219–231. DOI: 10.1007/s00709-013-0540-9.

Ma, F., Li, R., Guo, G., Nie, F. and Zhu, L. (2023) Introgression of QTL from *Aegilops tauschii* enhances yield-related traits in common wheat. *Crop Journal* 11(5), 1521–1532. DOI: 10.1016/j.cj.2023.05.001.

Mallick, N., Agarwal, P., Jha, S.K., Niranjana, M. and Vinod (2021) Marker-assisted breeding for rust management in wheat. *Indian Phytopathology* 74(2), 365–370. DOI: 10.1007/s42360-020-00317-9.

Mallick, N., Jha, S.K., Agarwal, P., Kumar, S., Mall, A. *et al.* (2022a) Marker-assisted transfer of leaf and stripe rust resistance from *Triticum turgidum* var. *durum* cv. Trinakria to wheat variety HD2932. *Frontiers in Genetics* 13, 941287. DOI: 10.3389/fgene.2022.941287.

Mallick, N., Jha, S.K., Agarwal, P., Mall, A., Kumar, S. *et al.* (2022b) Marker-assisted improvement of bread wheat variety HD2967 for leaf and stripe rust resistance. *Plants* 11(9), 1152. DOI: 10.3390/plants11091152.

Maruthasalam, S., Kalpana, K., Kumar, K.K., Loganathan, M., Poovannan, K. *et al.* (2007) Pyramiding transgenic resistance in elite Indica rice cultivars against the sheath blight and bacterial blight. *Plant Cell Reports* 26, 791–804. DOI: 10.1007/s00299-006-0292-5.

McIntosh, R.A., Yamazaki, Y., Dubcovsky, J., Rogers, J., Morris, C. *et al.* (2013) Catalogue of gene symbols for wheat. In: *Proceedings of the 12th International Wheat Genetics Symposium, 8–13 September*. Yokohama, Japan. Available at: https://shigen.nig.ac.jp/wheat/komugi/genes/download.jsp

McIntosh, R.A., Dubcovsky, J., Rogers, W.J., Xia, X.C. and Raupp, W.J. (2020) Catalogue of gene symbols for wheat: 2020 supplement. *Annual Wheat Newsletter* 66, 98. Available at: https://wheat.pw.usda.gov/GG3/wgc.

Mehrotra, M., Singh, A.K., Sanyal, I., Altosaar, I. and Amla, D.V. (2011) Pyramiding of modified *cry1Ab* and *cry1Ac* genes of *Bacillus thuringiensis* in transgenic chickpea (*Cicer arietinum* L.) for improved resistance to pod borer insect *Helicoverpa armigera*. *Euphytica* 182, 87–102. DOI: 10.1007/s10681-011-0501-3.

Mhetre, V.B., Patel, V.B., Singh, S.K., Mishra, G.P., Verma, M.K. *et al.* (2022) Unraveling the pathways influencing the berry color and firmness of grapevine cv. Flame Seedless treated with bioregulators using biochemical and RNA-seq analysis under semi-arid subtropics. *Food Chemistry: Molecular Sciences* 5, 100116. DOI: 10.1016/j.fochms.2022.100116.

Mitra, D., Rodriguez, A.M.D., Cota, F.I.P., Khoshru, B., Panneerselvam, P. et al. (2021) Amelioration of thermal stress in crops by plant growth-promoting rhizobacteria. *Physiological and Molecular Plant Pathology* 115: 101679. DOI: 10.1016/j.pmpp.2021.101679.

Mundt, C.C. (2014) Durable resistance: a key to sustainable management of pathogens and pests. *Infection, Genetics and Evolution* 27, 446–455. DOI: 10.1016/j.meegid.2014.01.011.

Mundt, C.C. (2018) Pyramiding for resistance durability: theory and practice. *Phytopathology* 108(7), 792–802. DOI: 10.1094/PHYTO-12-17-0426-RVW.

Muthurajan, R. and Balasubramanian, P. (2010) Pyramiding genes for enhancing tolerance to abiotic and biotic stresses. In: Jain, S. and Brar, D. (eds) *Molecular Techniques in Crop Improvement*. Springer, Dordrecht, Netherlands, pp. 163–184. Available at: http://dx.doi.org/10.1007/978-90-481-2967-6_7

Mwobobia, E.G., Sichangi, A.W. and Thiong'o, K.B. (2020) Characterization of wheat production using earth-based observations: a case study of Meru County, Kenya. *Modeling Earth Systems and Environment* 6(1), 13–25. DOI: 10.1007/s40808-019-00699-4.

Najera, V.A., Twyman, R.M., Christou, P. and Zhu, C. (2019) Applications of multiplex genome editing in higher plants. *Current Opinion in Biotechnology* 59, 93–102. DOI: 10.1016/j.copbio.2019.02.015.

Narayanan, N.N., Baisakh, N., Vera, C.M., Gnanamanickam, S.S., Datta, K. *et al.* (2002) Molecular breeding for the development of blast and bacterial blight resistance in rice cv.IR50. *Crop Science* 42, 2072–2079. DOI: 10.2135/cropsci2002.2072.

Nelson, R.R. (1978) Genetics of horizontal resistance to plant diseases. *Annual Review of Phytopathology* 16(1), 359–378. DOI: 10.1146/annurev.py.16.090178.002043.

Nguyen, T.X., Nguyen, T., Alameldin, H., Goheen, B., Loescher, W. *et al.* (2013) Transgene pyramiding of the *HVA1* and *mtlD* in T3 maize (*Zea mays* L.) plants confers drought and salt tolerance, along with an increase in crop biomass. *International Journal of Agronomy* 203, 598163. DOI: 10.1155/2013/598163.

Nicoletto, C., Maucieri, C., Zanin, G., Vianello, F. and Sambo, P. (2019) Vegetables quality and biotic stress. In: Ansari, R.A. and Mahmood, I. (eds) *Plant Health Under Biotic Stress: Organic Strategies*, Vol. 1. Springer Nature Singapore, pp. 107–128. DOI: 10.1007/978-981-13-6043-5_6.

Nymagoud, S., Tyagi, S., Chandra, A.K., Agarwal, P., Niranjana, M. *et al.* (2022) Development and characterisation of leaf rust resistant *Triticum timopheevii*-derived introgression lines in hexaploid wheat. *Indian Journal of Genetics and Plant Breeding* 82(4), 381–390. Available at: https://www.isgpb.org/journal/index.php/IJGPB/article/view/3802

Olweny, C.O., Opolot, O., Asea, G., Isabirye, D. and Smith, J. (2020) Performance of promiscuous soybean varieties under different management practices in uganda. *Agronomy* 10(5), 713. DOI: 10.3390/agronomy10050713.

Ookawa, T., Hobo, T., Yano, M., Murata, K., Ando, T. *et al.* (2010) New approach for rice improvement using a pleiotropic QTL gene for lodging resistance and yield. *Nature Communications* 1(1), 132. DOI: 10.1038/ncomms1132.

Pancaldi, F. and Trindade, L.M. (2020) Marginal lands to grow novel bio-based crops: a plant breeding perspective. *Frontiers in Plant Science* 11, 496406. DOI: 10.3389/fpls.2020.00227.

Pandey, P., Irulappan, V., Bagavathiannan, M.V. and Senthil, K.M. (2017) Impact of combined abiotic and biotic stresses on plant growth and avenues for crop improvement by exploiting physio-morphological traits. *Frontiers in Plant Science* 8, 537. DOI: 10.3389/fpls.2017.00537.

Pandurangan, S., Workman, C., Nilsen, K. and Kumar, S. (2021) Introduction to marker-assisted selection in wheat breeding. In: Gosal, S.S. and Wani, S.H. (eds) *Accelerated Breeding of Cereal Crops*. Springer, New York, USA, pp. 77–117. DOI: 10.1007/978-1-0716-1526-3_3.

Panigrahi, J., Mishra, R.R., Sahu, A.R., Rath, S.C. and Kole, C.R. (2013) Marker-assisted breeding for stress resistance in crop plants. In: Al-Khayri, J., Jain, S. and Johnson, D. (eds) *Advances in Plant Breeding Strategies: Breeding, Biotechnology and Molecular Tools*. Springer, Cham, Switzerland, pp. 1–23. DOI: 10.1007/978-81-322-0807-5_16.

Pehlivan, N., Sun, L., Jarrett, P., Yang, X., Mishra, N, *et al.* (2016) Co-overexpressing a plasma membrane and a vacuolar membrane sodium/proton antiporter significantly improves salt tolerance in transgenic Arabidopsis plants. *Plant and Cell Physiology* 57, 1069–1084. DOI: 10.1093/pcp/pcw055.

Peng, A., Zhang, J., Zou, X., He, Y., Xu, L. *et al.* (2021) Pyramiding the antimicrobial *PR1aCB* and *AATCB* genes in 'T orange (*Citrus sinensis Osbeck*)) to enhance citrus canker resistance. *Transgenic Research* 30(5), 635–647. DOI: 10.1007/s11248-021-00245-y.

Peterson, R.K. and Higley, L.G. (2001) Illuminating the black box: the relationship between injury and yield. In: Peterson, R.K.D. and Higley, L.G. (eds) *Biotic Stress and Yield Loss*. CRC Press, Boca Raton, Florida, pp. 15–26.

Pham, L.V. and Smith, C. (2014) Drivers of agricultural sustainability in developing countries: A review. *Environment Systems and Decisions* 34, 326–341. DOI: 10.1007/s10669-014-9494-5.

Pikunova, A., Goryunova, S., Goryunov, D., Golyaeva, O., Dolzhikova, M. *et al.* (2022) Genetic diversity and pedigree analysis of red currant germplasm. *Plants* 11(13), 1623. DOI: 10.3390/plants11131623.

Pilet-Nayel, M.L., Moury, B., Caffier, V., Montarry, J., Kerlan, M.C. *et al.* (2017) Quantitative resistance to plant pathogens in pyramiding strategies for durable crop protection. *Frontiers in Plant Science* 8, 295209. DOI: 10.3389/fpls.2017.01838.

Piquerez, S.J.M., Harvey, S.E., Beynon, J.L. and Ntoukakis, V. (2014) Improving crop disease resistance: lessons from research on Arabidopsis and tomato. *Frontiers in Plant Science* 5, 671. DOI: 10.3389/fpls.2014.00671.

Poland, J., Endelman, J., Dawson, J., Rutkoski, J., Wu, S. *et al.* (2012) Genomic selection in wheat breeding using genotyping-by-sequencing. *Plant Genome* 5(3), 103–113. DOI: 10.3835/plantgenome2012.06.0006.

Pradhan, M., Bastia, D., Samal, K.C., Dash, M. and Sahoo, J.P. (2023) Pyramiding resistance genes for bacterial leaf blight (*Xanthomonas oryzae* pv. *oryzae*) into the popular rice variety, Pratikshya through marker assisted backcrossing. *Molecular Biology Reports* 50(11), 9047–9060. DOI: 10.1007/s11033-023-08805-7.

Prasanna, H.C., Sinha, D.P., Rai, G.K., Krishna, R., Kashyap, S.P. *et al.* (2015) Pyramiding *Ty-2* and *Ty-3* genes for resistance tomonopartite and bipartite tomato leaf curl viruses of India. *Plant Pathology* 64, 256–264. DOI: 10.1111/ppa.12267.

Pruthvi, V., Narasimhan, R. and Nataraja, K.N. (2014) Simultaneous expression of abiotic stress responsive transcription factors, *AtDREB2A, AtHB7* and *AtABF3* atdreb2a, athb7 and atabf3 improves salinity and drought tolerance in peanut (*Arachis hypogaea* L.). *PLOS ONE* 9, e111152. DOI: 10.1371/journal.pone.0111152.

Pulindala, S., Bommisetty, R., Withanawasam, D.M., Somagutta, S., Kommana, M. *et al.* (2022) Enhancement of yield through marker-assisted pyramiding of yield contributing genes in rice (Oryza sativa L.) var BPT5204. *Euphytica* 218(10), 148. DOI: 10.1007/s10681-022-03104-x.

Puspito, A.N., Rao, A.Q., Hafeez, M.N., Iqbal, M.S., Bajwa, K.S. *et al.* (2015) Transformation and evaluation of *Cry1Ac+ Cry2A* and *GT* gene in *Gossypium hirsutum* L. *Frontiers in Plant Science* 6, 943. DOI: 10.3389/fpls.2015.00943.

Ragel, P., Raddatz, N., Leidi, E.O., Quintero, F.J. and Pardo, J.M. (2019) Regulation of $K^+$ nutrition in plants. *Frontiers in Plant Science* 10, 440547. DOI: 10.3389/fpls.2019.00281.

Ramalingam, J., Raveendra, C., Savitha, P., Vidya, V., Chaithra, T.L. *et al.* (2020) Gene pyramiding for achieving enhanced resistance to bacterial blight, blast, and sheath blight diseases in rice. *Frontiers in Plant Science* 11, e591457. DOI: 10.3389/fpls.2020.591457.

Ramu, V.S., Swetha, T.N., Sheela, S.H., Babitha, C.K., Rohini, S. *et al.* (2016) Simultaneous expression of regulatory genes associated with specific drought-adaptive traits improves drought adaptation in peanut. *Plant Biotechnology Journal* 14, 1008–1020. DOI: 10.1111/pbi.12461.

Rana, M., Sood, A., Hussain, W., Kaldate, R., Sharma, T.R. *et al.* (2019) Gene pyramiding and multiple character breeding. In: Singh, M. (ed.) *Lentils*. Academic Press, Cambridge, MA, USA, pp. 83–124. DOI: 10.1016/B978-0-12-813522-8.00006-6.

Rani, K., Raghu, B.R., Jha, S.K., Agarwal, P., Mallick, N. *et al.* (2020) A novel leaf rust resistance gene introgressed from *Aegilops markgrafii* maps on chromosome arm 2AS of wheat. *Theoretical and Applied Genetics* 133, 2685–2694. DOI: 10.1007/s00122-020-03625-w.

Reddy, I.N.B.L., Kim, B.K., Yoon, I.S., Kim, K.H. and Kwon, T.R. (2017) Salt tolerance in rice: focus on mechanisms and approaches. *Rice Science* 24(3), 123–144. DOI: 10.1016/j.rsci.2016.09.004.

Ribaut, J.-M., De Vicente, M.C. and Delannay, X. (2010) Molecular breeding in developing countries: challenges and perspectives. *Current Opinion in Plant Biology* 13(2), 213–218. DOI: 10.1016/j.pbi.2009.12.011.

Ruengphayak, S., Chaichumpoo, E., Phromphan, S., Kamolsukyunyong, W., Sukhaket, W. *et al.* (2015) Pseudo-backcrossing design for rapidly pyramiding multiple traits into a preferential rice variety. *Rice* 8, 7. DOI: 10.1186/s12284-014-0035-0.

Sanchez, A.C., Brar, D.S., Huang, N., Li, Z. and Khush, G.S. (2000) Sequence tagged site marker-assisted selection for three bacterial blight resistance genes in rice. *Crop Science* 40, 792–797. DOI: 10.2135/cropsci2000.403792x.

Sandhu, N., Yadaw, R. B., Chaudhary, B., Prasai, H., Iftekharuddaula, K. *et al.* (2019a) Evaluating the performance of rice genotypes for improving yield and adaptability under direct seeded aerobic cultivation conditions. *Frontiers in Plant Science* 10: 159. DOI: 10.3389/fpls.2019.00159.

Sandhu, N., Dixit, S., Swamy, B.P.M., Raman, A., Kumar, S. *et al.* (2019b) Marker assisted breeding to develop multiple stress tolerant varieties for flood and drought prone areas. *Rice* 12, 8. DOI: 10.1186/s12284-019-0269-y.

Schneider, A. (2002) Mapping of a nodulation loci Sym9 and Sym10 of pea. *Theoretical and Applied Genetics* 104, 1312–1316. DOI: 10.1007/s00122-002-0896-2.

Shahwar, D., Khan, Z. and Park, Y. (2024) Molecular markers for marker-assisted breeding for biotic and abiotic stress in melon (*Cucumis melo* L.): a review. *International Journal of Molecular Sciences* 25(12), 6307. DOI: 10.3390/ijms25126307.

Shamsudin, N.A.A., Swamy, B.M., Ratnam, W., Sta. Cruz, M.T., Sandhu, N. *et al.* (2016a) Pyramiding of drought yield qtls into a high quality Malaysian rice cultivar MRQ74 improves yield under reproductive stage drought. *Rice* 9, 2113. DOI: 10.1186/s12284-016-0093-6.

Shamsudin, N.A.A., Swamy, B.P.M., Ratnam, W., Sta. Cruz, Ma.T. and Raman, A. (2016b) Marker assisted pyramiding of drought yield QTLs into a popular malaysian rice cultivar, MR219. *BMC Genetics* 17, 30. DOI: 10.1186/s12863-016-0334-0.

Shi, A., Chen, P., Zheng, C., Hou, A. and Zhu, S. (2006) Gene pyramiding for soybean mosaic virus resistance using microsatellite markers. In: *International Meeting of the ASA-CSSA-SSSA, 12–16 November 2006*.

Shi, A., Chen, P., Li, D., Zheng, C., Zhang, B. *et al.* (2009) Pyramiding multiple genes for resistance to soybean mosaic virus in soybean using molecular markers. *Molecular Breeding* 23, 113–124. DOI: 10.1007/s11032-008-9219-x.

Shinada, H., Iwata, N., Sato, T. and Fujino, K. (2014) QTL pyramiding for improving of cold tolerance at fertilization stage in rice. *Breeding Science* 63, 483–488.

Simmonds, J., Scott, P., Leverington-Waite, M., Turner, A.S., Brinton, J. *et al.* (2014) Identification and independent validation of a stable yield and thousand grain weight QTL on chromosome 6A of hexaploid wheat (*Triticum aestivum* L.). *BMC Plant Biology* 14, 191 1–13. DOI: 10.1186/s12870-014-0191-9.

Singh, A.K., Singh, A., Singh, V.K., Gopala, K.S., Ellur, R.K. *et al.* (2012) Pyramiding genes for bacterial blight and blast resistance into an elite basmati rice restorer line (PRR78) through marker-assisted selection. In: Xie, F. and Hardy, B. (eds) *Public – Private Partnership for Hybrid Rice. Proceedings of the 6th International Hybrid Rice Symposium*, International Rice Research Institute, pp. 261–272 (Vol. 10).

Singh, R.P., Hodson, D.P., Huerta-Espino, J., Jin, Y., Bhavani, S. *et al.* (2011) The emergence of Ug99 races of the stem rust fungus is a threat to world wheat production. *Annual Review of Phytopathology* 49, 465–481. DOI: 10.1146/annurev-phyto-072910-095423.

Singh, R.P., Singh, P.K., Rutkoski, J., Hodson, D.P., He, X. *et al.* (2016) Disease impact on wheat yield potential and prospects of genetic control. *Annual Review of Phytopathology* 54, 303–322. DOI: 10.1146/annurev-phyto-080615-095835.

Singh, S., Sidhu, J.S., Huang, N., Vikal, Y., Li, Z. *et al.* (2001) Pyramiding three bacterial blight resistance genes (*Xa5*, *Xa13* and *Xa21*) using marker-assisted selection into indica rice cultivar PR106. *Theoretical and Applied Genetics* 102, 1011–1015. DOI: 10.1007/s001220000495.

Sliwka, J., Jakuczun, H., Chmielarz, M., Hara-Skrzypiec, A., Tomczyńska, I. et al. (2012) Late blight resistance gene from *Solanum ruiz-ceballosii* is located on potato chromosome X and linked to violet flower colour. *BMC Genetics* 13: 11. DOI: 10.1186/1471-2156-13-11.

Song, L., Wang, R., Yang, X., Zhang, A. and Liu, D. (2023) Molecular markers and their applications in marker-assisted selection (MAS) in bread wheat (*Triticum aestivum* L.). *Agriculture* 13(3), 642. DOI: 10.3390/agriculture13030642.

Stefańczyk, E., Plich, J., Janiszewska, M., Smyda-Dajmund, P., Sobkowiak, S. *et al.* (2020) Marker-assisted pyramiding of potato late blight resistance genes *Rpi-rzc1* and *Rpi-phu1* on di-and tetraploid levels. *Molecular Breeding* 40(9), 89. DOI: 10.1007/s11032-020-01169-x.

Steiner, B., Michel, S., Maccaferri, M., Lemmens, M., Tuberosa, R. *et al.* (2019) Exploring and exploiting the genetic variation of Fusarium head blight resistance for genomic-assisted breeding in the elite durum wheat gene pool. *Theoretical and Applied Genetics* 132, 969–988. DOI: 10.1007/s00122-018-3253-9.

Sun, C., Dong, Z., Zhao, L., Ren, Y., Zhang, N. *et al.* (2020) The wheat 660K SNP array demonstrates great potential for marker-assisted selection in polyploid wheat. *Plant Biotechnology Journal* 18(6), 1354–1360. DOI: 10.1111/pbi.13361.

Sun, Y., Ma, L., Ma, J., Li, B., Zhu, Y. *et al.* (2022) Combined application of plant growth-promoting bacteria and iron oxide nanoparticles ameliorates the toxic effects of arsenic in Ajwain (*Trachyspermum ammi* L. *Frontiers in Plant Science* 13, e1098755. DOI: 10.3389/fpls.2022.1098755.

Suresh, S. and Malathi, D. (2013) Gene pyramiding for biotic stress tolerance in crop plants. *Weekly Science Research Journal* 1, 1–14.

Suryavanshi, P. and Singh, Y. (2020) Assessment of Unnat PBW 343 (rust resistant wheat variety) in Mohali, Punjab. *Indian Journal of Pure and Applied Bioscience* 8(3), 64–69. Available at: http://dx.doi.org/10.18782/2582-2845.8102

Su, X., Chu, Y., Li, H., Hou, Y., Zhang, B. *et al.* (2011) Expression of multiple resistance genes enhances tolerance to environmental stressors in transgenic poplar (*Populus* × *euramericana* 'Guariento'). *PLOS ONE* 6, e24614. DOI: 10.1371/journal.pone.0024614.

Tamanna, T., Islam, M., Chaity, A.R., Shams, S.N.U., Rasel, Md. 2023) Water relation, gas exchange characteristics and yield performance of selected mungbean genotypes under low soil moisture condition. *Agronomy*, 13, 1068. DOI: 10.3390/agronomy13041068.

Tan, M.Y.A., Hutten, R.C.B., Celis, C., Park, T.H., Niks, R.E. *et al.* (2008) The $R_{pi\text{-}mcd1}$ locus from solanum microdontum involved in resistance to *Phytophthora infestans*, causing a delay in infection, maps on potato chromosome 4 in a cluster of NBSLRR genes. *Molecular Plant – Microbe Interactions* 21, 909–918. DOI: 10.1094/MPMI-21-7-0909.

Tan, M.Y.A., Hutten, R.C.B., Visser, R.G. and Eck, H.J. (2010) The effect of pyramiding *Phytophthora infestans* resistance genes $R_{Pi\text{-}mcd1}$ and $R_{Pi\text{-}ber}$ in potato *Theoretical and Applied Genetics* 12, 17–25. DOI: 10.1007/s00122-010-1295-8.

Thitisaksakul, M., Tananuwong, K., Shoemaker, C.F., Chun, A., Tanadul, O.U.M. *et al.* (2015) Effects of timing and severity of salinity stress on rice (*Oryza sativa* L.) yield, grain composition, and starch functionality. *Journal of Agricultural and Food Chemistry* 63(8), 2296–2304. DOI: 10.1021/jf503948p.

Thomson, S.V. and Pfeiffer, D.G. (2010) Strategies for disease resistance breeding in tomato. *Acta Horticulturae* 871, 181–188.

Tiwari, S., SI, K., Kumar, V., Singh, B., Rao, A.R. *et al.* (2016) Mapping QTLs for salt tolerance in rice (*Oryza sativa* L.) by bulked segregant analysis of recombinant inbred lines using 50K SNP chip. *PLOS ONE* 11(4), e0153610. DOI: 10.1371/journal.pone.0153610.

Tura, H., Edwards, J., Gahlaut, V., Garcia, M. and Sznajder, B. (2020) QTL analysis and fine mapping of a QTL for yield-related traits in wheat grown in dry and hot environments. *Theoretical and Applied Genetics* 133(1), 239–257. DOI: 10.1007/s00122-019-03454-6.

Van Loon, L.C. and Strien, E.A. (1999) The families of pathogenesis-related proteins, their activities, and comparative analysis of PR-1 type proteins. *Physiological and Molecular Plant Pathology* 55(2), 85–97. DOI: 10.1006/pmpp.1999.0213.

Varshney, R.K, Mohan, S.M., Gaur, P.M., Gangarao, N.V.P.R., Pandey, M.K. *et al.* (2013) Achievements and prospects of genomics-assisted breeding in three legume crops of the semi-arid tropics. *Biotechnology Advances* 31(8), 1120–1134. DOI: 10.1016/j.biotechadv.2013.01.001.

Venkatesh, B., Vennapusa, A.R., Kumar, N.J., Jayamma, N., Reddy, B.M. *et al.* (2022) Co-expression of stress-responsive regulatory genes, *MuNAC4*, *MuWRKY3* and *MuMYB96* associated with resistant-traits improves drought adaptation in transgenic groundnut (*Arachis hypogaea* L.) plants. *Frontiers in Plant Science* 13, e1055851. DOI: 10.3389/fpls.2022.1055851.

Verlaan, M.G., Hutton, S.F., Ibrahem, R.M., Kormelink, R., Visser, R.G. *et al.* (2013) The tomato yellow leaf curl virus resistance genes *Ty-1* and *Ty-3* are allelic and code for DFDGD-class RNA-dependent RNA polymerases. *PLOS Genetics* 9, 1003399. DOI: 10.1371%2Fjournal.pgen.1003399.

Verma, S., Nizam, S. and Verma, P.K. (2013) Biotic and abiotic stress signaling in plants. In: Sarwat, M., Ahmad, A. and Abdin, M.Z. (eds) *Stress Signaling in Plants: Genomics and Proteomics Perspective*, Vol. 1. Springer Science, New York, USA, pp. 25–49. Available at: http://dx.doi.org/10.1007/978-1-4614-6372-6_2

Vivitha, P., Raveendran, M. and Vijayalakshmi, D. (2017) Introgression of QTLs controlling spikelet fertility maintains membrane integrity and grain yield in improved white Ponni derived progenies exposed to heat stress. *Rice Science* 24, 32–40. DOI: 10.1016/j.rsci.2016.05.006.

Vleeshouwers, V.G.A.A., Rietman, H., Krenek, P., Champouret, N., Young, C. *et al.* (2008) Effector genomics accelerates discovery and functional profiling of potato disease resistance and *Phytophthora infestans* avirulence genes. *PLOS ONE* 3(8), e2875. DOI: 10.1371/journal.pone.0002875.

Waddington, S.R., Li, X., Dixon, J., Hyman, G. and De Vicente, M.C. (2010) Getting the focus right: production constraints for six major food crops in Asian and African farming systems. *Food Security* 2, 27–48. Available at: http://dx.doi.org/10.1007/s12571-010-0053-8

Wang, A., Yu, X., Mao, Y., Liu, Y., Liu, G. *et al.* (2015) Overexpression of a small heat-shock-protein gene enhances tolerance to abiotic stresses in rice. *Plant Breeding* 134(4), 384–393. DOI: 10.1111/pbr.12289.

Wang, D., Chen, S., Huang, Z. and Lin, J. (2022) Identification and mapping of genetic locus conferring resistance to multiple plant viruses in soybean. *Theoretical and Applied Genetics* 135(9), 293–3305. DOI: 10.1007/s00122-022-04187-9.

Wang, J., Wang, J., Hu, M., Wu, S., Qi, J. *et al.* (2016) The amino acid substitution in the effector domain of the resistant allele of the blast resistance gene *Pi54* activates it. *Plant Physiology* 172(2), 901–911. DOI: 10.1104/pp.16.00717.

Wei, A., He, C., Li, B., Li, N. and Zhang, J. (2011) The pyramid of transgenes *TsVP* and *BetA* effectively enhances the drought tolerance of maize plants. *Plant Biotechnology Journal* 9, 216–229. DOI: 10.1111/j.1467-7652.2010.00548.x.

Werner, K., Friedt, W. and Ordon, F. (2005) Strategies for pyramiding resistance genes against the barley yellow mosaic virus complex (*BaMMV*, *BaYMV*, *BaYMV-2*). *Molecular Breeding* 16, 45–55. DOI: 10.1007/s11032-005-3445-2.

Widyasari, K., Alazem, M. and Kim, K.H. (2020) Soybean resistance to soybean mosaic virus. *Plants* 9(2), 219. DOI: 10.3390/plants9020219.

Williamson, V.M. and Kumar, A. (2006) Nematode resistance in plants: the battle underground. *Trends in Genetics* 22(7), 396–403. DOI: 10.1016/j.tig.2006.05.003.

Xu, P. and Cai, W. (2014) *RAN1* is involved in plant cold resistance and development in rice (*Oryza sativa*). *Journal of Experimental Botany* 65(12), 3277–3287. DOI: 10.1093%2Fjxb%2Feru178.

Womack, E.D., Williams, W.P., Smith, J.S., Warburton, M.L. and Bhattramakki, D. (2020) Mapping quantitative trait loci for resistance to fall armyworm (Lepidoptera: Noctuidae) leaf-feeding damage in maize inbred Mp705. *Journal of Economic Entomology* 113(2), 956–963. DOI: 10.1093/jee/toz357.

Yang, Q., Chen, Z.Z., Zhou, X.F., Yin, H.B., Li, X. *et al.* (2009) Overexpression of *SOS* (salt overly sensitive) genes increases salt tolerance in transgenic Arabidopsis. *Molecular Plant* 2(1), 22–31. DOI: 10.1093/mp/ssn058.

Yang, T., Zhang, S., Zhao, J., Liu, Q., Huang, Z. *et al.* (2016) Identification and pyramiding of QTLs for cold tolerance at the bud bursting and the seedling stages by use of single segment substitution lines in rice (*Oryza sativa* L. *Molecular Breeding* 36, 96. DOI: 10.1007/s11032-016-0520-9.

Yang, Y., Su, Q., Li, Y., Cheng, Z. and Song, Y. (2023) Fine mapping of a major QTL *qHYF_B06* for peanut yield. *Crop Journal* 11(5), 1533–1540. DOI: 10.1016/j.cj.2023.04.004.

Yang, Z., Huang, D., Tang, W., Zheng, Y., Liang, K. *et al.* (2013) Mapping of quantitative trait loci underlying cold tolerance in rice seedlings via high-throughput sequencing of pooled extremes. *PLOS ONE* 8, e68433. DOI: 10.1371/journal.pone.0068433.

Ye, C., Ishimaru, T., Lambio, L., Li, L., Long, Y. *et al.* (2022) Marker-assisted pyramiding of QTLs for heat tolerance and escape upgrades heat resilience in rice (*Oryza sativa* L.). *Theoretical and Applied Genetics* 135, 1345–1354. DOI: 10.1007/s00122-022-04035-w.

Ye, G. and Smith, K.F. (2009) Marker-assisted gene pyramiding for cultivar development. In: Janick, J. (ed.) *Plant Breeding Reviews*, Vol. 33. John Wiley & Sons, Hoboken, New Jersey, USA, pp. 219–256. DOI: 10.1002/9780470535486.ch5.

Yusop, M.R., Sukaimi, J., Amiruddin, M.D., Jalloh, M., Swaray, S. *et al.* (2020) Genetic improvement of oil palm through recurrent selection. *The Oil Palm Genome* 35–46. Available at: http://dx.doi.org/10.1007/978-3-030-22549-0_4

Zeigler, R.S., Tohme, J., Nelson, J., Levy, M. and Correa, F. (1994) Linking blast population analysis to resistance breeding: a proposed strategy for durable resistance. In: Zeigler, R.S., Leong, S. and Teng, P.S. (eds) *Rice Blast Disease*. CAB International, Wallingford, UK, pp. 267–292.

Zhang, B., Chi, D., Hiebert, C., Fetch, T., McCallum, B. *et al.* (2019a) Pyramiding stem rust resistance genes to race TTKSK (Ug99) in wheat. *Canadian Journal of Plant Pathology* 41, 443–449. DOI: 10.1080/07060661.2019.1596983.

Zhang, J., Fengler, K.A., Van Hemert, J.L., Gupta, R. and Mongar, N. (2019b) Identification and characterization of a novel stay-green QTL that increases yield in maize. *Plant Biotechnology Journal* 17(12), 2272–2285. DOI: 10.1111/pbi.13139.

Zhang, Y., Deng, G., Fan, W., Yuan, L., Wang, H. *et al.* (2019c) *NHX1* and *eIF4A1*-stacked transgenic sweet potato shows enhanced tolerance to drought stress. *Plant Cell Reports* 38, 1427–1438. DOI: 10.1007/s00299-019-02454-6.

Zhang, Y., Wang, Y., Wang, E., Wu, X., Zheng, Q. *et al.* (2021) *SlPHL1*, a MYB-CC transcription factor identified from tomato, positively regulates the phosphate starvation response. *Physiologia Plantarum* 173, 1063–1077. DOI: 10.1111/ppl.13503.

Zhao, C., Zhang, H., Song, C., Zhu, J.-K. and Shabala, S. (2020) Mechanisms of plant responses and adaptation to soil salinity. *The Innovation* 1(1), 100017. Available at: http://dx.doi.org/10.1016/j.xinn.2020.100017

Zhao, G., Liu, Y., Li, L., Che, R., Douglass, M. *et al.* (2024) Gene pyramiding for boosted plant growth and broad abiotic stress tolerance. *Plant Biotechnology Journal* 22, 678–697. DOI: 10.1111/pbi.14216.

Zheng, Z., Liang, S., Wenbin, H., Bin, Z., Shuhua, T. *et al.* (2021) Breeding high-grain quality and blast resistant rice variety using combination of traditional breeding and marker-assisted selection. *Rice Science* 28(5), 422. DOI: 10.1016/j.rsci.2021.07.002.

Zhou, S., Chen, X., Zhang, X. and Li, Y. (2008) Improved salt tolerance in tobacco plants by co-transformation of a betaine synthesis gene *BADH* and a vacuolar $Na^+/H^+$ antiporter gene senhx1 *SeNHX1*. *Biotechnology Letters* 30, 369–376. DOI: 10.1007/s10529-007-9548-6.

Zhu, S., Li, Y., Vossen, J.H., Visser, R.G.F. and Jacobsen, E. (2012) Functional stacking of three resistance genes against *phytophthora infestans* in potato. *Transgenic Research* 21, 89–99. DOI: 10.1007%2Fs11248-011-9510-1.

Zhu, Z., Xu, X., Cao, B., Chen, C. and Chen, Q. (2015) Pyramiding of *AtEDT1/HDG11* and *Cry2Aa2* cry2aa2 into pepper (*Capsicum annuum* L.) enhances drought tolerance and insect resistance without yield decrease. *Plant Cell Tissue and Organ Culture* 120, 919–932. DOI: 10.1007/s11240-014-0600-7.

# 4 Genome Wide Selection for Accelerating the Efficiency of Molecular Breeding for Crop Improvement

**Sadhan Debnath[1*], Kirti Devi[2] and Ng. Tombisana Meetei[3]**

*[1]National Institute of Plant Genome Research (NIPGR), Aruna Asaf Ali Marg, New Delhi, India; [2]National Agri-food Biotechnology Institute (NABI), Mohali, Punjab, India; [3]College of Post Graduate Studies in Agricultural Sciences (CPGSAS), Central Agricultural University (Imphal), Umiam, Meghalaya, India*

**Abstract**

In the pursuit of addressing global food security challenges, the integration of advanced genomics technologies with traditional breeding methods has become imperative. Molecular breeding uses molecular marker data for enhancing the potency and productiveness of various breeding programmes, including proper designing and implementation of breeding programmes and also improving selection efficiency. Genome-wide selection has emerged as a powerful strategy, leveraging high-throughput phenotyping and genotyping data to accelerate the speed of crop improvement. It involves the simultaneous assessment of thousands to millions of genetic markers distributed across the entire genome, allowing for the prediction of breeding value of an individual. The integration of this approach into molecular breeding pipelines enables the identification of superior genotypes at an early stage, reducing the breeding cycle duration and resource requirements. The synthesis of genomics data with advanced statistical and regression models propels the field towards a future where crop improvement is not only faster but also more precise and tailored to meet the dynamic demands of an accelerating population. This chapter mainly explores the principles, methodologies and applications of GWS in molecular breeding, with the primary aim on enhancing efficiency in crop improvement programmes. It also underscores the significance of large-scale genotyping platforms, such as SNP genotyping and next-generation sequencing in providing comprehensive genomic information for accurate prediction models. It also explains the incorporation of multi-omics data, including transcriptomics and metabolomics, to enhance the precision of genomic predictions and unravel complex genotype – phenotype relationships.

**Keywords:** genomics, genomic selection, genome-wide association studies, marker-assisted selection, integrated genomic selection, single-nucleotide polymorphisms, genome-wide selection

**Abbreviations**

ANN – Artificial neural network, AI – Artificial intelligence, BLUP – Best linear unbiased prediction, BRR – Bayesian ridge regression, CSA – Climate-smart agriculture, CSC – Climate-smart crop, GAB – Genomics-assisted breeding, GEBV – Genomic estimated breeding values, G × E – Genotype–environment interaction, GHGs – Greenhouse gases, GHI – Global Hunger Index, GS – Genomic selection, GWAS – Genome-wide association studies, IGS – Integrated genomic selection, LASSO – Least absolute shrinkage and selection operator, LSR – Least-squares regression, MABC – Marker-assisted backcrossing, MARS – Marker-assisted recurrent selection,

*Corresponding author: debnathsadhan123@gmail.com





DOI: 10.1079/9781800626638.0004

MAS – Marker-assisted selection, MRCE – Multivariate regression with covariance estimation, MTGS – Multi-trait genomic selection, QTL – Quantitative trait locus, RF – random forest, RR – Ridge regression, SNP – single-nucleotide polymorphism, STGS – Single trait genomics selection, SVM – support vector machine, TPEs – Target population of environments.

## 4.1 Introduction

Global population estimates indicate that within the next 35 years, there will be an additional 2 billion people on Earth, bringing the total population from 8 billion in 2022 to 10 billion by 2057. Providing adequate food for an extra 2 billion people by 2057 – a quarter of the current global population – presents a challenge to global food security. Moreover, as per United Nations* UN anticipation, India will be the most populous country, with an estimated population of 1.67 billion by 2050 (UN Population Division, 2022). According to anticipations, food production must increase by at least 60% by 2050 due to the impact of climate change on the production system (WHO/FAO, 2015). It is expected that these challenges will be exacerbated by an increase of 1–29% in food prices by 2050. Moreover, there will be a 55–60% rise in food demand from 2010 levels (World Economic Forum, 2020). According to the Global Hunger Index (GHI) report (2023), there has been an increase in the ubiquity of undernourishment, one of the factors used in the estimation of GHI scores, rising from 7.5% in 2017 to 9.2% in 2022. During this time, the global population of undernourished individuals rose from 572 million to 735 million (FAO, 2023). Following the start of the COVID-19 epidemic, an additional 150 million individuals have experienced food insecurity. Sustainable food production and nutrition are becoming increasingly important given the rate at which food insecurity is increasing. According to the Census Report (2023), 45 million children were categorized as wasted (too thin for height) and 149 million young children under 5 years of age were classed as stunted (too short for age). Furthermore, it was noted that 37 million kids worldwide were overweight or obese. Undernourishment accounted for 45% of child deaths in middle- and low-income countries, where juvenile obesity rates were also increasing (Fact Sheet – Malnutrition, 2024). Most victims of food security are under 5 years old. Clearly, more food must be produced, alongside other essential improvements.

In addition to increasing food demand and population growth, climate change is exacerbating food uncertainty by endangering crop production and agricultural sustainability. The physical characteristics of several agricultural crop plants have been affected by climate change, which has a negative impact on crop production and productivity. Throughout the next 5 years, there will be record-breaking global temperatures due to the combination of an El Niño phenomenon and greenhouse gasses that trap heat. There is a 98% possibility that the next 5-year period will be the warmest (WMO, 2023). As a result, the environment, water management, food security and health will all suffer. Several stresses including heatwaves, droughts, cold weather and flooding are examples of how climate change is manifesting itself. Abiotic stresses also significantly affect a plant's capacity to grow and produce. In their natural environments, plants face a variety of climatic challenges or abiotic stresses, such as dry spells, high temperatures, UV-B radiation, changes in light intensity, heavy metals, flooding and other chemical elements. To avert major harm to global food security in the next two to three decades, creative techniques that use genetic enhancement to improve the production and productivity of staple food crops are essential (Voss-Fels *et al.*, 2019). Therefore, climate smart high-yielding cultivars are needed to overcome these obstacles.

## 4.2 Future Food Security: Adoption of Climate-Smart Crops (CSC)

A continuously rising global population and rapid climate change pose huge threats and risks to the agricultural ecosystem and our environment. Climate change has impacted agricultural crop productivity and is predicted to cause a 30%

decrease in the yield of staple crops (Habib-ur-Rahman *et al.*, 2022). Plant breeding has been instrumental in fulfilling rising food demands since crop domestication. However, it has not yet shown to be adequate and has been superseded by cutting-edge integrated genomics techniques and contemporary molecular breeding protocols. But, as the consequences of climate change are intensifying day by day, it is vital to produce climate-smart crop (CSC) varieties or climate-resilient crops, to implement climate-smart agriculture (CSA) (Rashid *et al.*, 2022). These CSCs have been designed with the essential characters to tolerate dynamic and multidimensional stresses with higher crop production and efficient regulation of plant biomass (Rai, 2022). As a result, CSA is imperative to address the problems associated with climate change and the global rising food demand. A comprehensive strategy for managing landscapes that includes crops, livestock, forests and fisheries, CSA aims to mitigate the effects of climate change while maintaining food security.

Increasing productivity, improving tolerance to biotic and abiotic stresses, and lowering emission patterns to safeguard the agri-ecosystem are the three objectives that CSA is aiming to simultaneously achieve. Climate-smart agriculture can increase crop yield, lower the probability of "crop failure" and lessen the adverse consequences of climate change, among other methods, to improve food security (Ali *et al.*, 2023). Growing crop plants by following CSA can aid in sustainable development, since crop biomass has the potential to both create and remove greenhouse gases (GHGs). The development and successful adoption of CSCs that offer tolerance against various biotic and abiotic stress are openly supported by important international organizations, such as the World Bank, International Monetary Fund (IMF), Food and Agriculture Organization (FAO), and International Center for Agricultural Research in the Dry Areas (ICARDA) (www.worldbank.org;).

## 4.3 From Conventional Breeding to Marker-Assisted Selection (MAS)

Through conventional breeding, it has been possible to successfully improve the genetic makeup of crops through phenotypic selection. Conventional breeding techniques have produced high-yielding, nutrient-rich crops that can be mechanically harvested to meet the global population's expanding need for food. However, currently the rate of yield improvement for some important food crops, such as rice (*Oryza sativa*), wheat (*Triticum aestivum*) and maize (*Zea mays*), are not high enough to meet future demand (Hickey *et al.*, 2019). The agricultural production has increased annually by 0.8–1.2% due to the significant contribution made by modern crop-breeding and management practices. But by the 2050s, genetic advancement will not be able to keep up with the demands of the world's expected population. Using conventional breeding methods to achieve the necessary pace of genetic progress is difficult, especially considering the degrading effects of climate change on soil and the reduction of useable water resources. Furthermore, it is difficult to improve yields and agronomic traits using conventional breeding methods because they are largely genetically complex and greatly influenced by outside factors (Krishnappa *et al.*, 2021). Conventional methods of hybridization and selection need a lot of effort, land and time to generate climate-resilient cultivars, which limits the genetic gain that may be achieved.

But, recently, advancement in the field of genomics and molecular breeding has revealed numerous important genes and quantitative trait loci (QTLs) that confer resistance to various stresses, and they have been employed in marker-assisted selection (MAS). According to Fernando and Grossman (1989), MAS is an indirect selection procedure in which candidates for a particular character of interest are chosen depending on the established molecular markers associated with it. This technology allows the addition and removal of genetic material or altering the genetic materials at specific locations in the genome by replacing an existing sequence with a customized DNA segment (Shahwar *et al.*, 2024). When selecting individuals for plant breeding, this approach has been more efficient than the conventional phenotype-based selection procedure, in terms of selection accuracy. However, MAS is only really practicable if the desired trait of interest is linked to one or few key genes, because MAS can enhance qualitative traits regulated by a small number of genes with

**Table 4.1.** List of improved varieties developed using MAS in various crops.

| Varieties developed | Crop | Reference |
|---|---|---|
| Improved Pusa Basmati1 | Rice | Gopalakrishnan *et al.*, 2008 |
| Pusa Basmati 1728 | Rice | Singh *et al.*, 2017a |
| Pusa Basmati 1637 | Rice | Singh *et al.*, 2017b |
| Pusa Samba 1850 | Rice | Krishnan *et al.*, 2019 |
| Improved Pusa RH10 | Rice | |
| Improved Samba Mahsuri | Rice | Madhavi *et al.*, 2016 |
| Swarna-Sub1 | Rice | Neeraja *et al.*, 2007 |
| Sambha Mahsuri-Sub1 | Rice | |
| HUW510 | Wheat | Vasistha *et al.*, 2017 |
| Vivek QPM 9 | Maize | |
| HHB67 | Pearl millet | Rai *et al.*, 2008 |
| HHB 67-2 | Pearl millet | |
| C214 | Chickpea | Varshney *et al.*, 2014a |
| JTN5503 and DS880 | Soybean | Arelli *et al.*, 2006; Arelli *et al.*, 2009 |
| JL24 and TAG24 | Groundnut | Varshney *et al.*, 2014b |

significant impacts (Bernardo, 2008). Some of the successful examples of varieties developed using MAS are listed in Table 4.1.

## 4.4 Genomic Selection

Marker-assisted selection has little effect on enhancing quantitative qualities that are influenced by multiple genes. As the most stress tolerance-related traits are quantitative traits (i.e. polygenic characters controlled by several hundreds of minor genes), MAS is irrelevant or impractical. To address this problem, a novel method of selection, known as genomic selection (GS), was proposed. It is a potent method that selects individuals for the subsequent breeding cycle depending on their genomic estimated breeding values (GEBVs), which are derived from genome-wide markers. This method estimates the individual impacts of every marker and uses the cumulative marker effects to get the GEBV of each individual. Using gene-sequencing methods, GS examines diversity within a crop population. Lande and Thompson (1990) first proposed GS in 2000, and then Meuwissen *et al.* (2001) popularized it. The net genetic merit of an individual is determined by analysing the impact of numerous markers dispersed across the genome – to select for such features, GS can help with the selection process (Meuwissen *et al.*, 2001). As a subset of marker-assisted breeding (MAB), genomic selection ensures that each quantitative trait locus (QTL) is in linkage disequilibrium (LD) with a minimum of one genetic marker. This guarantees a successful selection of targeted characteristics. Because of its ability to maximize genetic gains, decrease phenotyping, shorten the breeding cycle and enhance selection efficiency, GS has been widely used in animal breeding programmes worldwide during the past 20 years. Furthermore, the possibility of incorporating GS into breeding crops is being investigated, given the encouraging results of the early study of GS for enhancing yield performance, abiotic and biotic stress tolerance, and quality in cereal crops such as wheat, maize and rice. Two effective data science methods for examining marker-trait associations (MTAs) for the fundamental comprehension of simple and complicated adaptive and functional traits are GS and genome-wide association studies (GWAS) (Varshney *et al.*, 2021). Both strategies effectively shorten the breeding cycle and quicken the pace of genetic gain in crops.

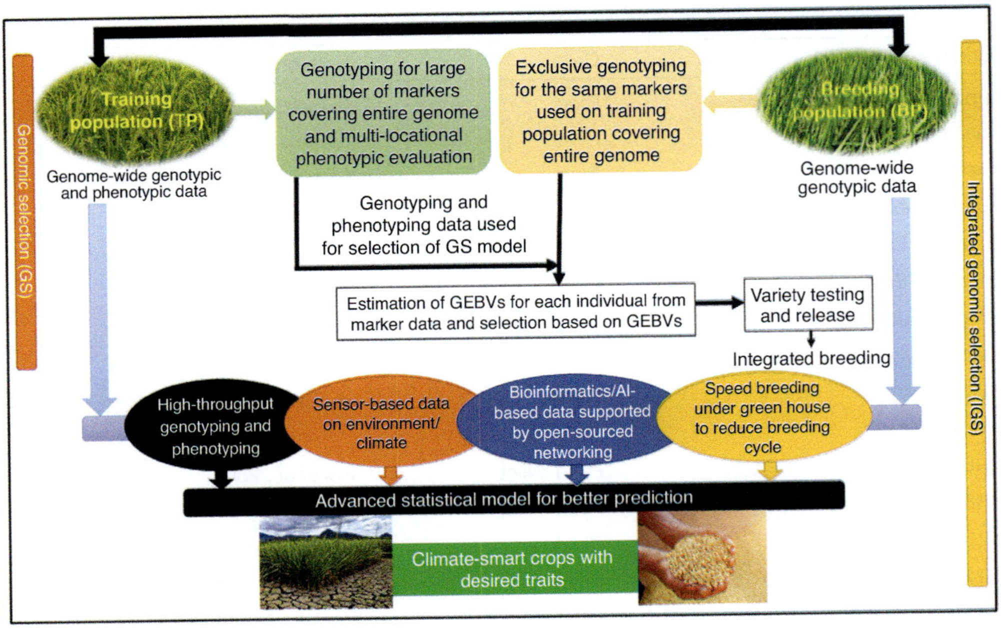


**Fig. 4.1.** General methodology and steps of genomic selection (GS) and integrated genomic selection (IGS).

## 4.5 General Methods Involved in Genomic Selection

Two distinct but closely related populations, namely a training population and a breeding population, are the foundation of the GS technique (Heffner *et al.*, 2009) (Fig. 4.1). The training population is mainly used to estimate the marker-related effects required for estimating the GEBVs of an individual or line in a breeding population, as well as to train the GS model, whereas, in a breeding population, the individuals undergo genomic selection in order to isolate superior lines for use as new varieties or parents of newly improved hybrids. Figure 4.1 also explains the basic GS process.

The following steps are followed in a regular GS method:

1. Establishing a training population appropriate for the relevant breeding population is the initial stage in a GS programme.
2. A huge number of molecular markers uniformly dispersed at an appropriate density throughout the entire genome are genotyped for the individuals/lines in the training population.
3. The training population's individuals/lines undergo comprehensive phenotypic assessment for the targeted character or traits in replicated trials conducted over the locations and years.
4. Model training, which involves calculating the variables of the GS model using the marker genotype and phenotypic data, is carried out. The calculations of the GS model variables are retained for future implementation in the breeding population.
5. The same markers that were utilized to calculate the model variables in the training population are also utilized to evaluate the breeding population. However, the breeding population is not phenotypically evaluated.
6. The genotypic data obtained by using molecular markers and the predicted marker-related effects from the training population are used to determine the GEBVs of the individuals/lines in the breeding population.
7. Depending on the estimated GEBV, the higher or superior individuals/lines are selected from the breeding population.

## 4.6 Training Population

In the genomic selection, the fundamental step is to create a training population, or a group of individuals possessing both phenotypic and genotypic information. This population is then utilized to develop a model in which the genotype is utilized as a predictor and the phenotype as a responder. Afterwards, the data from the created model are utilized to calculate the GEBV of the breeding population, i.e. the individuals/lines with solely genotypic data. It is imperative that the training population accurately represents the breeding population. It ought to increase the percentage of trait variance connected to the markers. This can be accomplished by incorporating individuals or lines with different GEBVs into the population. There should be little collinearity between the markers in the training population. Recombination breaks down the collinearity of markers, hence the individuals or lines that make up a particular training population should have gone through multiple series of recombination. As higher collinearity tends to degrade the accuracy of prediction of a particular GS model, low collinearity between markers is required. Lastly, the training population should appropriately display the genetic variability available in the breeding population. This can be accomplished by choosing lines/individual from the breeding population and incorporating them based on some kind of clustering, and involving them in the training population (Jannink *et al.*, 2010).

Choosing the makeup of the training population is still arguably one of the trickiest decisions in the GS process. The parents or the immediate ancestors of the breeding population in question should make up a sizable portion of the training population. On the other hand, it's possible that the breeding population and the training population are unconnected. Given this, it is expected that the two populations will differ in terms of marker and/or QTL allele frequencies, as well as the impact of their respective genetic origins. According to recent studies, for high GEBV accuracy, members of the breeding population should be included in the training population (Rutkoski *et al.*, 2016). A training population usually contains factual data, or it could be an actual population made-up of existing individuals. Several real training populations can be generated for plants, including biparental crossings, doubled haploid testcrosses, and intermated inbred lines. Ideally, a new training population should be generated for each of the breeding population. Because the breeding population and the training population would be directly associated, this strategy will produce GEBV predictions with a high degree of accuracy. Consequently, the two populations will have similar QTL effects, moderate allele frequencies and genetic backgrounds. However, this will require a precise phenotypic assessment in the target environments of a distinct training population for every breeding population. It is ideal for the breeding and training populations to share a similar genetic makeup and be closely connected.

GEBV forecast accuracy rises approximately linearly with population size. As a result, the training population needs to be sufficiently large, especially if it is not connected to the breeding population tightly. Furthermore, as the number of the successful breeding population increases, the size of the training population should also increase. In addition, compared to traits with higher heritability, traits with lower heritability should be trained on a larger population. When genetic diversity is higher, breeding populations are smaller, characteristics are less heritable and there are more QTLs controlling a trait, the ratio of training to breeding populations should generally be higher. Lastly, compared to cross-pollinated species, the proper training population size for self-pollinated species would be less.

## 4.7 Factors Affecting the Accuracy of GEBV Estimates

To predict the GEBVs of individuals or lines in GS, a genotyped and phenotyped training population is employed. A number of variables may affect the accuracy of the genomic prediction, including the size of the breeding and training populations, the genetic variability of the breeding population, the heritability of the principal trait, the impact of genotype and environment (G × E) interactions, the density/frequency of markers, and the genetic relatedness between the breeding and training populations or selection candidates (Howard *et al.*, 2014). Moreover, the estimates of GEBV are significantly impacted by variations in epistatic interactions and gene frequencies. The effect of epistasis, which is crucial, particularly in cross-pollinated plants, is ignored by the majority of models used to estimate GEBV (Heffner *et al.*, 2009). The Pearson correlation between GEBV and true breeding value estimations can be used to determine how accurate GEBV estimates are. This accuracy metric is directly related to gain from selection based on GEBV. Therefore:

$$R = ir\Sigma A \quad (4.1)$$

where, R means response to selection, i means intensity of selection, r means accuracy of the GEBV estimates, and ΣA means square root of additive genetic variance of the true breeding value estimates. Hence, the following elements would influence the effectiveness of a GS:

- the methodology used for calculation of marker effects;
- the polygenic effect term based on kinship;
- the procedure of phenotypic assessment of the training population;
- the types of markers and marker density; and
- the heritability of the character and the number of QTLs involved.

## 4.8 Various Statistical Models for GS

Improved statistical models that utilize genetic data to increase prediction accuracy are essential for the success of GS-enabled breeding operations. Implementing GS requires careful model selection, and numerous genomic prediction models, both parametric and non-parametric, are available. Choosing the right candidates for GS begins with a basic linear model, usually known as ordinary least-squares regression or least-squares regression:

$$Y = 1_n\mu + X\beta + \varepsilon \quad (4.2)$$

Here, $Y = n \times 1$ vectors of observations, $\mu$ = mean, X = design matrix of order $n \times p$ (where each row represents the genotype/individuals/lines (n) and each column corresponds to the marker (p)), $\beta = p \times 1$ vectors of marker effects, $\varepsilon = n \times 1$ vectors of random residual effects, and $\varepsilon \sim N(0, \sigma^2_e)$.

The fact that there are more markers (p) than observations (n), i.e. genotype/individuals/lines, in linear models with thousands of genome-wide markers is a significant issue. This leads to the overparameterization issue (large 'p' and small 'n' problem ($p$>>n)). To solve the large 'p' and small 'n' problem, an alternate approach is to use a subset of significant markers. For GS, Meuwissen *et al.* (2001) modified the least-squares regression method, where they applied least-squares regression analysis on individual marker separately with the following model:

$$Y = X_j\beta_j + \varepsilon \quad (4.3)$$

where $X_j$ means the j[th] column of the design matrix of the markers and $\beta_j$ indicates the genetic effect of the j[th] marker.

The log likelihood of the model is used to pick markers that have significant impacts, and those markers are then used to estimate breeding values. It must be acknowledged, none the less, that selection depending on the subset of markers may eliminate some important information. Ridge regression (RR), which is a penalized regression-based technique, is thus a productive way to solve the over-parameterization issue in linear models (Meuwissen *et al.*, 2001). Simultaneously, it resolves multi-collinearity issues (correlated predictors, such as single-nucleotide polymorphisms (SNPs) or markers). RR uses ℓ2 penalized least squares to solve the regression issue by uniformly shrinking the coefficients of connected predictors towards zero. Another kind of penalized regression, very much related to RR, is the least absolute shrinkage and selection operator (LASSO) (Tibshirani, 1996; Usai *et al.*, 2009). It employs the ℓ1 penalized least-squares principal to produce a sparse solution. A version of LASSO that is resistant to extremely high correlations between the predictors is the elastic net (ENET) (Friedman *et al.*, 2010).

Not all traits exhibit equal variance contributions from each marker, as assumed by the RR model. Consequently, it is important to speculate the variation of the markers based on the genetic makeup of the character. To achieve this goal, a number of Bayesian models with the assumption of a prior marker effect distribution have been put forth. Moreover, conclusions regarding the parameters of the model are put up from the rear distributions of the marker effects. For genomic estimation, there are a number of variations of Bayesian models, which include Bayes A, Bayes B, π, and Dπ (Meuwissen *et al.*, 2001; Habier *et al.*, 2011), Bayesian LASSO and Bayesian ridge regression (BRR) including various derivatives. Besides the models which depend on markers, the best linear unbiased prediction (BLUP) is another most widely used genomic prediction method (Henderson *et al.*, 1959). The BLUP model is one of the most acceptable and extensively utilized parametric genomic prediction models. The estimation of the marker effects is done by a whole-genome regression approach based on mixed models, which has already been effectively utilized to predict various complex characters (Habier *et al.*, 2009, 2013). In general, only traits with additive genetic architectures were shown to have efficient parametric model performance. Using parametric models becomes difficult for features that are strongly influenced by non-additive or epistatic interactions (Moore and Williams, 2009).

Many BLUP variations are available for this kind of application, such as GBLUP (genomic BLUP), ssGBLUP (single-step GBLUP), RRBLUP (ridge regression BLUP) and rrGBLUP (GBLUP with linear ridge kernel regression); among these, GBLUP is the most commonly utilized. Unlike the traditional BLUP, which determines the GEBV of the individuals/lines or using pedigree relationships, GBLUP involves genetic relationships determined based on markers (Meuwissen *et al.*, 2001).

However, these models perform poorly when there are epistatic interactions and their performance drastically degrades. For this reason, Gianola *et al.* (2006) employed semiparametric and non-parametric techniques to represent complicated genetic architecture. The modelling of both epistatic and additive effects for genomic selection was then done using several statistical techniques (Cai *et al.*, 2011; Legarra and Reverter, 2018). Numerous non-parametric techniques have been investigated in connection with genomic selection, such as the reproductive kernel Hilbert space (Gianola *et al.*, 2006), artificial neural network (ANN) (Gianola *et al.*, 2011), random forest (RF) (Holliday *et al.*, 2012), and support vector machine (SVM) (Maenhout *et al.*, 2007; Long *et al.*, 2011); among these, NN, SVM and RF are mainly based on a machine learning approach.

When there is pleiotropy, i.e. one gene connected to many traits, the performance of these single trait genomics selection (STGS) based approaches may be greatly impacted, and the accuracy of the genomic predictions may also suffer. These days, with the abundance of traits available to us, multi-trait genomic selection (MTGS) based techniques could yield a greater prediction accuracy and a more accurate GEBV. Various MTGS based approaches have been used with regard to GS, namely the Bayesian multi-trait model (Jia and Jannink, 2012; Cheng *et al.*, 2018), multivariate mixed model approach (Jia and Jannink, 2012; Klápště *et al.*, 2020), conditional Gaussian graphical model (Chiquet *et al.*, 2017) and multivariate regression with covariance estimation (MRCE) (Rothman *et al.*, 2010). In addition, Jia and Jannink (2012) proposed three multivariate linear models: GBLUP, Bayes A and Bayes Cπ, and compared with univariate models.

## 4.9 Statistical Tools for Implementing GS

The need for analysis of a huge number of markers, and the related expenditure for producing them, are the primary barriers to GS adoption. However, the economic considerations have been somewhat eliminated by recent advancements in high-throughput next-generation and third generation DNA sequencing technologies. A variety of software tools, including STGS, MTGS, RRBLUP and BreadWheat GS (BWGS) are available for the incorporation and assessment of vast volumes of data produced by high-throughput sequencing and phenotyping. One of the original GS techniques was RR. A number of packages and tools have been created for the assessment of genomic prediction and GS implementation, some of them are explained in Table 4.2.

**Table 4.2.** List of statistical tools for efficient implementation of GS.

| Statistical tools | Remarks |
|---|---|
| GMStool | • A GWAS-based tool, which uses genome-wide marker data<br>• Looks for sufficient numbers of markers using suitable statistical and deep learning/machine learning-based models<br>• Identifies the top 100 SNP markers, which have lowest *P* values in the GWAS<br>• Algorithm is mainly divided into three general steps: data preparation > marker selection > final prediction<br>• R-based tool, available freely at: https://github.com/JaeYoonKim72/GMStool (accessed 6 September 2024) |
| solGS | • An open-sourced tool that is Linux based<br>• Three available approaches for prediction: trial approach, trait-based approach and custom lists approach<br>• Designed to store a huge amount of phenotypic, genotypic and experimental data<br>• Uses two R packages: 'nlme' for data preprocessing and 'rrBLUP' for statistical modelling<br>• Algorithm is essentially divided into two steps: training of the prediction model and obtaining GEBV |
| rrBLUP | • One of the most used R packages for genomic prediction in both animal and plant breeding programmes<br>• The R package is based on a mixed linear model (BLUP)<br>• The package estimates the marker effects from training datasets for estimation of the GEBV of individuals<br>• The most widely used function of this package is the *mixed.solve* function, which is a linear mixed model equation<br>• An open-source package, which can be accessed at https://CRAN.R-project.org/package=rrBLUP (accessed 6 September 2024) |
| BWGS | • BreedWheat genomic selection is an integrated pipeline that is based on R<br>• Mainly consists of three modules: (i) imputation of missing data; (ii) reduction of dimension by reducing number of markers to increase the speed of computation on large datasets; and (iii) GEBV estimation<br>• Consists of nearly 15 parametric and non-parametric statistical models, which could be used for estimation of GEBVs for a wide range of genetic architectures<br>• Two main functions are: *bwgs.cv* (used for dimension reduction, missing value imputation and cross validation) and *bwgs.predict* (used for estimation of GEBV and model calibration)<br>• Freely available at https://CRAN.R-project.org/package=BWGS (accessed 6 September 2024) |

*Continued*

**Table 4.2.** Continued

| Statistical tools | Remarks |
|---|---|
| lme4GS | • An R-based package, and an extension of the lme4 R package, which is the standard package mainly used for fitting linear mixed models<br>• Also considered as an extension of the RRBLUP package<br>• Available freely via the GitHub repository at https://github.com/perpdgo/lme4GS (accessed 6 September 2024) |
| GSelection | • An R-based package that is comprised of a set of functions for selecting important markers and estimating GEBV by using an integrated model framework<br>• Was developed by integration of the best performing model from both additive and non-additive genetic models<br>• Available freely at: https://CRAN.R-project.org/package=GSelection (accessed 6 September 2024) |
| STGS | • An R-based package mainly developed for genomic predictions by estimating marker effects and GEBV<br>• Performs GS based on a single trait, hence 'single-trait genomic selection' (STGS)<br>• A comprehensive package for a single-step solution for GS, and based on widely used statistical methods like LASSO, RR, BLUP, ANN, SVM and RF<br>• Available freely through the CRAN server at: https://cran.r-project.org/web/packages/STGS/index.html (accessed 24 September 2024) |
| MTGS | • An R-based package mainly developed for genomic predictions by estimating marker effects based on information available on multiple traits<br>• A comprehensive package for a single-step solution for GS; based on some of the widely used MTGS methods, such as multivariate LASSO (MLASSO), MRCE, and kernelized multivariate LASSO (KMLASSO)<br>• Available freely through the CRAN server at: www.r-project.org/ (accessed 24 September 2024) |
| GenSel | • Mainly used for the estimation of breeding values of animals for any trait of interest based on molecular marker<br>• Works through a command line interface (MAC or Linux) or as a user-friendly tool<br>• Bayesian algorithm is mainly used for the estimation of marker effects from training population and GEBV for breeding population<br>• Can be accessed through the GitHub repository at: https://github.com/austin-putz/GenSel (accessed 6 September 2024) |
| BGLR | • An extension of the BLR package<br>• Offers versatility with regard to prior density distribution and is utilized in the implementation of several Bayesian models<br>• In the public domain and available freely via the CRAN mirror at: https://CRAN.R-project.org/package=BGLR (accessed 6 September 2024) |

*Continued*

## 4.10 Benefits of GS Over Conventional Breeding and Marker-Assisted Selection (MAS)

The fundamental traits that serve as the foundation for the overall plant breeding and management strategy are the phenotypic behaviours of plants. Morphological or phenotype-based (marker) selection strategies are the foundation of traditional plant breeding techniques (Wang *et al.*, 2018). Because phenotype is influenced by several loci in the genome and the outcome of G × E interactions, accurate phenotypic prediction is challenging. Due to their susceptibility to environmental influences and the growth cycle, phenotypic selection techniques are indirect and ineffective (Wang *et al.*, 2015). In contemporary plant breeding systems, phenotypic selection gave way to a variety of alternative forms of selection techniques, including cytological, biochemical and DNA/molecular markers, in an effort to increase crop output. However, modern molecular MAS-based methods were created to address the issues with the above techniques (Manzo-Sánchez *et al.*, 2015). In 1986, Beckmann and Soller coined the phrase 'marker-assisted selection', by which molecular markers are utilized for the indirect selection of superior genotypes. Genomic selection, which is a variation of MAS that utilizes SNP marker data, and the markers cover the entire genome of an organism. However, the efficiency, throughput and accuracy of conventional selection, marker-assisted selection and genomics selection vary considerably (Table 4.3) (Fig. 4.2).

## 4.11 Integrated Genomic Selection for Producing Climate-Smart Crops

It is now well acknowledged that climate change poses a serious threat. The global food demand – supply mismatch has widened because of a decline in the yield of important food crops (Ray *et al.*, 2012). As a result, it is now essential – and a top research goal – to develop crops resistant to biotic and abiotic stresses (Steinwand and Ronald, 2020). The foundation of climate-smart agriculture is climate-smart crops, which are resilient to harsh climatic circumstances. Additionally, by increasing output, they can effectively meet the rising demand for food. As the MAS was found to be ineffective in acquiring all the advantageous and commercially important alleles throughout crop improvement, plant breeding has acknowledged and embraced genomic selection as a highly powerful breeding tool for quick and efficient selection of superior genotype with minimal breeding period (Anilkumar *et al.*, 2022).

Better selection intensity through precise and high-throughput phenotyping can provide an advancement in the genetic gain as per the Lush equation (Lush, 1943), as does having a large genetic basis representing various ecogeography in breeding programmes. By combining GS with high-throughput genotyping and phenotyping, machine learning or deep learning based approaches, and speed-breeding techniques, more advancements were achieved. Integrated genomic selection (IGS) is the collective term for this kind of integration (Sandhu *et al.*, 2022). When genome sequences are available, it allows breeders to implement GS in bigger breeding programmes with a variety of useful applications. A new age in plant breeding and a genetic revolution have been brought about by the recent advancements in sequencing technologies. Genomic selection is one of the flexible plant breeding selection techniques, which is based on DNA sequencing technologies. Results from integrated genomic selection demonstrate improved GS accuracy and a large genetic gain/unit time with less effort.

When combined with a phenological model, GS in Asian rice (*O. sativa* L.) showed increased prediction accuracy in predicting the emergence date of untested genotypes under untested conditions (Onogi *et al.*, 2016). For varietal protection and nutritive characteristics, sensor-based phenomics employed in the IGS strategy beyond biomass aided in the growth of perennial ryegrass (*Lolium perenne* L.) breeding (Malmberg *et al.*, 2023). Similarly, moderately accurate early selection can be achieved by greenhouse phenotyping combined with GS. As demonstrated in a population of *Populus deltoides*, it can be used in connection with reproductive strategies that reduce breeding time and aid in accelerating genetic gains (Alves *et al.*, 2020)

**Table 4.3.** Comparison between marker-assisted selection (MAS) and genomic selection (GS) (Jannink *et al.*, 2002; Sandhu *et al.*, 2022; Parveen *et al.*, 2023).

| Character | MAS | GS |
|---|---|---|
| Number of markers involved | The trait is indirectly selected by using a genetic marker linked to the genomic loci governing the character | A variation of MAS and based on GEBVs calculated by a trained GS model utilizing the effects of all the markers |
| Nature of trait | Effective for major QTL traits with significant impacts and oligogenic characteristics | Effective for polygenic characteristics (major and minor QTLs) that have both major and minor effects |
| Prerequisite | QTL mapping and confirmation of molecular markers associated with the trait | Model selection and training a suitable GS model in a TP utilizing phenotype and genotype data |
| Approach | Targeted approach where selected markers linked to a few validated major QTLs are used | Holistic approach whereby all the markers used in training a GS model are used to implement GS in the BP |
| Nature of population | Any crop population if the QTL is validated, which is very unlikely. Comparatively less successful in enhancing quantitative traits | Feasible on a BP that is a subset of the TP or associated with it. Incredibly successful at enhancing quantitative traits |
| Implementation | To supplement any of the traditional breeding techniques, such as MA-Pedigree, MA-Recurrent selection, or MA-Backcross | More suitably applied in accordance with developing breeding |
| Genetic gain | There is little genetic gain per unit of time, and QTL identification and validation take a lot of time | Since all QTLs with significant and modest impacts are taken into account, the genetic gain per unit of time is relatively high |
| Disadvantage | In quantitative traits, linkage drag, background noise and environmental instability may be observed | Factors influencing prediction accuracy |
| Suitability | Numerous chromosomes, heterozygosity, high polyploidy, complex genome, and low/medium-density markers | Enhances heritability assessment, anticipates the breeding potential of particular lines, covers the entire genome, is preferred in purebred breeding across numerous animal species, and improves breeding efficiency and prediction |

BP – Breeding population, TP – Training population.

When environmental data are projected using climate change models, or obtained from historical weather databases, they can provide useful metadata about phenology when applied to crop models. By assisting with the investigation of the target population of environments (TPEs), these data can offer improved insights into TPEs that exist today and in the future. This method can also aid in the better design of phenotypic testing (Heslot *et al.*, 2014). There has also been

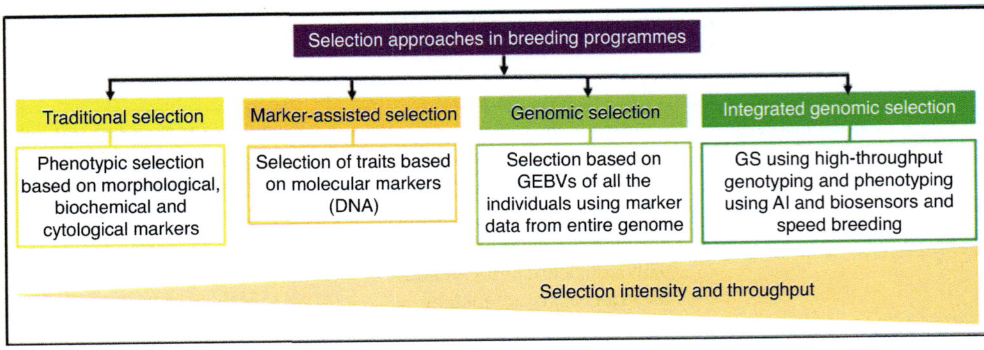


**Fig. 4.2.** A comparison of various selection approaches involved in plant breeding programmes.

assessment of the possibilities of GS and combining omics data for evaluating wheat diseases.

Additionally, researchers have stressed that sophisticated plant breeding programmes cannot simply adopt GS techniques from the breeding of dairy cattle. Since genomic selection uses additive genetic effects to forecast genetic outcomes, it can improve crop enhancement and genetic advancement by including a variety of yet-undiscovered sources of genetic diversity as non-additive effects. This increases the significance of IGS. It is challenging to enhance grain production in breeding programmes and crop improvement, particularly wheat breeding, because of the genotype and environment interaction (Yang *et al.*, 2007)

However, it was discovered that IGS was an effective tool for making decisions when detecting genotypes (Fiddy *et al.*, 2006). It could be used in conjunction with breeders' observations, expertise in germplasm, and experience both inside and outside of a breeding programme to maximize yield, as was noted for the simultaneous selection that produced superior wheat varieties for grain output, protein content, and dough rheological traits associated with baking quality. With a large hybrid maize population, the target-oriented priority (TOP) machine learning approach has been developed to implement an integrated multi-trait breeding strategy. The approach was found to be robust and dependable in the study, and it is expected to provide support in breeding decisions throughout a comprehensive search for germplasm demonstrating high yield and resilience to climate fluctuation (Michel *et al.*, 2019).

Huge resources including transcriptomes, proteomes, and genome sequence data have become available due to advancements in genomics techniques. These resources have accelerated the process of identifying target genes that mitigate the consequences of climate change (Varshney *et al.*, 2018). This informational sequence also results in the identification of several nucleotide-level mutant loci that may be linked to complicated traits, such as yield in general and under various stress conditions, which are otherwise highly challenging to interpret. IGS has become a significant technique that may be used to simulate crop yield under various environmental conditions and make quick and efficient selections to solve production difficulties in a world where climate change is occurring.

## 4.12 Integration of Artificial Intelligence, Speed Breeding and Genomics-Assisted Breeding (GAB) for Accelerated Crop Improvement

A set of methods called genomics-assisted breeding (GAB) is used to improve crop plants (Vishwakarma *et al.*, 2022). Three methods, namely marker-assisted recurrent selection (MARS), marker-assisted backcrossing (MABC) and marker-assisted selection (MAS), require knowledge of the genes linked to certain markers controlling particular traits of interest (Zargar *et al.*, 2015). These methods are helpful for breeding variations in traits controlled by major or key genes (Bhat and Yu, 2021), like higher nutritional quality and stress tolerance (Vishwakarma *et al.*, 2022). However, for complex quantitative characters like yield, quality and stress tolerance, phenotypic selection, which is mostly driven by minute gene variations, may be superior to classic GAB approaches (Bhat and Yu, 2021). On the other hand, the more recent GS method appears to have potential for breeding complex characters (Bhat *et al.*, 2016) using marker data covering an entire genome with at least one marker associated with each gene regulating the trait variation. With greater accuracy and precision than conventional techniques, GAB's environment-independent selection methodology has the potential to hasten the production of superior cultivars (Varshney *et al.*, 2021). Even though GAB is an essential strategy for creating better crop cultivars resistant to climate change, proper phenotyping data is necessary to identify marker – trait correlations using linkage or GWAS analysis and to train GS models for more exact genomic predictions. Accurate phenotypic data collection, large-scale gene identification, and genomic prediction are essential for the success of AI applications in GAB. To develop improved agricultural cultivars more quickly, the speed-breeding approach shortens breeding cycles. It has demonstrated considerable promise to swiftly and accurately generate superior agricultural cultivars by combining speed-breeding methods with different GAB methodologies, such as MAS and GS. Figure. 4.3 clearly explains

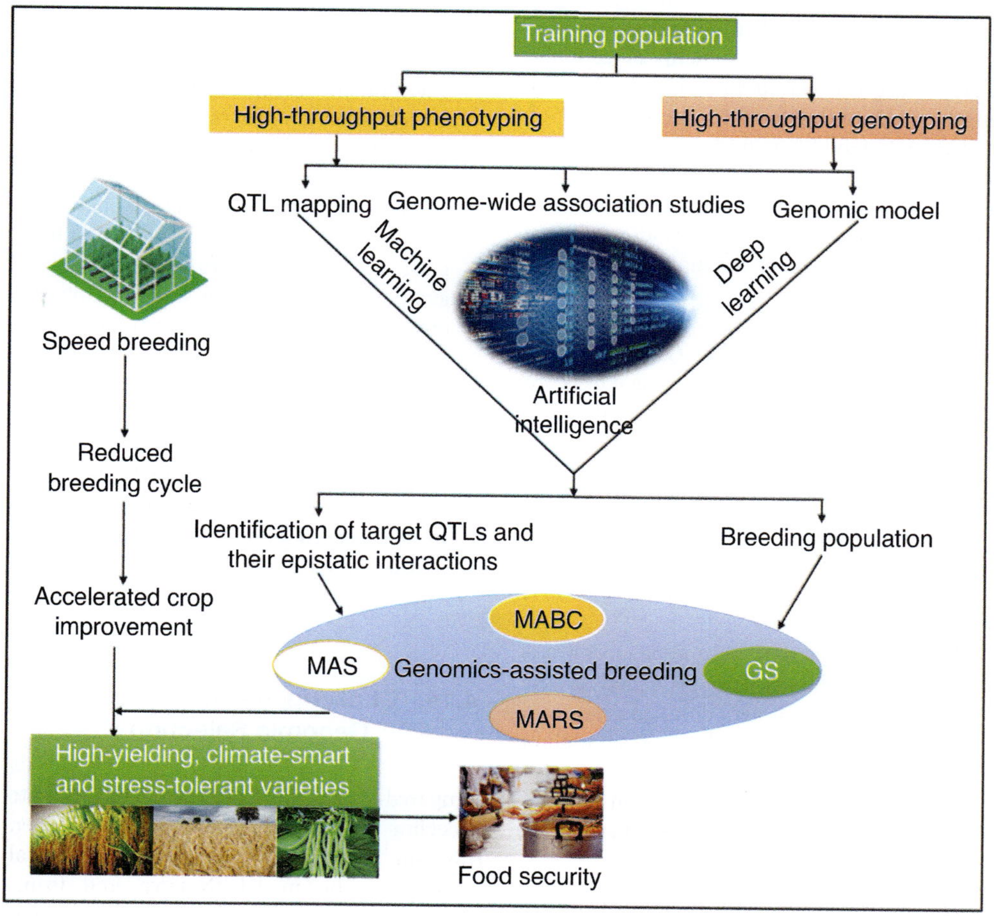


**Fig. 4.3.** A model for accelerated crop improvement using AI, speed breeding and genomics-assisted breeding.

how AI, speed breeding, and GAB could be integrated for accelerated crop improvement.

## 4.13 Advantages of Genomic Selection

According to Heffner *et al.* (2010), the main benefit of using GS is that it can significantly shorten the breeding cycle compared to traditional breeding and minimize the cost of extensive phenotyping. This leads to an acceleration of genetic gains and ensures food and nutritional security. Because of the abundance of single nucleotide polymorphisms (SNPs) found by genome sequencing, and the advancement of novel methods for effectively genotyping large numbers of SNPs, this strategy is now more feasible (Goddard and Hayes, 2007). Among the many benefits of genomic selection are the elimination of the need for lengthy field testing and the acceleration of the transmission of genetic gain at a lower cost and with shorter generation intervals. Potential benefits also include the ability to assess additive and non-additive genetic variations (Li and Dungey, 2018). Some of the major benefits are as follows:

1. There is no need for QTL discovery, mapping, etc. because the marker effects are estimated from the training population and used directly for GS in the relevant breeding population.
2. It is evident from both simulation and empirical research that GS outperforms phenotypic selection in terms of gains made per unit of time (Bernardo and Yu, 2007; Spindel *et al.*, 2015).
3. The phenotype data on one or more traits can be combined with or added to the genome-wide marker data to create a selection index. According to simulation studies, this combined selection index strategy of GS improves selection efficacy, especially for traits with low heritability (Heffner *et al.*, 2010).
4. Compared to selection based on breeding values inferred from phenotype data, GS would typically lower the rate of inbreeding and the loss of genetic variation without lowering selection gains. This might be especially crucial for species that exhibit higher inbreeding depression.
5. In GS, phenotyping is not necessary for each selection cycle in the breeding population. This drastically shortens the breeding cycle, especially for perennial plants.
6. Rapid crop improvement is made possible by genomic selection, which eliminates the need for a thorough examination of each locus
7. The selection cycle is shortened because GS does not call for phenotypic selection, which necessitates assessing testcross performance of the plants under selection. Compared to phenotypic selection, GS is anticipated to enable the commercialization of better genotypes in perennial species at substantially shorter time intervals (Wong and Bernardo, 2008).
8. Based on GEBVs of selected lines from target environment, breeders may select parents for hybridization programmes from among those lines that have not been evaluated in the target environment.
9. GS can incorporate data on trait phenotype and marker genotype that have been gathered throughout time in a range of evaluation programmes covering diverse environments into GEBV estimations of the different individuals/lines. As a result, GEBV estimate would be possible even for traits for which they have never been tested.
10. Production of superior genetic gains per unit of time and cost, as well as an increase in both speed and accuracy of the GS-assisted breeding process.

## 4.14 Challenges and Limitations of Genomic Selection

Compared to pedigree-based selection, the rate of decline in selection response in GS is higher. This can be reduced by including additional markers in the model (Nakaya and Isobe, 2012). On the other hand, GS implementation is more expensive than a conventional breeding programme. However, the regular use of GS into crop-breeding initiatives requires that GS models be improved to take into consideration non-additive effects and genotype–environment interaction (GEI), in addition to lowering related expenses. One of the major challenges is that there is currently insufficient evidence to support GS's practical utility; the plant breeding community has not yet completely adopted GS. Indeed, most conversations about its value are statistical analyses and simulations that plant breeders find difficult to understand. Also, as GS has been primarily examined in simulation studies, its potential worth should be judged cautiously. To prove the efficacy of GS in breeding scenarios, evaluation in these conditions is desperately needed. Additive genetic variances based simulation models have been used to assess the accuracy of GEBV estimates. These models do not consider epistatic effects, which is unrealistic (Heffner *et al.*, 2009).

The other challenge with GS is that the selection response decreases more quickly than it does under pedigree selection. This can be reduced by consistently adding additional markers for the prediction of GEBVs. By giving more weight to the low-frequency advantageous alleles, especially early in the GS programme, it is also possible to improve the long-term response under GS (Nakaya and Isobe, 2012). When up to three generations of plants are grown annually in off-season greenhouse facilities, GS is more productive per unit of time than phenotypic selection. In situations when such facilities are not available, the value and cost-effectiveness of GS would be questioned.

Breeding programmes become much more expensive since every generation of selection requires the genotyping of a huge number of marker loci. Future research indicates that the application of marker data is expected to receive more attention than phenotypic data. However, to achieve such a shift, a single-marker data point's cost would need to be just 1/5000 of the total cost of phenotyping an entry. Additionally, large-scale infrastructure and other resources would be needed to implement GS, which might be outside the budget of the majority of medium-sized public sector breeding projects, especially in developing nations. GS also requires a lot of preparation and execution, and breeders would need to change the way they approach their breeding programmes.

## 4.15 Future Prospects

The branch of molecular markers is constantly evolving, with the development of novel marker types, such as epigenetic variation, copy-number variation and variation at genomic sequence level due to natural mutation, as well as genotyping systems. To increase the accuracy of GEBV estimates, it would be necessary to establish suitable statistical tools and methods using AI and machine learning for the proper exploitation of novel marker systems, which would generate qualitatively and quantitatively distinct information. To enable the practical application of GS in the future crop improvement and speed-breeding programmes, appropriate user-friendly software tools for applying these statistical advancements must be created concurrently. GS now has an R package accessible at www.r-project.org (accessed 9 September 2024). Furthermore, for the enormous volumes of data produced by plant breeding projects – on a variety of topics, such as the lines under evaluation – the test conditions, genotyping and phenotype information, etc. must be preserved in suitable formats in user-friendly and AI-enabled databases. Establishing suitable data management systems is necessary for the effective integration and use of data kept in different databases. A few of these public sector databases are currently in use, e.g. The Hordeum Toolbox of the Barley Coordinated Agricultural Project (https://barley.triticeaetoolbox.org/, accessed 24 September 2024) and the Canadian COOL-DUDE. Practical use of the GS technique will be substantially facilitated by linking these and other comparable instruments with GS and by developing suitable software packages for genomic selection (Heffner *et al.*, 2009).

Appropriate guidelines must be developed for the creation of adequate training populations, the addition of new lines to them, and the selection of the best population for a particular breeding programme or goal. The training population needs to be built with the least amount of phenotyping resource demand possible to enable reliable GEBV prediction over time. By adding new lines established in the breeding population and restricting the phenotyping to the new inclusions, the population can be updated on a regular basis. The association between the trait phenotype and the marker genotype may be captured in different ways by different methods for estimating marker effects. As a result, several of these techniques might be complementary to one another, and combining them could increase the accuracy of GEBV predictions. Parametric techniques, such as ridge regression and Bayes B, outperformed machine learning and non-parametric techniques in simulated research. Unexpectedly, the simple mean of all the techniques outperformed each one individually, indicating that a meta-predictor might end up being the most accurate. It is necessary to thoroughly investigate these problems by empirical research and modelling (Jannink *et al.*, 2010).

Compared to selection based on breeding values inferred from pedigree data, GS is predicted to reduce inbreeding and preserve more genetic diversity. None the less, GS does record some genetic relationship information, which raises the possibility of inbreeding. Therefore, it might be necessary to control the rate of inbreeding during GS, for instance by utilizing marker data. Changing the weight assigned to marker information based on allele frequency at each marker locus is one way to address this (Goddard, 2009).

Currently, available GS models typically only include additive gene effects. In the near future, GS models will likely adequately account for the dominant gene effects; nevertheless, it is still difficult to include epistatic interactions in a reasonable manner. The development of machine learning techniques and semiparametric GS models will make it possible to incorporate epistatic interactions into regression models. Based on existing research, incorporating epistatic gene effects significantly improves GS accuracy. Furthermore, the impact of genetic background on QTL expression is not considered in the current GS models. The impact of interactions like this, which exist in training and/or breeding populations, on the precision of GEBV estimations is essentially unknown. GS is probably going to make it easier to develop optimal genotypes and produce them through targeted GS. The main flaw in the GS method appears to be its lack of appreciation for knowledge of the biological processes that underlie the emergence of the relevant traits. This is because the GS algorithms ignore a number of genomics and genetics-related discoveries, such as the identification of QTLs and the genes they represent. Creating GS algorithms that leverage the discoveries in the disciplines of genetics and genomics could prove to be a lucrative endeavour. It seems that GS is where the next stage of MAS begins. Plant breeding strategies are likely to incorporate genetic diversity as long as the associated problems are sufficiently addressed and the talks about GS grow less theoretical and quantitative. Moreover, the gradual decrease in marker genotyping expenses will hasten the integration of GS into standard plant breeding initiatives. However, combining phenotypic, genotypic and environmental datasets with additional AI based algorithms utilizing omics datasets will enable the accurate selection of superior and stable agricultural varieties. Furthermore, the GS breeding cycle duration would be further decreased by combining speed breeding with AI and GS, facilitating the quick creation of reliable crop cultivars that perform better in a variety of mega-environments.

## References

Ali, H., Menza, M., Hagos, F. and Haileslassie, A. (2023) Impact of climate-smart agriculture adoption on food security and multidimensional poverty of rural farm households in the central rift valley of Ethiopia. *Agriculture and Food Security* 11, 62. DOI: 10.1186/s40066-022-00401-5.

Alves, F.C., Balmant, K.M., Resende, M.F.R., Kirst, M. and Los Campos, G. (2020) Accelerating forest tree breeding by integrating genomic selection and greenhouse phenotyping. *Plant Genome* 13, e20048. DOI: 10.1002/tpg2.20048.

Anilkumar, C., Sunitha, N.C., Harikrishna Devate, N.B. and Ramesh, S. (2022) Advances in integrated genomic selection for rapid genetic gain in crop improvement: A review. *Planta* 256, 87. DOI: 10.1007/s00425-022-03996-y.

Arelli, P.R., Young, L.D. and Mengistu, A. (2006) Registration of high yielding and multiple disease-resistant soybean germplasm JTN-5503. *Crop Sci* 46, 2723–2724. DOI: 10.2135/cropsci2005.12.0471crg.

Arelli, P.R., Young, L.D. and Concibido, V.C. (2009) Inheritance of resistance in soybean PI 567516C to LY1 nematode population infecting cv. Hartwig. *Euphytica* 165, 1–4. DOI: 10.1007/S10681-008-9760-Z.

Bernardo, R. (2008) Molecular markers and selection for complex traits in plants: Learning from the last 20 years. *Crop Science* 48, 1649–1664. DOI: 10.2135/CROPSCI2008.03.0131.

Bernardo, R. and Yu, J. (2007) Prospects for genomewide selection for quantitative traits in maize. *Crop Science* 47, 1082–1090. DOI: 10.2135/CROPSCI2006.11.0690.

Bhat, J.A. and Yu, D. (2021) High-throughput NGS-based genotyping and phenotyping: Role in genomics-assisted breeding for soybean improvement. *Legume Science* 3, e81. DOI: 10.1002/leg3.81.

Bhat, J.A., Ali, S., Salgotra, R.K., Mir, Z.A., Dutta, S. *et al.* (2016) Genomic selection in the era of next generation sequencing for complex traits in plant breeding. *Frontiers in Genetics* 7, 221. DOI: 10.3389/fgene.2016.00221.

Cai, X., Huang, A. and Xu, S. (2011) Fast empirical bayesian LASSO for multiple quantitative trait locus mapping. *BMC Bioinformatics* 12, 211–213. DOI: 10.1186/1471-2105-12-211/FIGURES/5.

Census Report (2023) Available at: https://censusindia.gov.in/census.website/ (accessed 18 November 2024).

Cheng, H., Kizilkaya, K., Zeng, J., Garrick, D. and Fernando, R. (2018) Genomic prediction from multiple-trait Bayesian regression methods using mixture priors. *Genetics* 209, 89–103. DOI: 10.1534/GENETICS.118.300650/-/DC1.

Chiquet, J., Mary-Huard, T., Robin, S. and Robin, S. (2017) Structured regularization for conditional Gaussian graphical models. *Computational Statistics* 27, 789–804. DOI: 10.1007/s11222-016-9654-1.

Fact Sheets—Malnutrition (2024) Available at: www.who.int/news-room/fact-sheets/detail/malnutrition (accessed 18 November 2024).

FAO (2023) Available at: www.fao.org/documents/card/en?details=cc3017en (accessed 18 November 2024).

Fernando, R. and Grossman, M. (1989) Marker assisted selection using best linear unbiased prediction. *Genetics Selection Evolution* 21, 467–477. DOI: 10./1297-9686-21-4-467.

Fiddy, S., Cattermole, D., Xie, D., Duan, X.Y. and Mott, R. (2006) An integrated system for genetic analysis. *BMC Bioinformatics* 7, 210. DOI: 10.1186/1471-2105-7-210.

Friedman, J., Hastie, T. and Tibshirani, R. (2010) Regularization paths for generalized linear models via coordinate descent. *Journal of Statistical Software* 33, 1–22. DOI: 10.18637/jss.v033.i01.

Gianola, D., Fernando, R.L. and Stella, A. (2006) Genomic-assisted prediction of genetic value with semiparametric procedures. *Genetics* 173, 1761–1776. DOI: 10.1534/GENETICS.105.049510.

Gianola, D., Okut, H., Weigel, K.A. and Rosa, G.J. (2011) Predicting complex quantitative traits with Bayesian neural networks: A case study with Jersey cows and wheat. *BMC Genetics* 12, 87. DOI: 10.1186/1471-2156-12-87.

Global Hunger Index (GHI) (2023) Available at: www.globalhungerindex.org/pdf/en/2023.pdf (accessed 18 November 2024).

Goddard, M. (2009) Genomic selection: Prediction of accuracy and maximisation of long term response. *Genetica* 136, 245–257. DOI: 10.1007/s10709-008-9308-0.

Goddard, M.E. and Hayes, B.J. (2007) Genomic selection. *Journal of Animal Breeding and Genetics* 124, 323–330. DOI: 10.1111/j.1439-0388.2007.00702.x.

Gopalakrishnan, S., Sharma, R.K., Anand Rajkumar, K., Joseph, M., Singh, V.P. *et al.* (2008) Integrating marker assisted background analysis with foreground selection for identification of superior bacterial blight resistant recombinants in Basmati rice. *Plant Breeding* 127, 131–139. DOI: 10.1111/J.1439-0523.2007.01458.

Habib-ur-Rahman, M., Ahmad, A., Raza, A., Hasnain, M.U., Alharby, H.F. *et al.* (2022) Impact of climate change on agricultural production: Issues, challenges, and opportunities in Asia. *Frontiers in Plant Science* 13, 925548. DOI: 10.3389/fpls.2022.925548.

Habier, D., Fernando, R.L. and Dekkers, J.C.M. (2009) Genomic selection using low-density marker panels. *Genetics* 182, 343–353. DOI: 10.1534/GENETICS.%20108.100289.

Habier, D., Fernando, R.L., Kizilkaya, K. and Garrick, D.J. (2011) Extension of the Bayesian alphabet for genomic selection. *BMC Bioinformatics* 12, 186–197. DOI: 10.1186/1471-2105-12-186/FIGURES/2.

Habier, D., Fernando, R.L. and Garrick, D.J. (2013) Genomic BLUP decoded: A look into the black box of genomic prediction. *Genetics* 194, 597–607. DOI: 10.1534/GENETICS.113.152207.

Heffner, E.L., Sorrells, M.E. and Jannink, J.L. (2009) Genomic selection for crop improvement. *Crop Science* 49, 1–12. DOI: 10.2135/CROPSCI2008.08.0512.

Heffner, E.L., Lorenz, A.J., Jannink, J.L. and Sorrells, M.E. (2010) Plant breeding with genomic selection: Gain per unit time and cost. *Crop Science* 50, 1681–1690. DOI: 10.2135/cropsci2009.11.0662.

Henderson, C.R., Kempthorne, O., Searle, S.R. and von Krosigk, C.M. (1959) The estimation of environmental and genetic trends from records subject to culling. *Biometrics* 15, 192. DOI: 10.2307/2527669.

Heslot, N., Akdemir, D., Sorrells, M.E. and Jannink, J.L. (2014) Integrating environmental covariates and crop modeling into the genomic selection framework to predict genotype by environment interactions. *Theoretical and Applied Genetics* 127, 463–480.

Hickey, L.T., Hafeez, A.N., Robinson, H., Jackson, S.A., Leal-Bertioli, S.C.M. *et al.* (2019) Breeding crops to feed 10 billion. *Nature Biotechnology* 37, 744–754.

Holliday, J.A., Wang, T. and Aitken, S. (2012) Predicting adaptive phenotypes from multilocus genotypes in Sitka spruce (*Picea sitchensis*) using random forest. *G3 (Bethesda)* 2, 1085–1093. DOI: 10.1534/g3.112.002733.

Howard, R., Carriquiry, A.L. and Beavis, W.D. (2014) Parametric and nonparametric statistical methods for genomic selection of traits with additive and epistatic genetic architectures. *G3 (Bethesda)* 4, 1027–1046. DOI: 10.1534/G3.114.010298/-/DC1.

Jannink, J.L. and Walsh, B. (2002) Association mapping in plant populations. In: Kang, M.S. (ed.) *Quantitative Genetics, Genomics and Plant Breeding*. CABI Publishing, Wallingford, UK, pp. 59–68.

Jannink, J.L., Lorenz, A.J. and Iwata, H. (2010) Genomic selection in plant breeding: From theory to practice. *Briefings in Functional Genomics* 9, 166–177. DOI: 10.1093/bfgp/elq001.

Jia, Y. and Jannink, J.-L. (2012) Multiple-trait genomic selection methods increase genetic value prediction accuracy. *Genetics* 192, 1513–1522. DOI: 10.1534/GENETICS.112.144246.

Klápště, J., Dungey, H.S., Telfer, E.J., Suontama, M., Graham, N.J. *et al.* (2020) Marker selection in multivariate genomic prediction improves accuracy of low heritability traits. *Frontiers in Genetics* 11, 499094. DOI: 10.3389/FGENE.2020.499094/FULL.

Krishnan, S.G., Singh, A.K., Rathour, R., Nagarajan, M., Bhowmick, P.K. *et al.* (2019) Rice variety Pusa Samba 1850. *Indian Journal of Genetics* 79, 109–110.

Krishnappa, G., Savadi, S., Tyagi, B.S., Singh, S.K., Mamrutha, H.M. *et al.* (2021) Integrated genomic selection for rapid improvement of crops. *Genomics* 113, 1070–1086. DOI: 10.1016/j.ygeno.2021.02.007.

Lande, R. and Thompson, R. (1990) Efficiency of marker-assisted selection in the improvement of quantitative traits. *Genetics* 124, 743–756. DOI: 10.1093/genetics/124.3.743.

Legarra, A. and Reverter, A. (2018) Semi-parametric estimates of population accuracy and bias of predictions of breeding values and future phenotypes using the LR method. *Genetics Selection Evolution* 50, 53–18. DOI: 10.1186/S12711-018-0426-6/FIGURES/3.

Li, Y. and Dungey, H.S. (2018) Expected benefit of genomic selection over forward selection in conifer breeding and deployment. *PLOS ONE* 13, e0208232. DOI: 10.1371/journal.pone.0208232.

Long, N., Gianola, D., Rosa, G.J.M. and Weigel, K.A. (2011) Application of support vector regression to genome-assisted prediction of quantitative traits. *Theoretical and Applied Genetics* 123, 1065–1074. DOI: 10.1007/S00122-011-1648-Y.

Lush, J.L. (1943) *Animal Breeding Plans*, 2nd edn. Bibliolife DBA of Bibilio Bazaar, Charleston, South Carolina.

Madhavi, K.R., Rambabu, R., Abhilash Kumar, V., Vijay Kumar, S., Aruna, J. *et al.* (2016) Marker assisted introgression of blast (*Pi-2* and *Pi-54*) genes in to the genetic background of elite, bacterial blight resistant indica rice variety, Improved Samba Mahsuri. *Euphytica* 212, 331–342. DOI: 10.1007/S10681-016-1784-1.

Maenhout, S., Baets, B., Haesaert, G. and Bockstaele, E. (2007) Support vector machine regression for the prediction of maize hybrid performance. *Theoretical and Applied Genetics* 115, 1003–1013. DOI: 10.1007/s00122-007-0627-9.

Malmberg, M.M., Smith, C., Thakur, P., Drayton, M.C., Wilson, J. *et al.* (2023) Developing an integrated genomic selection approach beyond biomass for varietal protection and nutritive traits in perennial ryegrass (*Lolium perenne* L.). *Theoretical and Applied Genetics* 136, 44.

Manzo-Sánchez, G., Buenrostro-Nava, M.T., Guzmán-González, S., Orozco-Santos, M., Youssef, M. *et al.* (2015) Genetic diversity in bananas and plantains (*Musa* spp). In: Ozcan, B., Caliskan, M., Oz, G.C., Kavakli, I.H. and Ozcan, B. (eds) *Molecular Approaches to Genetic Diversity*. IntechOpen, London, UK.

Meuwissen, T.H.E., Hayes, B.J. and Goddard, M.E. (2001) Prediction of total genetic value using genome-wide dense marker maps. *Genetics* 157, 1819–1829. DOI: 10.1093/genetics/157.4.1819.

Michel, S., Löschenberger, F., Ametz, C., Pachler, B., Sparry, E. *et al.* (2019) Combining grain yield, protein content and protein quality by multi-trait genomic selection in bread wheat. *Theoretical and Applied Genetics* 132, 2767–2780.

Moore, J.H. and Williams, S.M. (2009) Epistasis and its implications for personal genetics. *American Journal of Human Genetics* 85, 309–320. DOI: 10.1016/j.ajhg.2009.08.006.

Nakaya, A. and Isobe, S.N. (2012) Will genomic selection be a practical method for plant breeding? *Annals of Botany* 110, 1303–1316. DOI: 10.1093/aob/mcs109.

Neeraja, C.N., Maghirang-Rodriguez, R., Pamplona, A., Heuer, S., Collard, B.C.Y. *et al.* (2007) A marker-assisted backcross approach for developing submergence-tolerant rice cultivars. *Theoretical and Applied Genetics* 115, 767–776. DOI: 10.1007/s00122-007-0607-0.

Onogi, A., Watanabe, M., Mochizuki, T., Hayashi, T., Nakagawa, H. *et al.* (2016) Toward integration of genomic selection with crop modelling: The development of an integrated approach to predicting rice heading dates. *Theoretical and Applied Genetics* 129, 805–817. DOI: 10.1007/s00122-016-2667-5.

Parveen, R., Kumar, M., Singh, D., Shahani, M., Imam, Z. *et al.* (2023) Understanding the genomic selection for crop improvement: Current progress and future prospects. *Molecular Genetics & Genomic Medicine* 298, 813–821. DOI: 10.1007/s00438-023-02026-0.

Rai, K.K. (2022) Integrating speed breeding with artificial intelligence for developing climate-smart crops. *Molecular Biology Reports* 49, 11385–11402. DOI: 10.1007/s11033-022-07769-4.

Rai, K.N., Hash, C.T., Singh, A.K. and Velu, G. (2008) Adaptation and quality traits of a germplasm-derived commercial seed parent of pearl millet. *Plant Genetic Resources Newsletter* 154, 20–24.

Rashid, S., Bin Mushtaq, M., Farooq, I. and Khan, Z. (2022) Climate smart crops for food security. In: Harris, S. (ed.) *The Nature, Causes, Effects and Mitigation of Climate Change on the Environment*. IntechOpen, London, UK.

Ray, D.K., Ramankutty, N., Mueller, N.D., West, P.C. and Foley, J.A. (2012) Recent patterns of crop yield growth and stagnation. *Nature Communications* 3, 1293. DOI: 10.1038/ncomms2296.

Report from UN Population Division (2022) Available at: www.un.org/development/desa/pd/content/World-Population-Prospects-2022 (accessed 19 November 2024).

Rothman, A.J., Levina, E. and Zhu, J. (2010) Sparse multivariate regression with covariance estimation. *Journal of Computational and Graphical Statistics* 19, 947–962. DOI: 10.1198/JCGS.2010.09188.

Rutkoski, J., Poland, J., Mondal, S., Autrique, E., Pérez, L.G. *et al.* (2016) Canopy temperature and vegetation indices from high-throughput phenotyping improve accuracy of pedigree and genomic selection for grain yield in wheat. *G3 (Bethesda)* 6, 2799–2808. DOI: 10.1534/g3.116.032888.

Sandhu, K.S., Shiv, A., Kaur, G., Meena, M.R., Raja, A.K. *et al.* (2022) Integrated approach in genomic selection to accelerate genetic gain in sugarcane. *Plants* 11, 2139. DOI: 10.3390/plants11162139.

Shahwar, D., Khan, Z. and Park, Y. (2024) Molecular markers for marker-assisted breeding for biotic and abiotic stress in melon (*Cucumis melo* L.): A review. *International Journal of Molecular Sciences* 25, 6307. DOI: 10.3390/ijms25126307.

Singh, A.K., Gopala Krishnan, S., Ellur, R.K., Bhowmick, P.K., Nagarajan, M. *et al.* (2017a) Notification of Basmati rice variety, Pusa Basmati 1728. *Indian Journal of Genetics* 77, 584.

Singh, A.K., Gopala Krishnan, S., Nagarajan, M., Bhowmick, P.K., Ellur, R.K. *et al.* (2017b) Notification of Basmati rice variety Pusa Basmati 1637. *Indian Journal of Genetics* 77, 583–584.

Spindel, J., Begum, H., Akdemir, D., Virk, P., Collard, B. *et al.* (2015) Correction: Genomic selection and association mapping in rice (*Oryza sativa*): Effect of trait genetic architecture, training population composition, marker number and statistical model on accuracy of rice genomic selection in elite, tropical rice breeding lines. *PLOS Genetics* 11, e1005350. DOI: 10.1371/JOURNAL.PGEN.1005350.

Steinwand, M.A. and Ronald, P.C. (2020) Crop biotechnology and the future of food. *Natural Foods* 1, 273–283. DOI:10.1038/s43016-020-0072-3.

Tibshirani, R. (1996) Regression shrinkage and selection via the Lasso. *Journal of the Royal Statistical Society, Series B (Methodological)* 58, 267–288. DOI: 10.1111/J.2517-6161.1996.%20TB02080.X.

Usai, M.G., Goddard, M.E. and Hayes, B.J. (2009) LASSO with cross-validation for genomic selection. *Genetics Research International* 91, 427–436. DOI: 10.1017/S0016672309990334.

Varshney, R.K., Mohan, S.M., Gaur, P.M., Chamarthi, S.K., Singh, V.K. *et al.* (2014a) Marker-assisted backcrossing to introgress resistance to Fusarium wilt race 1 and Ascochyta blight in C 214, an elite cultivar of chickpea. *Plant Genome* 7, 35. DOI: 10.3835/plantgenome2013.10.0035.

Varshney, R.K., Pandey, M.K., Janila, P., Nigam, S.N., Sudini, H. *et al.* (2014b) Marker-assisted introgression of a QTL region to improve rust resistance in three elite and popular varieties of peanut (Arachis hypogaea L). *Theoretical and Applied Genetics* 127, 1771–1781. DOI: 10.1007/S00122-014-2338-3.

Varshney, R.K., Singh, V.K., Kumar, A., Powell, W. and Sorrells, M.E. (2018). Can genomics deliver climate-change ready crops? *Current Opinion in Plant Biology* 45, 205–211. DOI: 10.1016/J.PBI.2018.03.007.

Varshney, R.K., Bohra, A., Yu, J., Graner, A., Zhang, Q. *et al.* (2021) Designing future crops: Genomics-assisted breeding comes of age. *Trends in Plant Science* 6, 631–649. DOI: 10.1016/j.tplants.2021.03.010.

Vasistha, N.K., Balasubramaniam, A., Mishra, V.K., Srinivasa, J., Chand, R. and Joshi, A.K. (2017). Molecular Introgression of leaf rust resistance gene *Lr*34 validates enhanced effect on resistance to spot blotch in spring wheat. *Euphytica* 213, 262. DOI: 10.1007/s10681-017-2051-9.

Vishwakarma, M.K., Yadav, P.S., Rai, V.P., Kumar, U. and Joshi, A.K. (2022) Molecular markers and genomics assisted breeding for improving crop plants. In: Samuel, J., Kumar, A. and Singh, J. (eds) *Relationship between Microbes and the Environment for Sustainable Ecosystem Services*, Vol. 1. Elsevier, Amsterdam, Netherlands, pp. 303–334.

Voss-Fels, K.P., Stahl, A. and Hickey, L.T. (2019) Modern crop breeding for future food security. *BMC Biology* 17, 18. DOI: 10.1186/s12915-019-0638-4.

Wang, X., Yang, Z. and Xu, C.A. (2015) A comparison of genomic selection methods for breeding value prediction. *Science Bulletin* 60, 925–935. DOI: 10.1007/s11434-015-0791-2.

Wang, X., Xu, Y., Hu, Z. and Xu, C. (2018) Genomic selection methods for crop improvement: Current status and prospects. *Crop Journal* 6, 330–340. DOI: 10.1016/j.cj.2018.03.001.

WHO/FAO (2015) Available at: www.fao.org/3/i8304e/i8304e.pdf (accessed 19 November 2024).

WMO (2023) Global Temperatures Set to Reach New Records in Next Five Years. [Press release.] Available at: https://wmo.int/news/media-centre/global-temperatures-set-reach-new-records-next-five-years (accessed 4 September 2024).

Wong, C.K. and Bernardo, R. (2008) Genome-wide selection in oil palm: Increasing selection gain per unit time and cost with small populations. *Theoretical and Applied Genetics* 116, 815–824. DOI: 10.1007/s00122-008-0715-5.

World Economic Forum (2020) This Is Why Food Security Matters Now More than Ever. Available at: www.weforum.org/agenda/2020/11/foodsecurity-why-it-matters/ (accessed 19 November 2024).

Yang, D.L., Jing, R.L., Chang, X.P. and Li, W. (2007) Identification of quantitative trait loci and environmental interactions for accumulation and remobilization of water-soluble carbohydrates in wheat (*Triticum aestivum* L.) stems. *Genetics* 176, 571–584. DOI: 10.1534/genetics.106.068361.

Zargar, S.M., Raatz, B., Sonah, H., Bhat, J.A., Dar, Z.A. *et al.* (2015) Recent advances in molecular marker techniques: Insight into QTL mapping, GWAS and genomic selection in plants. *Journal of Crop Science and Biotechnology* 5, 293–308. DOI: 10.1007/s12892-015-0037-5.

# 5 Enhancing Crop Disease Resistance and Improvement through Molecular Approaches: A Focus on CRISPR-Cas9 Gene Editing

**Aliya Firdaus[1], Syed Khalida Izhar[1], Mohd Hadi Yunus[2], Zeba Khan[3]* and Mohd Yunus Khalil Ansari[4]**

*[1]Biocontrol Research Laboratory, Integral Information and Research Centre-4 (IIRC-4), Integral University, Lucknow, UP – 226026, India; [2]Department of Horticultural Bioscience, Pusan National University, Miryang 50463, Republic of Korea; [3]Department of Botany, Raja Mahendra Pratap Singh State University, Aligarh – 202002, India; [4]Department of Botany, Aligarh Muslim University, Aligarh – 202001, India*

## Abstract

The demand for sustainable and resilient agricultural practices is growing alongside the global population. Leading this innovation is CRISPR-Cas9 gene editing, offering a cutting-edge solution to bolster disease resistance in food crops. This chapter examines the application of CRISPR-Cas9's application in enhancing disease resistance across several vital food crops. Renowned for its precision, the CRISPR-Cas9 system allows targeted modifications of specific genes responsible for susceptibility to pathogens. By elucidating molecular mechanisms, researchers aim to fortify crop immunity and reduce reliance on chemical interventions, thus promoting environmentally sustainable farming. It also explores the potential impact of CRISPR-edited crops on global food security, promising increased yields despite heightened susceptibility to diseases and a more stable agricultural ecosystem. However, it discusses ethical considerations, regulatory challenges, and public perceptions regarding genetically modified organisms, underscoring the importance of responsible and transparent deployment of CRISPR-Cas9 technology in agriculture. In conclusion, this chapter synthesizes recent advancements in the use of CRISPR-Cas9 for enhancing disease resistance in food crops, highlighting its transformative potential for precision gene editing in tackling the challenges of feeding a growing population, while ensuring sustainability and resilience against evolving plant pathogens.

**Keywords:** CRISPR-Cas9, gene editing, food security, regulatory challenges, sustainable agriculture

## Abbreviations

CRISPR-Cas9 – Clustered regularly interspaced short palindromic repeats/CRISPR-associated protein 9, DSBs – Double-strand breaks, GM – Genetically modified, GMO – Genetically modified organism, gRNA – Guide RNA, HDR – Homology-directed repair indel – Insertion/deletion, MAS – Marker-assisted selection, NHEJ – Non-homologous end joining, PAM – Protospacer adjacent motif, PAMP – pathogen-associated molecular pattern, PCR – Polymerase chain reaction, PEG – polyethylene glycol, PTI – PAMP-triggered immunity, R-gene – resistance gene, RNAi – RNA interference, sgRNA – Single guide RNA, TALENs – Transcription activator-like effector nucleases, TYLCV – tomato yellow leaf curl virus, ZFNs – Zinc finger nucleases.

*Corresponding author: khanzeba02@gmail.com


DOI: 10.1079/9781800626638.0005

## 5.1 Introduction

Plant diseases have a major influence on crop production because they reduce yields, lower quality and raise production costs. Pathogens such as nematodes, bacteria, viruses and fungi provide unique difficulties for farmers. Fungal infections, such as late blight, powdery mildew and rusts, can cause severe damage to leaf areas, leading to reduced photosynthesis and inhibited growth. *Ralstonia solanacearum* is a bacterial pathogen that causes bacterial wilt. The pathogen can rapidly propagate via soil and water, resulting in the devastation of many crops. Tobacco mosaic virus and other viral infections have the potential to reduce the size of fruits, hinder their growth and cause abnormalities that render the food unfit for sale. Nematode infections often hinder root development, leading to reduced nutrient absorption and overall plant distress (Strange and Scott, 2005). These diseases have a substantial economic impact. Farmers face significant expenses in disease management, which include the expenditures of fungicides, bactericides and resistant seed varieties. Infected plants often lead to decreased crop yields. Imposing limitations and prohibitions on market access due to disease can lead to increased damages, resulting in global food price increases and financial difficulties for farmers (Savary *et al.*, 2019).

Pathogens can inflict direct harm on crops, leading to diminished plant vitality, slowed development and, eventually, lower agricultural production. Pathogens, including fungi, bacteria, viruses and nematodes, can infect crops at different phases of growth, impacting physiological functions such as photosynthesis, nutrient absorption and water uptake. Severe disease outbreaks due to pathogenic action have the potential to cause significant damage to crops, resulting in either partial or entire loss. This is especially true for crop varieties that are more vulnerable to diseases or when the climatic circumstances are favourable for the rapid growth and spread of pathogens. Losses can happen at every step of crop growth, starting from germination and seedling establishment, and continuing through blooming, fruit set and maturation.

Pathogens can undermine the quality of harvested crops by inducing symptoms such as decay, discoloration, abnormalities and diminished nutritional content. Post-harvest infections may worsen the decline in crop quality during storage, transportation and marketing, resulting in further losses. When crops are afflicted by diseases, they may fail to satisfy the desired standards for appearance, size, form or taste. Therefore, their marketability decreases, and their prices are reduced. Consumers can avoid buying contaminated produce, resulting in market refusal and financial setbacks for growers and wholesalers.

To effectively manage infections, it is often necessary to use fungicides, bactericides, nematicides or other chemical treatments. These treatments help restrict the spread of pathogens and minimize the damage to crops. However, higher input expenses related to disease control, such as labour, materials and equipment, might diminish the profitability of farmers. Additionally, extensive application of chemical pesticides to manage crop diseases can result in detrimental environmental consequences, such as the pollution of soil and water, the harmful impacts on non-target animals, and the disturbance of ecosystem dynamics. Sustainable disease management strategies, such as integrated pest management and biological control, strive to reduce environmental harm while ensuring efficient disease control.

Chronic or repeated disease incidence can result in the accumulation of soilborne pathogens, decreased soil fertility, and a heightened likelihood of future disease outbreaks in succeeding crop cycles. Repetitive cultivation of a single crop and a lack of genetic variation might worsen the vulnerability to diseases and increase the intensity of disease outbreaks as time goes on (Fig. 5.1).

## 5.2 Conventional Breeding Techniques for Disease-Resistant Plant Varieties

Conventional breeding is a time-honoured technique used to create novel plant varieties that

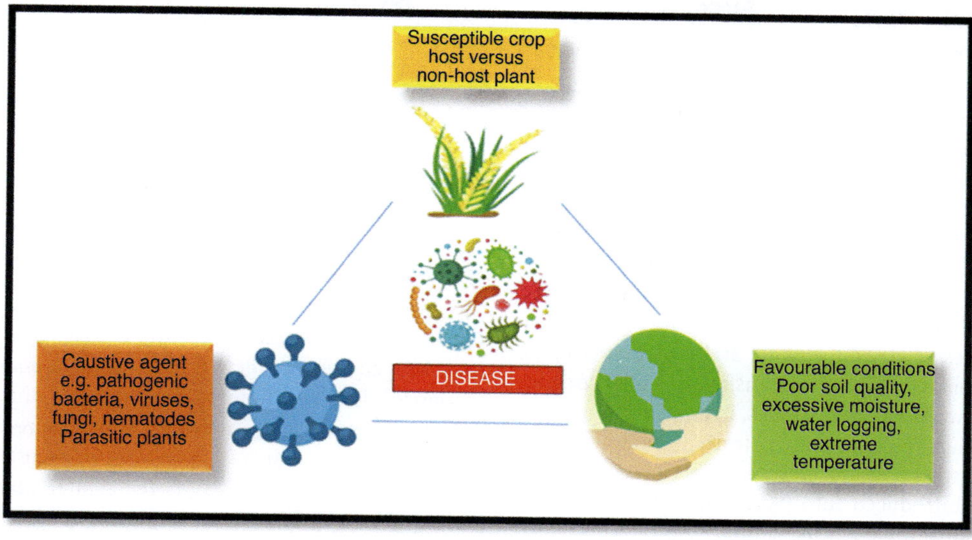


**Fig. 5.1.** Identifying, managing and preventing crop diseases.

possess desired characteristics, such as resistance to diseases (Acquaah, 2009). The process entails the deliberate selection and breeding of parent plants that possess advantageous attributes, resulting in progeny that exhibit the desired features.

Breeders initially find wild or produced types that have inherent resistance to illnesses. These resilient cultivars function as useful genetic reservoirs. They hybridize resistant plants with commercially valuable, yet susceptible, kinds. This approach integrates the targeted disease resistance feature with other favourable traits, such as high productivity, quality or flexibility (Nepolean *et al.*, 2018). The resulting offspring, referred to as hybrids or progeny, are carefully examined to determine their resistance to the specific illness being targeted. This involves exposing the plants to the pathogen and monitoring their reaction. Plants that demonstrate resistance are chosen for additional breeding. In some cases, breeders engage in backcrossing, a process by which they repeatedly mate the resistant offspring with the susceptible parent to preserve the desired qualities while eliminating undesired characteristics inherited from the resistant parent (Begna, 2021). The selected lines undergo rigorous testing in many contexts to verify the stability and efficacy of the resistance trait across varied situations. After the development and successful testing of a new variety that has enhanced resistance to diseases, there is a process of regulatory approval before it is made available to farmers for commercial production.

Conventional methods have a crucial role in generating new genetic variations, preserving natural germplasm, facilitating sexual hybridization between different parental lines, and inducing mutations. Over the past three decades, the International Rice Research Institute (IRRI) has used traditional breeding techniques to successfully produce a wide variety of elite cultivars that possess a diverse set of genes conferring resistance to many diseases (Miah *et al.*, 2013). Conventional breeding programmes utilize several strategies, including pedigree analysis, backcrossing, recurrent selection and mutant breeding. The pedigree approach is extensively employed in rice enhancement. If the resistance to insects and illnesses in rice is controlled by significant genes, the pedigree technique is an excellent approach for developing resistance. The breeding strategy of backcross breeding has been implemented in South and South-east Asia (Hasan *et al.*, 2015) to enhance the resistance to blast in premier cultivars such as KDML105, Basmati and Manawthukha (Chan-In *et al.*, 2020).

## 5.3 Biotechnological Advances in Enhancing Plant Disease Resistance

Biotechnological technologies provide advanced solutions for combating agricultural diseases by offering more accurate and efficient means to enhance plant resistance against pathogens. Biotechnology provides multiple strategies to address illnesses in crops. Genetic engineering is the process of introducing or modifying genes in a plant's genome to provide it with resistance to illnesses (Mushtaq *et al.*, 2019). This technique enables the accurate transfer of resistance genes between different species, irrespective of genetic compatibility. As an illustration, genes that code for toxins produced by bacteria or viruses can be introduced into plants to confer resistance against those diseases.

### 5.3.1 Marker-assisted selection (MAS)

Marker-assisted selection (MAS), is a method that uses molecular markers associated with specific features, like disease resistance genes, to speed up the traditional breeding process. It facilitates systematic improvements in breeding cycles, reduces reliance on conventional phenotypic evaluations, and ensures precise genetic modifications (Shahwar *et al.*, 2024). The technique leverages molecular markers that are linked to genes of interest, facilitating the identification and selection of plants with these traits without the need for phenotypic screening. One notable example of MAS in plant disease resistance is its application in rice breeding. According to a study by Ashkani *et al.* (2015), the integration of MAS in breeding programmes has resulted in rice cultivars with durable resistance, enhancing food security and agricultural sustainability. MAS has significantly accelerated the breeding process and improved the efficiency of developing disease-resistant plant varieties. Breeders can speed up breeding for disease resistance by selectively choosing plants that possess certain markers linked to resistance, thus eliminating the need for laborious phenotypic screening (Osei *et al.*, 2018). Although MAS has improved the efficiency of crossing and selection, it is constrained when dealing with crops that have complex genetics (Shahwar *et al.*, 2023).

### 5.3.2 RNA interference (RNAi)

RNA interference (RNAi) is a process by which the activity of certain genes is suppressed by introducing double-stranded RNA molecules that are complementary to the target gene sequence. This technology can suppress the activity of genes that are crucial for the virulence or reproduction of pathogens, therefore providing the plant with resistance against the disease (Ghag, 2017). This mechanism involves the degradation of target mRNA molecules, thereby preventing the production of specific proteins that may contribute to disease susceptibility. The application of RNAi in plant disease resistance has shown promising results across various crops and pathogens. For example, Jia *et al.* (2017) used RNAi technology to develop transgenic papaya resistant to papaya ringspot virus.

### 5.3.3 Gene editing (CRISPR-Cas9)

Gene editing methods, like CRISPR-Cas9, allow for accurate adjustments to the plant genome, including the insertion, removal or modification of specific DNA regions (Noman *et al.*, 2016). This technology enables the insertion or deletion of genetic material, or the modification of genetic sequences at precise locations within the genome, by substituting an existing sequence with a tailored DNA segment (Khan and Shahwar, 2024). This method can be employed to manipulate plants to improve their ability to withstand diseases. This can be achieved by modifying genes that are linked to susceptibility or by introducing new genes that provide resistance. For example, studies have demonstrated successful editing of genes like *TaEDR1* in wheat and various resistance genes in tomatoes, leading to enhanced resistance without the need for traditional chemical treatments (Nekrasov *et al.*, 2017; Zhang *et al.*, 2019). Beyond the immediate benefits of enhanced disease resistance, CRISPR-edited crops also hold promise for reducing environmental impact. By reducing reliance on chemical pesticides, these crops contribute to sustainable agricultural practices and mitigate potential environmental harm associated with pesticide use (Zhang *et al.*, 2024).

### 5.3.4 Transgenic approaches

Transgenic plants are plants that have undergone genetic modification by the incorporation of genes from other organisms (Low *et al.*, 2018). Transgenic methods can be employed to introduce genes from other plant species, as well as from non-plant organisms, to provide resistance against diseases. This genetic modification enables plants to acquire new traits, such as resistance to diseases that may otherwise cause significant crop losses. For instance, in soybeans, genes from bacteria have been introduced to enhance resistance against specific fungal pathogens (Mansoor *et al.*, 2017). Similarly, in potatoes, genes from wild relatives have been incorporated to provide resistance to late blight, a devastating fungal disease (Slater *et al.*, 2013). These transgenic approaches not only offer targeted solutions to specific disease challenges but also contribute to sustainable agriculture by reducing the need for chemical pesticides. However, like CRISPR-Cas9 technology, the adoption and regulatory approval of transgenic plants vary globally and are subject to rigorous safety assessments (Waltz, 2016).

### 5.3.5 Biocontrol agents

Biotechnological techniques can also be utilized to create biocontrol agents, such as genetically modified microorganisms or viruses, that specifically attack plant infections. These agents can be genetically modified to synthesize antifungal chemicals or proteins that hinder the growth of pathogens, offering a sustainable and eco-friendly substitute for chemical pesticides (Junaid *et al.*, 2013). For example, certain Trichoderma species and *Bacillus* spp. produce antimicrobial compounds that inhibit pathogen growth, while rhizobacteria like *Pseudomonas fluorescens* can induce systemic resistance in plants against various pathogens (Pieterse *et al.*, 2014; Contreras-Cornejo *et al.*, 2016). Additionally, viral biocontrol strategies utilize RNA viruses to manage insect pests indirectly by reducing their populations and limiting vector transmission. Despite their potential, the efficacy of biocontrol agents can be influenced by environmental factors and compatibility with existing agricultural practices. The integration of biotechnological tools, such as genetic engineering and omics technologies, is advancing our understanding of biocontrol mechanisms and aiding the development of more effective strategies. These approaches underscore the importance of continued research to optimize biocontrol agent applications and promote their widespread adoption in sustainable crop protection strategies.

## 5.4 Advancements in Genome-Editing Technologies

Recently, the development of highly adaptable genome-editing tools has given researchers the capability to make changes quickly and cost-effectively to the genetic material of many cell types and organisms (Li *et al.*, 2021). The utilization of these genome editing techniques has fundamentally transformed the domain of molecular biology and biotechnology, facilitating meticulous alteration of the genetic code with unparalleled simplicity and effectiveness. They offer great potential for future advancements in various disciplines, such as basic research, medicine and agriculture. Genome editing techniques allow for accurate alterations to an organism's DNA, providing exceptional prospects for fundamental research, biotechnology and medicine. The core technologies that are most used now to facilitate genome editing are: (i) CRISPR-Cas9 (clustered regularly interspaced short palindromic repeats/CRISPR-associated protein 9); (ii) TALENs (transcription activator-like effector nucleases); (iii) ZFNs (zinc-finger nucleases); and (iv) base editors.

### 5.4.1 CRISPR-Cas9

CRISPR-Cas9 is an innovative method for modifying genomes that originate from the immune system of bacteria. The system comprises a guide RNA (gRNA) that instructs the Cas9 nuclease to target a particular DNA sequence, resulting in the creation of double-strand breaks (DSBs). The cell's DNA repair system can fix these breaks, resulting in gene knockout, knockin, or precise editing. In comparison to earlier genome editing methods, CRISPR-Cas9 is more user-friendly, adaptable and efficient. Research and biotechnological applications greatly benefit from its programmability, which enables exact targeting of specific genetic loci.

### 5.4.2 Transcription activator-like effector nucleases (TALENs)

DNA-binding domains from transcription activator-like effectors (TALENs) linked to nuclease domains from endonucleases like FokI are the building blocks of TALENs, which are synthetic nucleases. Inducing DSBs at specified genomic regions, TALENs enable targeted genome editing, much like CRISPR-Cas9 (Becker and Boch, 2021). TALENs are a great alternative to CRISPR-Cas9 because of their effectiveness and high specificity in genome editing. Thanks to their modular design, DNA-binding specificities can be easily adjusted, allowing for precise targeting of genomic loci.

### 5.4.3 Zinc Finger Nucleases (ZFNs)

Zinc finger nucleases (ZFNs) are designed nucleases that combine a nuclease domain with a zinc finger DNA-binding domain. The zinc finger domain is highly adaptable for DNA binding because of its numerous repeats, each of which binds to a distinct set of DNA triplets (Carroll, 2014). To modify the genome, ZFNs, like TALENs and CRISPR-Cas9, cause DSBs at specific locations. They have found use in therapeutic applications like gene knockout studies and gene therapy, as well as in basic research. These ZFNs provide very targeted genome editing capabilities. Their adaptable structure permits fine-tuning of DNA-binding specificities, paving the way for precise editing of genomic regions.

### 5.4.4 Base editors

Genomics editing methods that allow for the direct conversion of one DNA base pair into another without producing DSBs are known as base editors. Enzymes like cytidine deaminase and adenine deaminase, which change the base pairs at certain genomic sites, are combined with a nuclease that is catalytically disabled to form these enzymes (Rees and Liu, 2018). In contrast to nucleases, which cause DSBs, base editors allow for efficient and accurate genome editing with fewer off-target consequences. Among their many uses, they have found a home in gene editing, disease modelling and agricultural advancement. Base editors make it possible to convert single nucleotides with pinpoint accuracy, enabling the repair of point mutations or the introduction of targeted alterations without causing double-strand breaks. They provide a useful resource for genome editing projects that call for exact modifications to nucleotides.

## 5.5 Fundamentals of CRISPR-Cas9 Gene Editing

The use of CRISPR-Cas9 technology has greatly improved agricultural genetic engineering by allowing for more efficient and precise targeted genome editing than ever before. An adaptable tool for editing genomes, CRISPR-Cas9 originated in the immune system of bacteria. A Cas9 nuclease enzyme is used in this method to target specific DNA regions using a single guide RNA (sgRNA). Gene knockout, knock-in, or precise sequence alterations can occur because of DNA repair mechanisms triggered by Cas9-induced DSBs at the target site (Ray and Raghavan, 2020). To edit the genome efficiently and selectively, the design of sgRNA is essential. The ubiquitous *Streptococcus pyogenes* Cas9 gene editing tool, together with specific sgRNAs, is engineered to complement target DNA sequences that are next to a protospacer adjacent motif (PAM), usually NGG (Zhou *et al.*, 2018). The optimization of sgRNA design and prediction of off-target effects are achieved using bioinformatics techniques. Protoplast transfection, particle bombardment (biolistics) and *Agrobacterium*-mediated transformation are some of the ways that Cas9 and sgRNA are introduced into plant cells. The specificity and efficiency of crop genome editing are affected by delivery techniques. Both homology-directed repair (HDR) and non-homologous end joining (NHEJ) are effective methods for repairing DSBs caused by Cas9 (Yang *et al.*, 2020). Gene disruption or knockout can occur because of insertion/deletion (indel) mutations, which are commonly caused by NHEJ. HDR allows for precise gene editing, including gene replacement or insertion; however, it does require a donor DNA template.

## 5.6 Mechanism of CRISPR-Cas9

The mechanism of CRISPR-Cas9 gene editing in crops involves several intricate steps, from the design of guide RNAs to the induction of targeted DNA modifications. The Cas9 nuclease and a gRNA are the two primary components of the CRISPR-Cas9 system (Fig. 5.2), which is used for genome editing in plants. To form the Cas9 – gRNA complex, a scaffold sequence is fused to the gRNA, which is designed to be complementary to the target DNA sequence. The target DNA sequence is usually 20 nucleotides long. When the Cas9 protein and gRNA assemble into a complex, the Cas9 protein is able to attach to specific DNA sequences by forming base pairing interactions. Following recognition of the target DNA sequence by the Cas9 – gRNA complex, DNA cleavage occurs at a location specified by the sequence specificity of the gRNA, where Cas9 produces a DSB. The DNA strands are cut by Cas9, which, depending on the strain, leaves behind blunt ends or overhangs.

Two major mechanisms, NHEJ and HDR, are utilized by plant cells to repair DNA damages after Cas9 production of DSBs. As a result of the insertion or deletion of nucleotides (indels) at the site of the DSB, NHEJ is the main repair process in plants. Knockout or disruption of genes can result from this mechanism. To fix the DSB with exact nucleotide sequence modifications, HDR uses a template DNA molecule, like an externally supplied homologous sequence. High-resolution gene editing allows for the precise replacement or insertion of sequences into genes.

## 5.7 Tools and Techniques for CRISPR-Cas9 Editing

### 5.7.1 Design tools for guide RNAs (sgRNA)

Designing sgRNAs targeting specific genes or genomic areas in crop genomes is made easier with the help of bioinformatics tools such

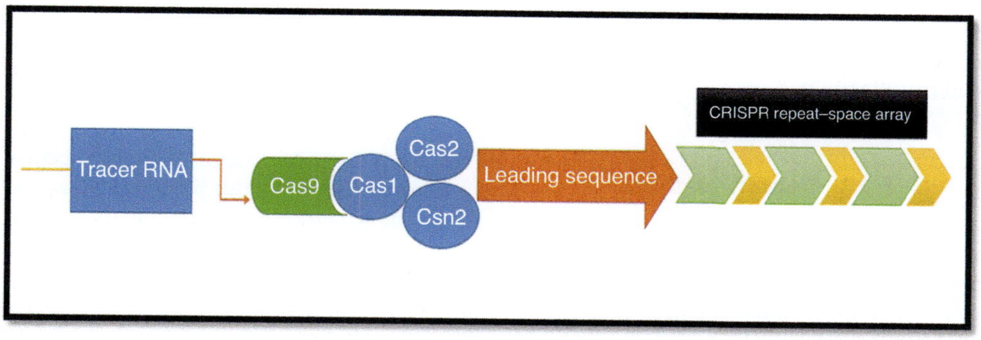


**Fig. 5.2.** Basic CRISPR-Cas9 structure. The CRISPR-Cas9 locus consists of three elements, namely a transactivating CRISPR RNA (tracrRNA) at the 5' end, a Cas protein-encoding gene, and a CRISPR locus at the 3' end, which include a lead sequence, a 23–50 bp repeat sequence, and a spacer sequence.

as CRISPR-PLANT, CRISPR-P and CRISPR direct (Arif *et al.*, 2024). By predicting possible off-target sites for a specific sgRNA, tools like CRISPOR and CCTop help in the selection of efficient and specific sgRNAs (Stemmer *et al.*, 2015). Off-target prediction tools, such as CRISPOR and CCTop, are essential for enhancing the precision and efficiency of CRISPR-Cas9 gene editing by identifying potential off-target sites for specific sgRNAs. CRISPOR is a widely used web-based tool that evaluates the specificity of sgRNAs by comparing their sequences against the target genome, providing a comprehensive list of potential off-target sites along with their likelihood of being edited. This tool employs various algorithms to assess off-target effects based on sequence homology and other genomic factors, ensuring that selected sgRNAs are both efficient and specific. Similarly, CCTop (CRISPR-Cas9 target online predictor) offers a robust platform for predicting off-target sites by scanning the entire genome and ranking potential off-targets based on their sequence similarity to the intended target (Manghwar *et al.*, 2020).

### 5.7.2 Cas9 expression constructs

Plant cells are frequently transfected with Cas9 expression constructs using plasmid vectors. In addition to selection markers for detecting transformed cells, these vectors include promoters that drive Cas9 expression (Ma *et al.*, 2016). The introduction of Cas9 constructs into plant cells is made easier by binary vectors that are compatible with *Agrobacterium tumefaciens*-mediated transformation (Azizi-Dargahlou and Pouresmaeil, 2023)

### 5.7.3 Delivery methods

To introduce Cas9 and sgRNA into the plant genome, *Agrobacterium* bacteria that carry these expression constructs infect plant tissues (Schmitz *et al.*, 2020). To directly transport CRISPR-Cas9 components into plant cells, a gene gun is used to shoot gold or tungsten particles coated with Cas9 – sgRNA complexes into plant tissues (Ghogare *et al.*, 2021). Electroporation and polyethylene glycol (PEG)-mediated transfection are two methods for introducing CRISPR-Cas9 components into plant protoplasts, which are isolated plant cells that have had their cell walls removed (John Martin *et al.*, 2022) (Fig. 5.3).

### 5.7.4 Genome editing detection and analysis

Detection of CRISPR-induced mutations can be achieved by polymerase chain reaction (PCR) amplification of specific genomic areas, followed by gel electrophoresis or sequencing. High-throughput investigation of CRISPR-induced mutations across the complete genome of edited plants is made possible by NGS techniques like amplicon sequencing or whole-genome sequencing. A T7 endonuclease I assay (T7EI) enables the identification of CRISPR-induced mutations using gel electrophoresis by cleaving heteroduplex DNA, which is generated by annealing wild-type and mutant DNA strands (Bhattacharya and Van Meir, 2019).

### 5.7.5 Validation of edited plants

Utilizing PCR, sequencing or other molecular methods, modified plants are examined for the existence of target mutations. Research on the impact of CRISPR-induced mutations on desired features, such as disease resistance, yield or nutritional content, is carried out through the phenotypic analysis of modified plants.

### 5.7.6 Plant – pathogen interactions

Plant – pathogen interactions involve molecular and biochemical exchanges between a plant and an invading pathogen, which can be broken down into several stages. Initially, plants use pattern recognition receptors to identify specific pathogen-associated molecular patterns (PAMPs), initiating PAMP-triggered immunity (PTI). To counteract PTI and manipulate the host environment, pathogens deliver effector proteins into plant cells. Plants have evolved resistance (R) genes coding for proteins that can

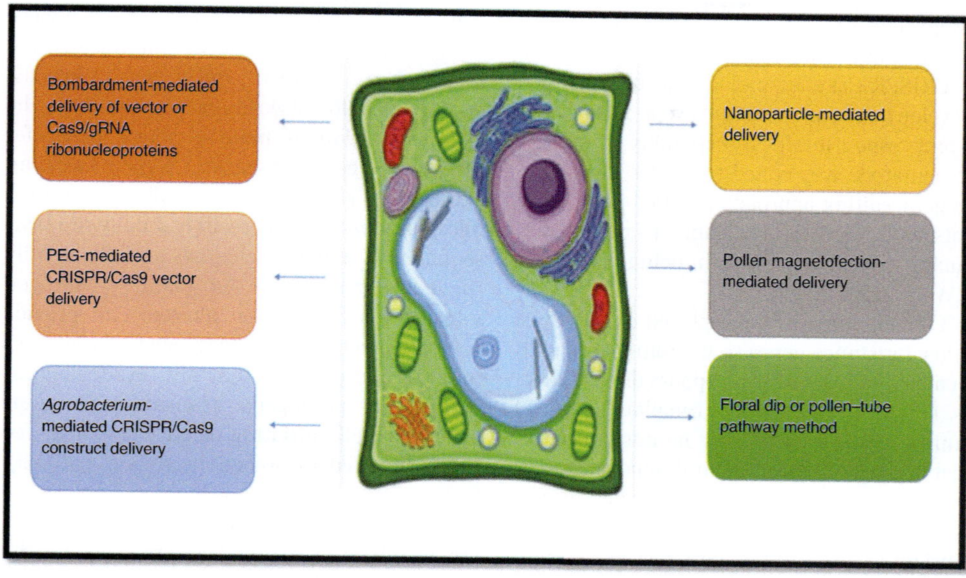


**Fig. 5.3.** Current and prospective methods of delivering CRISPR-Cas9 systems.

recognize these effectors, leading to a stronger and more targeted immune response known as effector-triggered immunity (Zvereva and Pooggin, 2012).

## 5.8 CRISPR Technology in Plant Pathogen Resistance

CRISPR, and its associated protein Cas9, offer a powerful tool for genome editing, which can be harnessed to enhance plant disease resistance. The advent of CRISPR technology allowed for the precise editing of plant genomes to increase pathogen resistance, which shook up the plant science industry. Using this gene-editing technology, which is based on the CRISPR-Cas9 system, researchers can precisely modify certain genes in plant DNA. CRISPR provides a potent method for developing disease-resistant crops by either disabling genes that make plants vulnerable to pathogen invasion or strengthening genes that activate defence mechanisms against pathogens.

CRISPR-Cas9 has been successfully used to confer resistance against various plant viruses. For example, researchers have used CRISPR-Cas9 to target and edit the viral genome of tomato yellow leaf curl virus (TYLCV) in tomato plants. This editing resulted in plants with enhanced resistance to TYLCV, showcasing the potential of CRISPR-Cas9 in combating viral diseases (Tashkandi *et al.*, 2018).

CRISPR-Cas9 has been employed to improve resistance to fungal pathogens in crops. In wheat, researchers have edited the *TaMLO* gene, which is involved in susceptibility to powdery mildew caused by *Blumeria graminis*. The edited wheat plants exhibited strong resistance to powdery mildew, demonstrating the effectiveness of CRISPR-Cas9 in enhancing fungal disease resistance (Wang *et al.*, 2014). Another example is the *MLO* gene found in wheat and barley, which may be specifically targeted for knockout. This has been demonstrated to provide resistance to powdery mildew without causing any major negative impacts on the plant's growth and productivity (Garcia-Ruiz *et al.*, 2021).

Gene editing using CRISPR-Cas9 has also been applied to increase resistance to bacterial pathogens. In rice, the *OsSWEET13* gene, which is targeted by the bacterial blight pathogen *Xanthomonas oryzae* pv. *oryzae*, was edited to disrupt its function. The resulting rice plants showed enhanced resistance to bacterial blight, highlighting the utility of CRISPR-Cas9 in addressing bacterial diseases (Zhou *et al.*, 2017).

CRISPR-Cas9 technology has been utilized to develop nematode-resistant crops. In soybean, the gene responsible for susceptibility to soybean cyst nematode was edited to confer resistance. This gene editing approach resulted in soybean plants with improved resistance to nematodes, reducing crop losses and enhancing productivity (Wang *et al.*, 2019).

CRISPR can also be employed to introduce or alter resistance genes that equip plants with the capacity to identify and oppose disease effectors. This strategy enhances the plant's natural immune responses and may be customized to target various diseases by simultaneously modifying numerous genes, a process called multiplexing. Nevertheless, the utilization of CRISPR in agriculture encounters obstacles like regulatory barriers, possible off-target consequences and the requirement for public approval, despite its promising capabilities. Moreover, the ever-changing nature of interactions between plants and diseases, characterized by the evolution of pathogens, necessitates continuous adaptations and improvements in techniques to combat resistance. However, CRISPR remains promising due to its ability to enhance crop resilience, decrease reliance on chemical pesticides and, potentially, enhance global food security as a component of integrated disease management systems.

CRISPR technology can be utilized to insert or alter resistance genes (R genes) in plants, thereby improving their capacity to detect and react to pathogen effectors. For instance, the introduction of a particular R gene that can identify a wide range of disease effectors can provide resistance against numerous pathogens. In addition to conventional knockouts, CRISPR base editors can perform accurate modifications to nucleotide sequences without inducing double-strand breaks. This method can be employed to optimize gene functioning to augment resistance (Yao *et al.*, 2023). Prime editing, a recently developed approach based on CRISPR, enables highly accurate modifications and controlled introduction of desired characteristics, making it suitable for correcting mutations and enhancing qualities.

CRISPR can simultaneously target several genes. This enables the modification of many genes associated with resistance pathways, hence enhancing the efficacy of the defence mechanism against infections. Targeted gene editing for disease resistance entails the exact alteration of plant genomes to confer resistance against diseases. CRISPR-Cas9, an extensively employed gene editing method, allows researchers to precisely modify the DNA sequence of a plant. Scientists initially identify the genes responsible for plant – pathogen interactions that, when altered, can provide resistance to the specific pathogen. These genes may encode proteins that are responsible for recognizing pathogens, transmitting signals inside the cell, initiating defence responses, or being targeted by infections as susceptibility factors.

Plant cells that have been transformed are examined to identify those that possess the appropriate genetic modifications. Confirmation of the intended mutations can be achieved by several methods, including PCR-based genotyping, sequencing or phenotypic testing. Plant cells that have been genetically modified with the required gene changes are, subsequently, regenerated into whole plants utilizing tissue culture procedures. These genetically modified plants include modified genes and can be further analysed to determine their resistance characteristics. The modified plants undergo pathogen challenge assays to evaluate their resistance to the specific pathogen. This process may entail the introduction of pathogens into plants and the subsequent observation of disease signs, pathogen proliferation or other pertinent factors. If the edited plants exhibit improved disease resistance in controlled laboratory and greenhouse settings, they may be subjected to field trials to assess their performance in natural environments. Before the commercial release of the modified plants, it may be necessary to obtain regulatory permission.

## 5.9 Application of CRISPR-Cas9 in Plant Breeding

### 5.9.1 Crop domestication

The process of domesticating wild plants for the benefit of humans can be traced back to the beginning of agriculture. The process of crop domestication and plant breeding has led to the development of high-yielding crops that are well-suited to local environmental conditions. To address the various problems posed by climate change, it will be necessary to utilize novel crop plants, wild cousins and orphan crops that are more suited to their environment. This will help to ensure that food production can meet the needs of a growing population. The commercial cultivation of wild crops is impeded by their low productivity and unappealing aesthetics. CRISPR-Cas9 technology is now being used to domesticate wild plants. Wild tomatoes with drought-tolerant traits have been domesticated using CRISPR-Cas9. However, they showed some deficiencies in fruit yield and appearance. Numerous established genes in domesticated crops can be used as a foundation for CRISPR-Cas9 editing. This editing technique allows for the efficient alteration of the genome of wild plants, transforming them into advantageous agronomic crops (Shen *et al.*, 2017).

### 5.9.2 Crop yield and quality improvement

CRISPR-Cas9 has the potential to create valuable crops that possess enhanced nutritional content and increased resistance to various factors, as stated by Rasheed *et al.* (2021). CRISPR-Cas9 can be used to delete or suppress genes that reduce food output and have a negative impact on food quality. Several crops have been improved using CRISPR-Cas9 technology. For example, the size, shape and structure of tomato plants can be modified by modifying its genes using CRISPR-Cas9 technology. This is typically achieved by targeting the promoter regions of genes that control these traits. For example, the *SlERF.B3* gene is known to influence fruit shape and, by editing its promoter region, researchers can develop tomatoes with desired shapes. Similarly, modifying genes like *FW2.2* and *LOCULE NUMBER* (*LC*) can influence fruit size and structure, leading to more marketable and high-yield tomato varieties (Rodríguez-Leal *et al.*, 2017).

CRISPR-Cas9 has been instrumental in improving yield-related traits in rice, for instance by increasing the number of grains, grain size, and dense erect panicles. Targeted editing of the *GS3*, *DEP1* and *Gn1a* genes, which negatively affect these traits, has led to significant improvements. For instance, knocking out the *GS3* gene, which regulates grain length and weight, results in longer and heavier grains. Similarly, *DEP1* and *Gn1a* are involved in panicle architecture and grain number, respectively, and their disruption leads to increased grain production (Li *et al.*, 2016).

The genes *GW2*, *GW5* and *TGW6* in rice are associated with reduced grain weight. By disrupting these genes using CRISPR-Cas9, significant enhancements in grain output have been achieved. For example, knockout of the *GW2* gene, which acts as a negative regulator of grain size, leads to increased grain width and weight. Similarly, mutations in *GW5* and *TGW6* have shown to positively influence grain filling and overall grain yield (Xu *et al.*, 2016). In wheat, the *TaGW2* gene is known to negatively affect seed size. CRISPR-Cas9-mediated disruption of this gene has resulted in increased seed size. By knocking out *TaGW2*, researchers observed larger and heavier seeds, which can contribute to higher yield per plant. This gene editing approach provides a practical solution to improving wheat productivity and meeting global food demands (Wang *et al.*, 2018). CRISPR-Cas9 technology has been used to enhance the nutritional value of tomatoes by targeting the *GDP-galactose phosphorylase 1* (*GGP1*) gene. This gene is involved in the biosynthesis of vitamin C (ascorbic acid). By editing the *GGP1* gene, researchers have developed tomato cultivars with significantly increased vitamin C content, thereby improving their nutritional profile and potential health benefits for consumers (Li *et al.*, 2018).

CRISPR-Cas9 may also be used to create transgenic lines in crops like *Brassica campestris* and Chinese kale, with exceptional accuracy. By employing CRISPR-Cas9, scientists can introduce or modify specific genes in these crops to enhance desirable traits, such as disease resistance, yield and nutritional content, providing a robust tool for crop improvement and agricultural sustainability (Yang *et al.*, 2017).

### 5.9.3 Development of disease-resistant varieties

Disease resistant plants have been developed using conventional breeding methods and transgenic approaches. CRISPR-Cas9 demonstrates significant potential for enhanced efficiency and reduced time to consumption. The utilization of the CRISPR-Cas9 approach enables us to deliberately modify a vulnerable gene, thus facilitating the evaluation of plants and pathogens in novel ways and on a previously impossible scale. CRISPR-Cas9 can be utilized to create disease-resistant crops by augmenting the crop's natural immunity or by introducing engineered immunity, hence boosting its market value (Eş *et al.*, 2019).

The *ERF transcription factor* gene (*OsERF922*) was knocked out using CRISPR-Cas9 to boost rice blast resistance in the japonica rice variety Kuiku131. Mutating the promoter region of the *CsLOBI* gene leads to increased resistance against citrus canker. By utilizing Cas9, the gene *TaMLO* and EDR1 homologues can be edited and targeted in wheat, leading to the improvement and strengthening of resistance against powdery mildew. The CRISPR-Cas system has greatly enhanced the potential for effective control of viral infections. Cas9 can be employed to modify the genome of viruses as a means of combating viral infection, rather than simply treating the sickness caused by the viruses. Utilizing the CRISPR-Cas9 system can generate resistance against tobacco mosaic virus and cucumber mosaic virus. FnCas9, produced from the bacterium *Francisella novicida*, halts the replication of the viral genome.

### 5.9.4 Management of abiotic stress

Managing abiotic stress is crucial, due to the increasing impact of climate change. The plant breeding system has been revolutionized by developing crops that exhibit tolerance to abiotic stressors. Molecular breeders have identified numerous genes responsible for abiotic stress tolerance, which have been integrated into crops. Additionally, CRISPR-Cas9 technology allows for precise modification of any sequence and the introduction of desired traits into crops.

An instance of increased drought responsiveness was seen in tomato plants, by using the CRISPR-Cas9 technology to create Slmapk3 mutants. CRISPR-Cas9 was employed to do targeted genome editing on high-quality rice cultivars, resulting in the generation of mutants that exhibited improved resistance to salinity. Recent progress in abiotic stress management with CRISPR-Cas9 includes the creation of drought-tolerant chickpea cultivars (Zsögön *et al.*, 2018) and the modification of the *OsAnn3* gene in rice to enhance cold stress resistance.

### 5.9.5 Nutritional enhancement

Enhancing the nutritional value of crops, such by as raising the amount of vitamins or minerals, can be made easier with CRISPR-Cas9. Scientists used CRISPR-Cas9 to delete the *FAD2-1A* and *FAD2-1B* genes in soybeans, which raised the plant's beneficial oleic acid level (Fu *et al.*, 2022). There is great hope that CRISPR-Cas9 technology can speed up and improve plant breeding efforts, resulting in new crop varieties with better attributes and smaller environmental impacts (Fig. 5.4) (Table 5.1).

## 5.10 Ethical and Regulatory Considerations

### 5.10.1 Ethical issues in plant gene editing

Utilizing CRISPR-Cas9 gene-editing technology to enhance disease resistance in food crops raises several ethical and intellectual considerations that merit careful examination.

#### *5.10.1.1 Unintended consequences*

Altering the genetic makeup of food crops using CRISPR-Cas9 may lead to unintended consequences, including off-target effects or unexpected changes in the crop's characteristics. These unintended changes could have

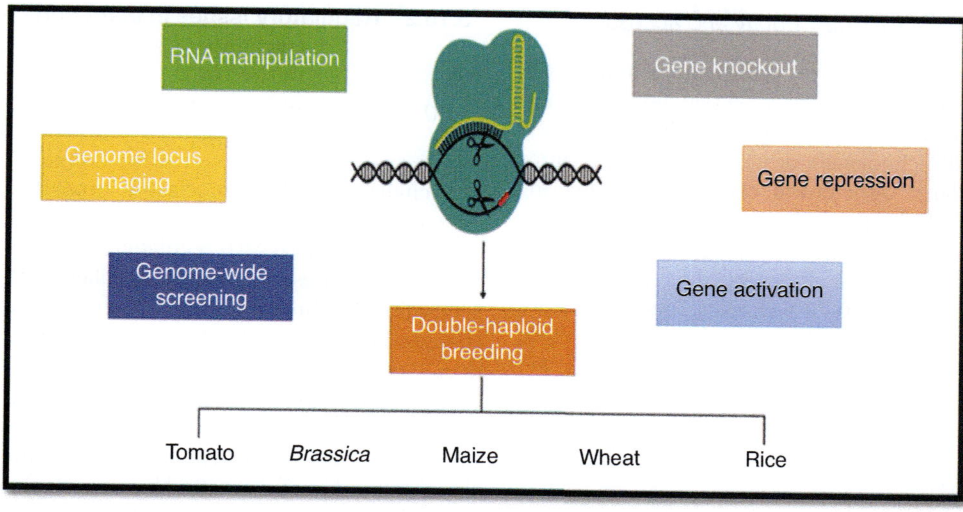


**Fig. 5.4.** Application of the CRISPR-Cas9 system in crop improvement.

**Table 5.1.** Numerous advances in the precise genetic editing of essential food crops have been made possible by CRISPR-Cas9. Various crops have yielded the following remarkable case studies and accomplishments.

| Crop | Achievements | Reference |
|---|---|---|
| Rice | Enhanced grain yield and biomass via *OsERF922* gene modification | Lou *et al.*, 2017 |
| Tomato | Delayed fruit senescence and enhanced quality via *SlLBD40* gene editing | Sharma *et al.*, 2023 |
| Wheat | Increased grain size and yield through *TaGW2* gene editing | Wang *et al.*, 2017 |
| Maize | Improved nutritional quality by editing *EPSPS* genes | Dong *et al.*, 2021 |
| Soyabean | Increased oleic acid content in oil via *FAD2* gene knockout | Haun *et al.*, 2014 |

unforeseen ecological or health impacts. For instance, unintentional gene edits could lead to the development of new allergens or toxins in the modified crops, which might pose health risks to consumers. Additionally, off-target effects could result in reduced crop viability or unintended harm to beneficial insects and soil microorganisms, thereby disrupting agricultural ecosystems (Esvelt and Gemmell, 2017)

#### 5.10.1.2 *Environmental concerns*

Introducing genetically modified (GM) crops into the environment could have ecological consequences, such as unintended effects on non-target organisms or disruption of ecosystems. For example, disease-resistant crops might outcompete wild relatives, leading to a loss of biodiversity. Furthermore, the spread of gene-edited traits through cross-pollination could inadvertently affect native plant populations. Ethical considerations involve weighing the potential benefits of disease-resistant crops against potential environmental risks. It is crucial to conduct thorough environmental impact assessments and develop strategies to mitigate potential negative effects on ecosystems (Snow *et al.*, 2005).

##### 5.10.1.3 Equity and access

There are concerns about equitable access to CRISPR-modified crops. Will small-scale farmers or farmers in developing countries have access to these technologies, or will they be monopolized by large agricultural corporations? Ensuring equitable access to genetically improved crops is an ethical imperative. It is essential to develop policies and frameworks that promote the sharing of CRISPR technology and its benefits, with marginalized and resource-poor farmers. This includes capacity-building initiatives, fair pricing strategies and collaborative research efforts, to address the specific needs of diverse farming communities (Cohen and Paarlberg, 2004)

##### 5.10.1.4 Consumer acceptance

There may be ethical concerns about consumer acceptance of GM foods. Transparency about the use of CRISPR technology in crop breeding and the clear labelling of GM products are essential for informed consumer choice. Public engagement and education are critical to address consumer concerns and building trust in CRISPR-modified crops. Ethical marketing practices, along with rigorous safety assessments and regulatory approvals, can help ensure that consumers are well informed and confident in the safety and benefits of GM foods (Frewer *et al.*, 2013).

##### 5.10.1.5 Long-term effects

Ethical considerations include the long-term effects of GM crops on agriculture, biodiversity and human health. Comprehensive risk assessments and ongoing monitoring are necessary to mitigate potential risks (Eş *et al.*, 2019). Long-term studies should evaluate the sustainability of CRISPR-edited crops, their impact on soil health and their interactions with other organisms in the ecosystem. Additionally, regulatory frameworks should be adaptive and responsive to new scientific findings, ensuring that the deployment of CRISPR technology in agriculture remains safe and beneficial over time (NASEM, 2016) .

### 5.10.2 Regulatory issues in plant gene editing

##### 5.10.2.1 Classification of CRISPR-edited organisms

One challenge is how to classify organisms edited using CRISPR-Cas9. Standard genetically modified organism (GMO) regulations may not adequately address organisms modified with gene-editing techniques, as they can produce changes indistinguishable from natural mutations. Regulators must determine whether CRISPR-edited crops fall under existing GMO regulations or require new regulatory frameworks. According to Borrelli *et al.* (2018), the classification dilemma is critical because it influences the approval process, labelling requirements and public perception of these crops.

##### 5.10.2.2 Regulatory harmonization

Regulatory frameworks for GMOs vary between countries, leading to inconsistencies in approval processes and market access. Achieving regulatory harmonization internationally is challenging but necessary to facilitate the global trade of CRISPR-edited crops and ensure food security. Harmonized regulations can help prevent trade barriers and promote the equitable distribution of CRISPR technology benefits across different regions. Establishing common standards and mutual recognition agreements among nations could streamline the regulatory process and foster international cooperation in biotechnology (Smyth, 2019).

##### 5.10.2.3 Risk assessment

Regulatory agencies must assess the safety of CRISPR-edited crops for human health and the environment. Assessing the potential risks associated with gene editing, such as off-target effects and unintended consequences, requires robust scientific evidence and standardized evaluation protocols. These assessments should include comprehensive genomic analyses to detect any unintended changes, along with long-term studies to monitor the effects of CRISPR-edited crops on ecosystems and human health. A rigorous and transparent risk assessment process is essential to gain public trust and ensure the safe implementation of gene-editing technologies in agriculture (Jones, 2015).

#### 5.10.2.4 *Transparency and labelling*

There is debate over whether CRISPR-edited crops should be subject to mandatory labelling requirements. Ensuring transparency in labelling allows consumers to make informed choices about the foods they purchase, but implementing labelling regulations presents logistical and regulatory challenges. Transparent labelling can help address public concerns about GM foods and build consumer trust. However, defining clear criteria for labelling and developing consistent standards across different regions are complex tasks that require careful consideration of both scientific and consumer perspectives (Shahriar *et al.*, 2021).

#### 5.10.2.5 *Intellectual property rights*

Gene-editing technologies like CRISPR are subject to intellectual property rights, which can complicate regulatory processes. Patent disputes and licensing agreements may affect the accessibility and affordability of CRISPR technology for agricultural research and development. Intellectual property issues can create barriers to innovation, particularly for small-scale researchers and developing countries. Addressing these challenges requires the development of fair and transparent licensing practices that encourage widespread access to CRISPR technology while respecting the rights of inventors and innovators (Lemley, 2017).

#### 5.10.2.6 *Public engagement*

Regulatory decisions regarding CRISPR-edited crops should consider public attitudes and values. Engaging stakeholders, including consumers, farmers, scientists and policymakers, in decision-making processes can enhance regulatory legitimacy and address public concerns about the safety and ethical implications of gene editing in agriculture. Public engagement efforts should aim to educate and inform stakeholders about the benefits and risks of CRISPR technology, fostering a dialogue that incorporates diverse viewpoints and values. This inclusive approach can help build consensus and support for the responsible use of gene-editing technologies in agriculture (Rasheed *et al.*, 2021).

## 5.11 Limitations and Future Prospects

The increasing availability of crop genome sequences and fungal/oomycete genomes, along with user-friendly CRISPR-Cas9 technologies that make use of free resources like plasmids and protocols, has accelerated the process of disease resistance in many crops. However, a few limitations will need to be taken into consideration (Fig. 5.5). For example, since agricultural plants have a longer life cycle than model plants, there are significant challenges with the frequency and accuracy of mutant generation. Recently, agroinfiltration was utilized to successfully demonstrate a disease-resistance phenotype in cocoa utilizing the CRISPR-Cas9 technology (Fister *et al.*, 2018). Hence, additional research into simple and fast ways to confirm the resistant phenotype using transient production of CRISPR-Cas9 components is necessary. The absence of selectable markers is one of the basic problems that restrict genome editing for oomycetes and fungi. The minimal inhibitory concentration of the newly created selectable marker (oxathiapiprolin-resistance gene) for *Phytophthora* spp. is significantly lower than that of geneticin (G418) or hygromycin B (Wang *et al.*, 2019); however, research on additional oomycetes is still in its early stages.

Before using CRISPR-Cas9 genome-editing tools on plant susceptibility genes, it is important to consider the potential negative consequences of choosing this technology. The susceptibility gene(s) are known to have pleiotropic effects and are associated with plant development and fitness (Tyagi *et al.*, 2021). This is also a common occurrence following the intentional enhancement of plant immunity by removing a negative regulator. Additional research and testing are required, even though these genes have not been shown to negatively affect plant development or health in previous studies (Pyott *et al.*, 2016). Another critical aspect to think

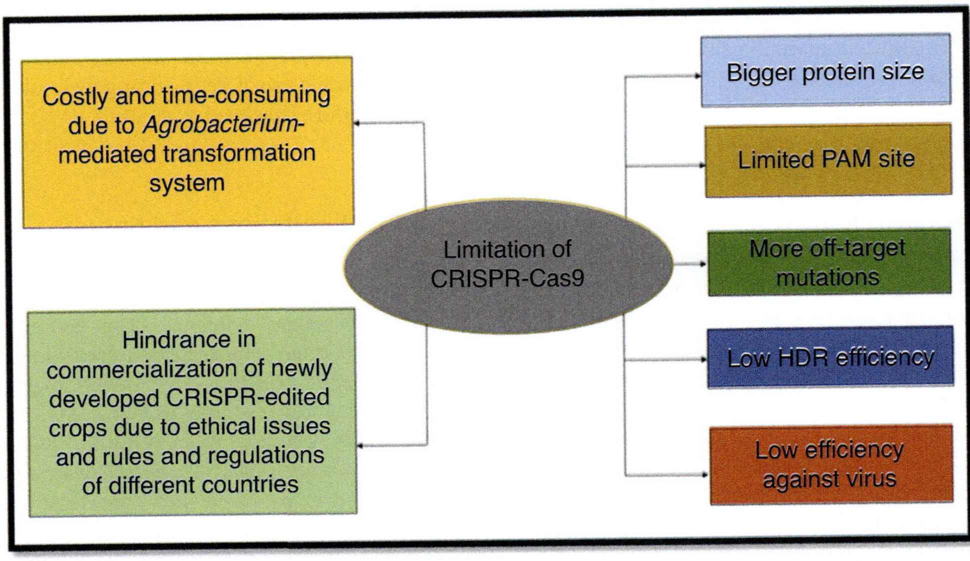


**Fig. 5.5.** Limitation of CRISPR-Cas9 system.

about is the possibility that the CRISPR-Cas9 system may introduce unintended mutations. The unintended consequence of modifying gene sequences in a way that affects protein structure and/or function is known as an off-target mutation. Researchers are currently re-editing CRISPR components including gRNA and Cas proteins to decrease off-target difficulties (Ahmad *et al.*, 2020).

When compared to conventional or molecular breeding techniques, genome editing using CRISPR-Cas9 could quickly create disease resistance. The fast evolution and genetic variety of oomycete and fungal populations mean that infections can eventually overcome disease resistance based on single-gene editing. These fast-developing CRISPR-Cas9 technologies have the potential to be strong tools for long-term disease management, through targeting several genes or combining genome editing of plants and fungi/oomycetes. Using combined genome editing, it is possible to show how a pathogen effector interacts with a host plant resistance gene. One way to suppress *Phytophthora* root and stem rot in soybean is to combine editing of the *Phytophthora sojae* effector *Avr* genes (Fang and Tyler, 2016) with editing of the soybean *Rps* genes (Nagy *et al.*, 2021). To further illustrate how CRISPR-Cas9 gene drive technology might be utilized for plant disease control, consider the eradication of malaria vector mosquitoes through targeted disruption of the double-sex gene (Paul *et al.*, 2021). Potentially useful for managing plant diseases transmitted sexually, this gene drive system offers a promising framework. Enhancing plant immunity and functioning as a biocontrol agent against fungal and oomycete pathogens can also be achieved by using the CRISPR-Cas9 system to engineer beneficial fungal species, including Trichoderma spp. Producing disease-resistant transgene-free crops will be an essential part of meeting global food demands in the future, and CRISPR-Cas9-based plant and pathogen genome editing will certainly be extensively used to improve plant disease resistance.

## 5.12 Conclusion

CRISPR-Cas9 is a powerful tool that is reshaping the future of plant breeding. Enabling precise genetic modifications, it opens possibilities for creating crop varieties with enhanced qualities and reduced vulnerabilities. Ethical considerations and regulatory norms must be carefully considered throughout the deployment of this technology. With its efficient and accurate methods of genetic editing, CRISPR-Cas9 technology is a game-changer in agricultural biotechnology and plant breeding. Many problems in agricultural production could be solved by this technology. For example, it could enhance crop resistance to diseases, increase tolerance to environmental stress, and improve the nutritional quality of food produced. With CRISPR-Cas9's capacity to alter specific regions of the genome, scientists can design crops with improved yields and a lesser impact on the environment.

Careful management of CRISPR-Cas9 deployment in agriculture is required to address regulatory and ethical concerns. Ethical concerns, such as unintended repercussions, environmental effects, fair access and consumer acceptability, demand careful deliberation. Categorization, risk assessment, transparency, labelling and intellectual property rights are some of the regulatory hurdles associated with CRISPR-edited species. The appropriate use of this technology requires actively involving stakeholders and making sure the public is part of the decision-making process.

However, there are still some challenges with CRISPR-Cas9 technology. These include the requirement for efficient delivery mechanisms and the possibility of off-target consequences. Research is currently underway to improve genome editing by addressing these constraints and making it more precise and effective. With the ability to transform plant breeding and aid in the fight for food security on a worldwide scale, CRISPR-Cas9 offers bright prospects for the future of agriculture. Scientists and breeders may use this technology to create crops that are more resistant to pests and have higher yields and better nutritional value, all while reducing the negative effects on the environment.

With its ongoing development, CRISPR-Cas9 will be pivotal in determining how agriculture is shaped in the years to come. We can create disease-resistant crop varieties that can thrive in different, and changing, environmental settings by integrating this technology with traditional breeding procedures and other biotechnological approaches. For sustainable agriculture practices and food security during global crises, this all-encompassing strategy for crop improvement is crucial.

## References


Acquaah, G. (2009) *Principles of Plant Genetics and Breeding*. John Wiley & Sons, Hoboken, New Jersey, USA.

Ahmad, S., Wei, X., Sheng, Z., Hu, P. and Tang, S. (2020) CRISPR/Cas9 for development of disease resistance in plants: Recent progress, limitations and future prospects. *Briefings in Functional Genomics* 19(1), 26–39.

Arif, A., Haider, R.A., Munir, A., Ghouri, M.Z., Khan, S.H. *et al.* (2024) Bioinformatics tools and databases in genome editing for plants. In: *CRISPRized Horticulture Crops*. Academic Press, pp. 51–66.

Ashkani, S., Rafii, M. Y., Shabanimofrad, M., Azadi, A., Ghasemzadeh, A. *et al.* (2015). Allele mining strategies: Principles and utilisation for blast resistance genes in rice (*Oryza sativa* L.) *Current Issues in Molecular Biology* 17, 57–74.

Azizi-Dargahlou, S. and Pouresmaeil, M. (2023) *Agrobacterium tumefaciens*-mediated plant transformation: A review. *Molecular Biotechnology* 66(7), 1563–1580.

Becker, S. and Boch, J. (2021) TALE and TALEN genome editing technologies. *Gene and Genome Editing* 2, 100007.

Begna, T. (2021) Conventional breeding methods widely used to improve self-pollinated crops. *International Journal of Research* 7(1), 1–16.

Bhattacharya, D. and Van Meir, E.G. (2019) A simple genotyping method to detect small CRISPR-Cas9 induced indels by agarose gel electrophoresis. *Scientific Reports* 9(1), 4437.

Borrelli, V.M., Brambilla, V., Rogowsky, P., Marocco, A. and Lanubile, A. (2018) The enhancement of plant disease resistance using CRISPR/Cas9 technology. *Frontiers in Plant Science* 9, 407423.

Carroll, D. (2014) Genome engineering with targetable nucleases. *Annual Review of Biochemistry* 83, 409–439.

Chan-In, P., Jamjod, S., Yimyam, N., Rerkasem, B. and Pusadee, T. (2020) Grain quality and allelic variation of the *Badh2* gene in Thai fragrant rice landraces. *Agronomy* 10(6), 779.

Cohen, J.I. and Paarlberg, R. (2004) Unlocking crop biotechnology in developing countries—A report from the field. *World Development* 32 (9), 1441–1466.

Contreras-Cornejo, H.A., Macías-Rodríguez, L., Cortés-Penagos, C. and López-Bucio, J. (2016) *Trichoderma* spp. improve growth of *Arabidopsis* seedlings under salt stress through enhanced root development, osmolite production, and $Na^+$ elimination through root exudates. *Molecular Plant – Microbe Interactions* 29(6), 412–423.

Dong, S., Qin, Y.L., Vakulskas, C.A., Collingwood, M.A., Marand, M. *et al.* (2021) Efficient targeted mutagenesis mediated by CRISPR-Cas12a ribonucleoprotein complexes in maize. *Frontiers in Genome Editing* 3, 670529.

Eş, I., Gavahian, M., Marti-Quijal, F.J., Lorenzo, J.M., Khaneghah, A.M. *et al.* (2019) The application of the CRISPR-Cas9 genome editing machinery in food and agricultural science: Current status, future perspectives, and associated challenges. *Biotechnology Advances* 37(3), 410–421.

Esvelt, K.M. and Gemmell, N.J. (2017) Conservation demands safe gene drive. *PLOS Biology* 15(11), e2003850.

Fang, Y. and Tyler, B.M. (2016) Efficient disruption and replacement of an effector gene in the oomycete Phytophthora sojae using CRISPR/Cas9. *Molecular Plant Pathology* 17(1), 127–139.

Fister, A.S., Landherr, L., Maximova, S.N. and Guiltinan, M.J. (2018) Transient expression of CRISPR/Cas9 machinery targeting *TcNPR3* enhances defense response in *Theobroma cacao*. *Frontiers in Plant Science* 9, 329023.

Frewer, L.J., van der Lans, I.A., Fischer, A.R., Reinders, M.J., Menozzi, D. *et al.* (2013) Public perceptions of agri-food applications of genetic modification: A systematic review and meta-analysis. *Trends in Food Science and Technology* 30(2), 142–152.

Fu, M., Chen, L., Cai, Y., Su, Q., Chen, Y. *et al.* (2022) CRISPR/Cas9-mediated mutagenesis of *GmFAD2-1A* and/or *GmFAD2-1B* to create high-oleic-acid soybean. *Agronomy* 12(12), 3218.

Garcia-Ruiz, H., Szurek, B. and Van den Ackerveken, G. (2021) Stop helping pathogens: Engineering plant susceptibility genes for durable resistance. *Current Opinion in Biotechnology* 70, 187–195.

Ghag, S.B. (2017) Host induced gene silencing, an emerging science to engineer crop resistance against harmful plant pathogens. *Physiological and Molecular Plant Pathology* 100, 242–254.

Ghogare, R., Ludwig, Y., Bueno, G.M., Slamet-Loedin, I.H. and Dhingra, A. (2021) Genome editing reagent delivery in plants. *Transgenic Research* 30 (4), 321–335.

Hasan, M.M., Rafii, M.Y., Ismail, M.R., Mahmood, M., Rahim, H.A. *et al.* (2015) Marker-assisted backcrossing: A useful method for rice improvement. *Biotechnology and Biotechnological Equipment 4* 29(2), 237–254.

Haun, W., Coffman, A., Clasen, B.M., Demorest, Z.L., Lowy, A. *et al.* (2014) Improved soybean oil quality by targeted mutagenesis of the fatty acid desaturase 2 gene family. *Plant Biotechnology* 12 (7), 934–940.

Jia, R., Zhao, H., Huang, J., Kong, H., Zhang, Y. *et al.* (2017) Use of RNAi technology to develop a PRSV-resistant transgenic papaya. *Scientific Reports* 7, 12636.

John Martin, J.J., Yarra, R., Wei, L. and Cao, H. (2022) Oil palm breeding in the modern era: Challenges and opportunities. *Plants* 11(11), 1395.

Jones, H.D. (2015) Regulatory uncertainty over genome editing. *Nature Plants* 1(4), 14011.

Junaid, J.M., Dar, N.A., Bhat, T.A., Bhat, A.H. and Bhat, M.A. (2013) Commercial biocontrol agents and their mechanism of action in the management of plant pathogens. *International Journal of Modern Plant & Animal Sciences* 1(2), 39–57.

Khan, Z. and Shahwar, D. (2024) Genome editing in plants via CRISPR/Cas9: A genomic scissor borrowed from bacterial immune system. In: Khan, Z., Shahwar, D. and Heikal, Y. (eds) *Genome Editing and Global Food Security*. Routledge, London, pp. 1–16.

Lemley, M.A. (2017) The surprising resilience of the patent system. *Texas Law Review* 95(1), 1–44.

Li, J., Li, Y., Chen, S. and Xie, G. (2018) Targeted mutagenesis of ascorbate oxidase gene in tomato results in enhanced ascorbate accumulation and tolerance to abiotic stress. *Plant Science* 267, 59–69.

Li, M., Li, X., Zhou, Z., Wu, P., Fang, M. *et al.* (2016) Reassessment of the four yield-related genes *Gn1a*, *DEP1*, *GS3*, and *IPA1* in rice using a CRISPR/Cas9 system. *Frontiers in Plant Science* 7, 377.

Li, S., Zhang, C., Li, J., Yan, L., Wang, N., & Xia, L. (2021). Present and future prospects for wheat improvement through genome editing and advanced technologies. *Plant Communications* 2, 100211.

Lou, D., Wang, H., Liang, G. and Yu, D. (2017) OsSAPK2 confers abscisic acid sensitivity and tolerance to drought stress in rice. *Frontiers in Plant Science* 8, 993.

Low, L.Y., Yang, S.K., Kok, D.X.A., Ong-Abdullah, J., Tan, N.P. *et al.* (2018) Transgenic plants: Gene constructs, vector and transformation method. In: Çelik, O. (ed.) *New Visions in Plant Science*. InTech Open, London, p. 41.

Ma, X., Zhu, Q., Chen, Y. and Liu, Y.G. (2016) CRISPR/Cas9 platforms for genome editing in plants: Developments and applications. *Molecular Plant* 9(7), 961–974.

Manghwar, H., Li, B., Ding, X., Hussain, A., Lindsey, K. *et al.* (2020) CRISPR/Cas systems in genome editing: Methodologies and tools for sgRNA design, off-target evaluation, and strategies to mitigate off-target effects. *Advanced Science* 7(6), 1902312.

Mansoor, S., Amin, I., Hussain, M., Zafar, Y. and Briddon, R.W. (2017) Engineering novel traits in plants through RNA interference. *Trends in Plant Science* 22(4), 344–346.

Miah, G., Rafii, M.Y., Ismail, M.R., Puteh, A.B., Rahim, H.A. *et al.* (2013) Blast resistance in rice: a review of conventional breeding to molecular approaches. *Molecular Biology Reports* 40 (3), 2369–2388.

Mushtaq, M., Sakina, A., Wani, S.H., Shikari, A.B., Tripathi, P. *et al.* (2019) Harnessing genome editing techniques to engineer disease resistance in plants. *Frontiers in Plant Science* 10, 550.

Nagy, E.D., Stevens, J.L., Yu, N., Hubmeier, C.S., LaFaver, N. *et al.* (2021) Novel disease resistance gene paralogs created by CRISPR/Cas9 in soy. *Plant Cell Reports* 40 (6), 1047–1058.

NASEM (National Academies of Sciences, Engineering, and Medicine) (2016) Genetically engineered crops: experiences and prospects. The National Academies Press, Washington, DC.

Nekrasov, V., Wang, C., Win, J., Lanz, C., Weigel, D. *et al.* (2017) Rapid generation of a transgene-free powdery mildew resistant tomato by genome deletion. *Scientific Reports* 7, (1) 482.

Nepolean, T., Kaul, J., Mukri, G. and Mittal, S. (2018) Genomics-enabled next-generation breeding approaches for developing system-specific drought-tolerant hybrids in maize. *Frontiers in Plant Science* 9, 325697.

Noman, A., Aqeel, M. and He, S. (2016) CRISPR-Cas9: Tool for qualitative and quantitative plant genome editing. *Frontiers in Plant Science* 7, 216074.

Osei, M.K., Prempeh, R., Adjebeng-Danquah, J., Opoku, J.A., Danquah, A. *et al.* (2018) Marker-assisted selection (MAS): A fast-track tool in tomato breeding. In: Nyaku, S.T. and Danquah, A. (eds) *Recent Advances in Tomato Breeding and Production*, pp. 93–113. InTech Open, London.

Paul, N.C., Park, S.W., Liu, H., Choi, S., Ma, J. *et al.* (2021) Plant and fungal genome editing to enhance plant disease resistance using the CRISPR/Cas9 system. *Frontiers in Plant Science* 12, 700925.

Pieterse, C.M., Zamioudis, C., Berendsen, R.L., Weller, D.M., Van Wees, S.C. *et al.* (2014) Induced systemic resistance by beneficial microbes. *Annual Review of Phytopathology* 52, 347–375.

Pyott, D.E., Sheehan, E. and Molnar, A. (2016) Engineering of CRISPR/Cas9-mediated potyvirus resistance in transgene-free *Arabidopsis* plants. *Molecular Plant Pathology* 17(8), 1276–1288.

Rasheed, A., Gill, R.A., Hassan, M.U., Mahmood, A., Qari, S. *et al.* (2021) A critical review: Recent advancements in the use of CRISPR/Cas9 technology to enhance crops and alleviate global food crises. *Current Issues in Molecular Biology* 43(3), 1950–1976.

Ray, U. and Raghavan, S.C. (2020) Modulation of DNA double-strand break repair as a strategy to improve precise genome editing. *Oncogene* 39(41), 6393–6405.

Rees, H.A. and Liu, D.R. (2018) Base editing: Precision chemistry on the genome and transcriptome of living cells. *Nature Reviews Genetics* 19, 770–788.

Rodríguez-Leal, D., Lemmon, Z.H., Man, J., Bartlett, M.E. and Lippman, Z.B. (2017) Engineering quantitative trait variation for crop improvement by genome editing. *Cell* 171(2), 470–480.

Savary, S., Willocquet, L., Pethybridge, S.J., Esker, P., McRoberts, N. *et al.* (2019) The global burden of pathogens and pests on major food crops. *Nature Ecology & Evolution* 3(3), 430–439.

Schmitz, D.J., Ali, Z., Wang, C., Aljedaani, F., Hooykaas, P.J. *et al.* (2020) CRISPR/Cas9 mutagenesis by translocation of Cas9 protein into plant cells via the Agrobacterium type IV secretion system. *Frontiers in Genome Editing* 2, 6.

Shahriar, S.A., Islam, M.N., Chun, C.N.W., Rahim, M.A., Paul, N.C. *et al.* (2021) Control of plant viral diseases by CRISPR/Cas9: Resistance mechanisms, strategies and challenges in food crops. *Plants* 10(7), 1264.

Shahwar, D., Khan, Z. and Park, Y. (2023) Molecular marker-assisted mapping, candidate gene identification, and breeding in melon (*Cucumis melo* L.): A review. *International Journal of Molecular Sciences* 24(20), 15490.

Shahwar, D., Khan, Z. and Park, Y. (2024) Molecular markers for marker-assisted breeding for biotic and abiotic stress in Melon (*Cucumis melo* L.): A review: *International Journal of Molecular Sciences* 25(12), 6307.

Sharma, P., Pandey, A., Malviya, R., Dey, S., Karmakar, S. *et al.* (2023) Genome editing for improving nutritional quality, post-harvest shelf life and stress tolerance of fruits, vegetables, and ornamentals. *Frontiers in Genome Editing* 5, 1094965.

Shen, C., Que, Z., Xia, Y., Tang, N., Li, D. *et al.* (2017) Knock out of the annexin gene *OsAnn3* via CRISPR/Cas9-mediated genome editing decreased cold tolerance in rice. *Journal of Plant Biology* 60, 539–547.

Slater, A.T., Cogan, N.O.I., Hayes, B.J., Schultz, L., Dale, M.F.B. *et al.* (2013) Improving breeding efficiency in potato using molecular and quantitative genetics. *Theoretical and Applied Genetics* 126(9), 2279–2292.

Smyth, S.J. (2019) Regulatory challenges of gene editing with CRISPR/Cas9: Aspects to consider. *Frontiers in Bioengineering and Biotechnology* 7, 164.

Snow, A.A., Andow, D.A., Gepts, P., Hallerman, E.M., Power, A. *et al.* (2005) Genetically engineered organisms and the environment: Current status and recommendations. *Ecological Applications* 15(2), 377–404.

Stemmer, M., Thumberger, T., del Sol Keyer, M., Wittbrodt, J. and Mateo, J.L. (2015) CCTop: An intuitive, flexible and reliable CRISPR/Cas9 target prediction tool. *PLOS ONE* 10(4), e0124633.

Strange, R.N. and Scott, P.R. (2005) Plant disease: A threat to global food security. *Annual Review of Phytopathology* 43, 83–116.

Tashkandi, M., Ali, Z., Aljedaani, F., Shami, A. and Mahfouz, M.M. (2018) Engineering resistance against tomato yellow leaf curl virus via the CRISPR/Cas9 system in tomato. *Plant Signaling & Behavior* 13(10), e1525996.

Tyagi, S., Kumar, R., Kumar, V., Won, S.Y. and Shukla, P. (2021) Engineering disease resistant plants through CRISPR-Cas9 technology. *GM Crops & Food* 12(1), 125–144.

Waltz, E. (2016) Gene-edited CRISPR mushroom escapes US regulation. *Nature Biotechnology* 34(9), 1003.

Wang, H.X., Li, M., Lee, C.M., Chakraborty, S., Kim, H.W. *et al.* (2017) CRISPR/Cas9-based genome editing for disease modeling and therapy: Challenges and opportunities for nonviral delivery. *Chemical Reviews* 117(15), 9874–9906. DOI: 10.1021/acs.chemrev.6b00799.

Wang, J., Meng, X., Hu, X., Sun, T., Li, J. *et al.* (2019) XCas9 expands the scope of genome editing with reduced efficiency in rice. *Plant Biotechnology Journal* 17(4), 709.

Wang, W., Simmonds, J., Pan, Q., Davidson, D., He, F. *et al.* (2018) Gene editing and mutagenesis reveal inter-cultivar differences and additivity in the contribution of *TaGW2* homoeologues to grain size and weight in wheat. *Theoretical and Applied Genetics* 131 (11), 2463–2475.

Wang, Y., Cheng, X., Shan, Q., Zhang, Y., Liu, J. *et al.* (2014) Simultaneous editing of three homoeoalleles in hexaploid bread wheat confers heritable resistance to powdery mildew. *Nature Biotechnology* 32(9), 947–951.

Xu, R., Yang, Y., Qin, R., Li, H., Qiu, C. *et al.* (2016) Rapid improvement of grain weight via highly efficient CRISPR/Cas9-mediated multiplex genome editing in rice. *Journal of Genetics and Genomics* 43(8), 529–532.

Yang, H., Wu, J.J., Tang, T., Liu, K.D. and Dai, C. (2017) CRISPR/Cas9-mediated genome editing efficiently creates specific mutations at multiple loci using one sgRNA in *Brassica napus*. *Scientific Reports* 7(1), 12215.

Yang, H., Ren, S., Yu, S., Pan, H., Li, T. *et al.* (2020) Methods favoring homology-directed repair choice in response to CRISPR/Cas9 induced-double strand breaks. *International Journal of Molecular Sciences* 21(18), 6461.

Yao, T., Yuan, G., Lu, H., Liu, Y., Zhang, J. *et al.* (2023) CRISPR/Cas9-based gene activation and base editing in Populus. *Horticulture Research* 10(6), uhad085.

Zhang, S., Zhu, H. (2024) Development and prospect of gene-edited fruits and vegetables. *Food Quality and Safety*, Volume 8, fyad045. DOI: 10.1093/fqsafe/fyad045.

Zhang, Y., Bai, Y., Wu, G., Zou, S., Chen, Y. *et al.* (2019) Simultaneous modification of three homoeologs of *TaEDR1* by genome editing enhances powdery mildew resistance in wheat. *Plant Journal* 98(3), 528–538.

Zhou, J., Deng, K., Cheng, Y., Zhong, Z., Tian, L. *et al.* (2017) CRISPR-Cas9 based genome editing reveals new insights into microRNA function and regulation in rice. *Frontiers in Plant Science* 8, 1598.

Zhou, W., Hu, L., Ying, L., Zhao, Z., Chu, P.K. *et al*. (2018) A CRISPR-Cas9-triggered strand displacement amplification method for ultrasensitive DNA detection. *Nature Communications* 9, 5012.
Zsögön, A., Čermák, T., Naves, E., Notini, M.M., Edel, K.H. *et al*. (2018) *De novo* domestication of wild tomato using genome editing. *Nature Biotechnology* 36, 1211–1216.
Zvereva, A.S. and Pooggin, M.M. (2012) Silencing and innate immunity in plant defense against viral and non-viral pathogens. *Viruses* 4(11), 2578–2597.

# 6 Artificial Intelligence (AI) and Data-Driven Plant Breeding

**Sadhan Debnath[1,2]*, Kirti Devi[3] and Sujan Majumder[4]**
*[1]National Institute of Plant Genome Research (NIPGR), Aruna Asaf Ali Marg, New Delhi, India; [2]College of Post Graduate Studies in Agricultural Sciences (CPGSAS), Central Agricultural University (Imphal), Umiam, Meghalaya, India; [3]National Agri-food Biotechnology Institute (NABI), Mohali, Punjab, India; [4]ICAR-Indian Institute of Vegetable Research (IIVR), Varanasi, India*

**Abstract**
Plant breeding is being used as an environmentally sound approach to increase food production and ensure food security. However, in order to boost crop yields and produce new varieties that are more resilient to various environmental stress factors and more nutritious, various high-throughput multi-omics strategies in plant breeding have been developed and implemented. These cutting-edge technologies are used to produce huge amounts of data about the plants' phenotype, biochemistry, physiology and genetic architecture, and could be used to manipulate the essential traits of plants, which are crucial for crop improvement. Therefore, in order to analyse and represent these enormous amounts of complicated data in an efficient way, plant breeders usually depend on advanced computational and bioinformatics tools and artificial intelligence (AI), including machine learning (ML) techniques. These technologies could also be used for plant phenotyping and genomic prediction for the efficient selection of superior plant types. Artificial intelligence refers to the ability of machines, in particular, computers, to simulate human intelligence processes. To write and operate machine learning algorithms, AI needs a foundation of sophisticated software and hardware components. The application of AI to plant breeding can provide breeders with highly productive and useful tools to accelerate the development of improved crop varieties and enhance the effectiveness of the breeding processes, which are crucial for addressing the difficulties that agriculture currently faces in the age of climate change. This chapter briefly explains how AI could revolutionize the field of future plant breeding while addressing the global food security and nutritional requirements.

**Keywords:** artificial intelligence, food security, genomics, genomic prediction, high-throughput phenomics, machine learning, plant breeding, simulation modelling

**Abbreviations**
ACID – Annotated Crop Image Dataset, AI – Artificial intelligence, ANN – Artificial neural network, BLUP – Best linear unbiased prediction, CGMs – Crop growth models, CNN – Convolutional neural network, CSC – Climate-smart crops, CT – Computed tomography, DL – Deep learning, DNNGP – Deep neural network for genomic prediction, G2P – Gene-to-phenotype, GAB – Genomics-assisted breeding, GS – Genomic selection, GWAS – Genome-wide association studies, MAS – Marker-assisted selection, ML – Machine learning, MLP – Multilayer perceptron, OPIA – Open Plant Image Archive.

*Corresponding author: debnathsadhan123@gmail.com

 *Plant Breeding Technology: Future Trends and Challenges* (ed. Zeba Khan)
DOI: 10.1079/9781800626638.0006

## 6.1 Introduction

An estimated world population of 10 billion by 2057 imposes tremendous pressure on the agriculture sector to fulfil the increasing demand for food (Yu *et al.*, 2020; UN, 2022). At the same time, global agricultural productivity has been impacted by the consequences of climate change, which have made it more difficult to improve crop yields. The geographical distribution of rainfall has altered because of intense sunshine, extreme high or low temperatures and an increased $CO_2$ level, making crops more vulnerable to insect pests and diseases.

It has been suggested that the biotic and abiotic stress factors caused by climate change are now a major factor influencing the growth and development of crop plants and agricultural production. A decline in agricultural productivity will negatively impact food security, leading to persistent malnutrition and micronutrient deficiencies for the world's expanding population. Hence, the development of improved cultivars that are stress tolerant and high yielding requires immediate attention. Therefore, it is now essential to develop climate-smart crops (CSC) using the next-generation crop breeding technology. The objective could be fulfilled by combining artificial intelligence (AI) with the classical or conventional breeding and genomics-assisted breeding (GAB), which includes both marker-assisted selection (MAS) and genomic selection (GS) (Breseghello and Coelho, 2013; Bhat and Yu, 2021).

Global food and nutrition security has been made possible by traditional plant breeding. Through the development of new plant varieties with desired characteristics, including high productivity, resistance to diseases and abiotic stress resistance, plant breeding plays a key role in the sustainability of agriculture and food and nutritional security. It has significantly increased crop yields, as seen through the outcome of the Green Revolution (Ahmar *et al.*, 2020). High-quality agronomic traits that help in yield enhancement and stress resistance can be found in the new varieties of crops developed through conventional breeding, which is achieved by identifying new genes and making use of genetic diversity.

However, the method of effectively utilizing genetic diversity and discovering novel genes through traditional breeding relies mostly on phenotypic selection, which may decrease the accuracy and precision in choosing superior genotypes, and may also make the procedure time consuming and labour intensive. Moreover, most desirable characteristics, like resilience to stress and yield-related traits, are complex traits that have strong genotype–environment interactions and low heritability, therefore requiring time and labour-intensive phenotyping operations (Karikari *et al.*, 2019; Hina *et al.*, 2020). For instance, developing a genetically homogeneous/stable line or variety takes nearly 6–7 years involving a series of steps that includes: selection of parents, hybridization programme, backcrossing, collecting and analysing data, and field assessment of individual characteristics. Therefore, conventional plant breeding is no longer adequate to ensure global food security (Athar and Ashraf, 2009; Bhat *et al.*, 2016).

Hence, to speed up the whole process and to obtain highly reliable genomic prediction, next-generation breeding technologies – which involve AI and machine learning (ML) or deep learning (DL) approaches, speed-breeding and high-throughput phenotyping and genotyping approaches – must be integrated into the ongoing molecular breeding and GAB programmes, to sustain the food and nutritional security of the rapidly rising global population (Fig. 6.1). Improved crop cultivars can be developed through conventional breeding within 8–10 years, whereas speed breeding can produce similar results within 3–5 years. Specialized laboratories are not necessary for speed breeding; all that is needed is the availability of optimal lighting and temperature management (Shivakumar *et al.*, 2018).

High-throughput phenotyping is an integral part of speed-breeding and integrated GS; however, the actual implementation of high-throughput phenotyping in crop breeding has been hampered by the lack of appropriate statistical tools for interpreting phenotypic data. In this context, by combining AI and phenomics, breeders may take advantage of large amounts of imaging data for practical breeding through non-invasive imaging (Casanova *et al.*, 2014) and deal with non-linear and epistatic interactions in GWAS and GS to fully utilize complicated trait improvement

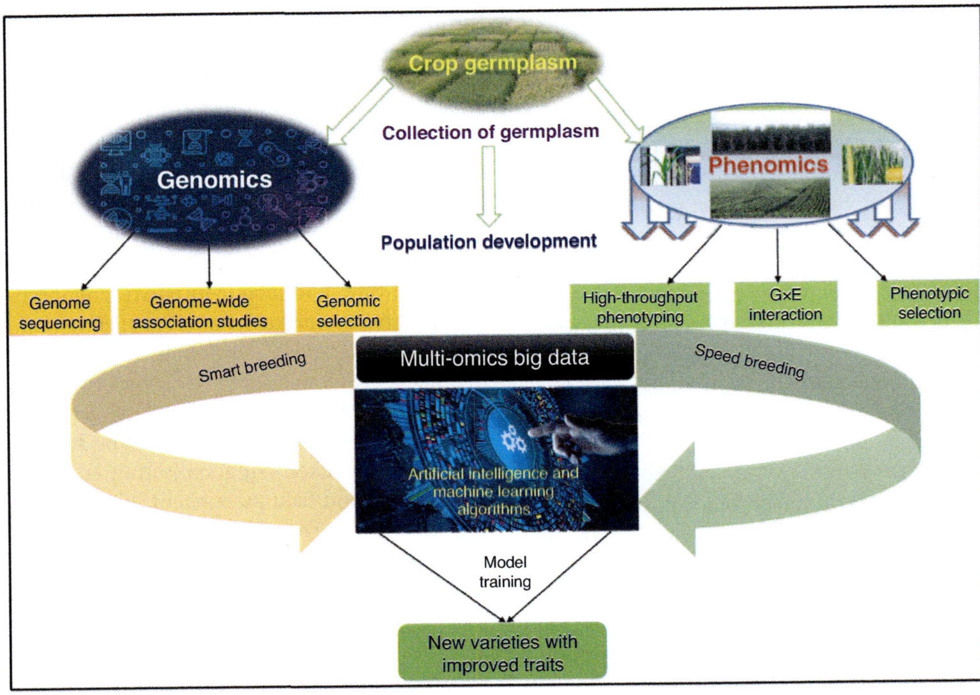


**Fig. 6.1.** An overview of the potential application of artificial intelligence in plant breeding.

in plant breeding programmes (Ansarifar and Wang, 2019; Yoosefzadeh-Najafabadi *et al.*, 2022a). Additionally, speed breeding has become increasingly popular, because it can reduce breeding cycles and increase crop generations (Watson *et al.*, 2018). In speed breeding, the life cycle of photo-insensitive crops is manipulated by using longer photoperiods under regulated environments. It is possible to create superior cultivars more quickly and accurately by combining speed breeding with other methods like AI and GAB. Recent developments in AI based approaches, such as ML and DL, provide opportunities to address long-standing limitations and difficulties in crop breeding, including phenotyping of plants, identification of genes and varietal selection (Xu *et al.*, 2022).

Genomics-assisted breeding has been shown to be a more feasible approach in this context for hastening the development of superior crop varieties (Mir *et al.*, 2019; van Bezouw *et al.*, 2019). The two main GAB approaches are MAS and GS (Bhat *et al.*, 2016). In MAS, major effect quantitative trait loci (QTLs) or genes linked to known markers are transferred from the donor parent to an elite recipient crop variety; whereas, in GS, a model that predicts genomic estimated breeding values (GEBVs) for selection candidates is trained using genome-wide marker data, considering that at least a single marker is in linkage disequilibrium with every gene controlling a complicated trait phenotype (Spindel *et al.*, 2015). Technological advancements in the field of AI, high-throughput genomics and sequencing, and high-throughput phenomics have paved the way for the practical implementation of GS in the breeding of crops. According to Xu *et al.*, 2022, plant breeding would be regulated by AI in four different ways: (i) AI-assisted breeding technologies will be important for agricultural field management, breeding method preparation, assessment and selections, as well as for theoretical investigations; (ii) robots with AI capabilities will engage with every step of collecting data, preservation, evaluation, collaborating and usage, greatly enhancing breeding systems that collect information; (iii) the historical expertise and pertinent information generated and acquired in breeding projects would be useful for AI systems; and (iv) big data and AI-powered breeding technologies will be able to create and anticipate a great deal using simulation of models and optimization.

## 6.2 Artificial Intelligence (AI)

The science of developing computer systems and machines to 'think and act like humans' is known as artificial intelligence (Warwick, 2013). It has gained popularity recently and is being used in many different fields and industries, such as the human sciences, medical treatment and the scientific study of plants and animals (Warwick, 2013; Pepe *et al.*, 2021a; Harfouche *et al.*, 2022). Artificial intelligence is capable of task automation, pattern recognition in massive datasets, insight generation, and prediction (Harfouche *et al.*, 2022). Artificial Intelligence tools like computer vision, robotics and ML can be applied to plant breeding to analyse vast volumes of data and recognize trends that help choose better plant varieties with desired traits (Chandra *et al.*, 2020). Another area of AI that focuses on creating algorithms that can recognize and identify things in pictures and videos is computer vision (Bhatt *et al.*, 2021). Computer vision algorithms are useful for applications like picture categorization, tracking of objects and recognition of faces (Voulodimos *et al.*, 2018). The creation of autonomous robots is the main goal of robotics in AI (Kunze *et al.*, 2018). For AI applications that need to perform manual activities like assembling components, navigating areas, or handling objects, robotics is crucial (Murphy, 2019).

Plant breeders can effectively monitor plant growth and development and select plants with desired features through the extensive applications of robotics and computer vision in plant breeding (Fig. 6.2). There have been a lot of sensors, robots and pipelines developed recently, for instance, the BoniRob V2 robot was developed to serve as a plant scout robot. It can measure properties of soil, characterize plants using multi-sensor applications based on their phenotypical and spectral characteristics, and provide a solution for weed control based on camera technology and image recognition (Bangert *et al.*, 2013). Another popular robot, Thorvald II, was created to allow for excellent customization for a range of settings, including open crop fields, greenhouses and polytunnels. This enables the robot to be quickly applied to the specified environment (Grimstad and From, 2017). Falk *et al.*, 2020 reported successful usage of robotics and computer vision in soybean-root phenotyping. The ARIA 2.0 robot was created to enable post-image capture automation from soybean roots and do away with image preprocessing procedures such image cropping (Falk *et al.*, 2020).

## 6.3 Machine Learning (ML)

The primary objectives of ML, a subfield of artificial intelligence, are to create programmes that can analyse data, spot undetectable trends, create prediction models, identify objects in photos, operate robotic devices and form conclusions (Libbrecht and Noble, 2015). It has been applied extensively to many dimensions of our lives, including communication and agriculture (Falk *et al.*, 2020; Jafari and Shahsavar, 2020). Machine learning algorithms have applications in the agricultural sector, including precision farming, automated irrigation, monitoring of plant development and crop production prediction. These algorithms are generally categorized as supervised learning, unsupervised learning and reinforcement learning (Hesami *et al.*, 2022). In supervised ML, predictions are made by the algorithm through training on a labelled dataset (Hesami and Jones, 2020). When a set of inputs and related outputs are provided to the model, it learns by modifying its internal parameters to yield the intended result. In plant breeding, the most popular and widely used category of ML algorithms is supervised learning, which is used to identify a particular genomic locus linked to a specific character (Yoosefzadeh-Najafabadi *et al.*, 2022b), make an early

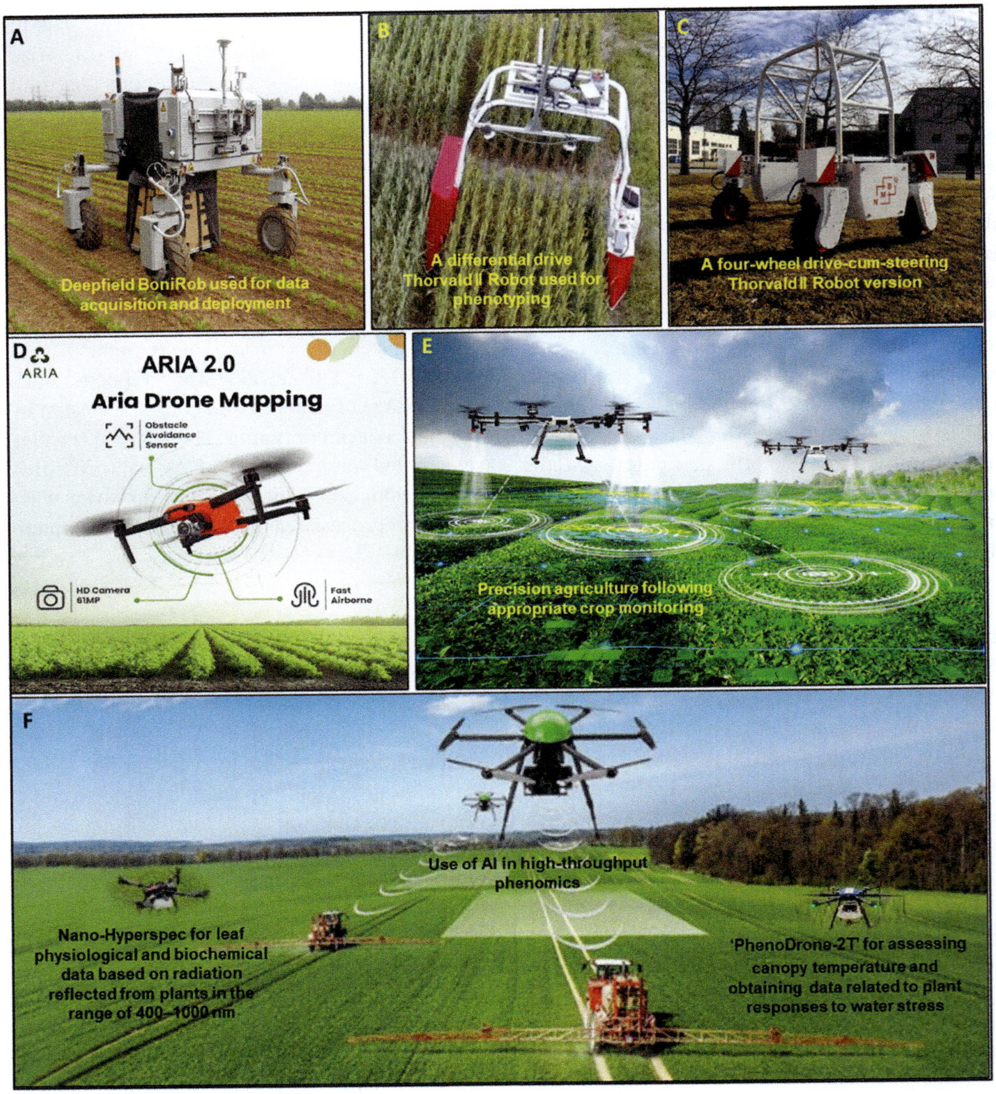


**Fig. 6.2.** The various applications of AI in plant breeding include a new version of BOSCH Deepfield BoniRob, suitable for data acquisition and deployment in plant breeding (A); a differential drive Thorvald II robot, useful for high-throughput phenomics (B); a four-wheel drive-steering Thorvald II robot for plant phenotyping (C); ARIA 2.0, useful for mapping and image based phenotyping (D); use of AI for crop monitoring in precision agriculture (E); and use of AI in high-throughput phenomics for plant morphological, physiological and biochemical data (F).

prediction of complex character, and choose superior genotypes through GS (Yoosefzadeh-Najafabadi *et al.*, 2022c).

If data is unlabelled, an algorithm that employs unsupervised learning analyses the data to identify trends and similarities on its own, inside the dataset (Mahesh, 2020). The way the model learns is by spotting patterns in the data, like groupings or clusters. Unsupervised learning is typically used in plant breeding to create kinship matrices, classify unstructured datasets and identify potential relationships between genotypes within a breeding population (Mahesh, 2020). Another ML algorithm is reinforcement learning, which exposes the model to an environment and gives it feedback in the form of incentives or punishments based on how it behaves (Lee *et al.*, 2012). By performing actions and modifying its settings to optimize the overall rewards earned, the model learns. Reinforcement learning is a very recent field in plant breeding and its usefulness needs to be further explored.

A number of essential considerations must be made in order for ML algorithms to successfully anticipate a particular complex trait in plant breeding. A few examples of factors involved in the collection of data (preliminary processing, obtaining features, training, evaluating, tuning of hyperparameter and deploying and tracking the model (Vabalas *et al.*, 2019; Sakib *et al.*, 2020) are discussed here in more detail:

1. The process of obtaining information for the training of models from various sources (genotypes, environments, etc.) in various formats, such as text, photos, numerical/categorial datasets and videos is known as data collecting (Xu *et al.*, 2022).
2. The cleaning, altering and arranging of data to make it more appropriate for ML algorithms is known as the pre-processing step (Vabalas *et al.*, 2019).
3. The process of extracting features or variables from the data so they can be represented in a way that is more appropriate for ML algorithms is known as feature extraction (Zhang *et al.*, 2021).
4. To fit models to the data, model training employs several ML methods (Niazian and Niedbała, 2020; Yoosefzadeh-Najafabadi *et al.*, 2021).
5. In the hyperparameter tuning process, tested ML algorithm parameters are adjusted once again for optimal performance (Hesami and Jones, 2020; Pepe *et al.*, 2021b).
6. The process of implementing a generated model in production, typically as an application, is summed up as model deployment (Derakhshan *et al.*, 2019).
7. The process of tracking a model's performance over time to make sure it stays correct is referred to as model monitoring (Yoosefzadeh-Najafabadi *et al.*, 2021).

## 6.4 Applications of AI in Plant Breeding

### 6.4.1 Genomic prediction

Genomic prediction is the process of forecasting a plant's phenotype using its genetic information. Methods that use AI, such as ML algorithms, have been applied to evaluate massive amounts of genomic data and make precise predictions about plant characteristics. This has allowed breeders to choose the most promising individuals for future breeding cycles. Genomic selection, also known as genomic prediction, is the key element of present-day crop-improvement methods. Genomic prediction forecasts the performance of possible breeding prospects based on genetic composition. Since the introduction of GS, modern plant breeding has made great progress. The method was first conceptualized and reported by Meuwissen *et al.* (2001) for the breeding of animals; thereafter, it has become an important method for determining the phenotypes of polygenic characteristics in crop plants. This strategy has been confirmed as successful in a number of crop species, including rice, wheat and maize, demonstrating its effectiveness in practical breeding activities (Montesinos-Lopez *et al.*, 2021; Wang *et al.*, 2023, Yan *et al.*, 2023).

However, accurate prediction is challenging when there are high-dimensional marker spaces where the number of genotypic markers is greater than the population size. Conventional GS models are often challenged by the highly dimensional nature of marker data and the complex interactions within and between genotypes and phenotypes. These models rely on statistical techniques like Bayesian and best linear unbiased prediction (BLUP) based approaches (e.g. ridge regression BLUP and genomic connection BLUP) (Danilevicz *et al.*, 2022). To manage the complexity of multi-dimensional genotypic data, programmes like DNNGP (a deep neural network for genomic prediction) and DeepGS (an R package that uses DL methods in predicting phenotypes from genotypes) use deep neural networks and techniques like convolution, sampling, dropping out and collaborative learning. DNNGP makes itself distinctive by employing a deep neural network that uses a multi-level hierarchical structure for flexible feature learning and merging multi-omics data from multiple plant species. Since it can manage breeding data of different sizes, improve prediction accuracy and compute more quickly than normal approaches, this model is found to be a highly important tool for genomic selection (Wang *et al.*, 2023).

Another noteworthy example of DL being utilized for genomic prediction in plant breeding is the recently created SoyDNGP model (Gao *et al.*, 2023). Conventional methods generally transform DNA sequences into binary data by 'one-hot encoding', resulting in large, sparse matrices that fall short of accurately capturing the biological importance of genomic sequences. On the other hand, SoyDNGP converts genomic variations into a format similar to an image, so that the data may be interpreted by a convolutional neural network (CNN) in a way that is comparable to an image recognition task. Because of its distinct representation, the SoyDNGP can identify complex patterns and traits within genomic data by employing many convolutional and pooling layers to determine the intricate linkages and interactions between distinct genetic locations (Feng *et al.*, 2024).

### 6.4.2 Image based phenotyping

High-throughput phenotyping of plant characteristics, such as leaf area, plant height and disease signs, using digital photos is made possible by AI based image analysis tools. The use of CNNs, in particular, has demonstrated remarkable accuracy in the extraction of phenotypic information from images. This has made it possible for breeding programmes to evaluate phenotypic information quickly and non-destructively. Plant phenomics, a cutting-edge discipline that uses high-throughput phenotyping technologies for the rapid and accurate acquisition of a wide variety of phenotypic data, has been widely employed to find desirable characteristics. These modern technologies offer numerous notable advantages, including non-intrusiveness and the ability to swiftly and accurately gather multi-dimensional phenotypic data from plants, resulting in the speedy accumulation of images and i-traits. Plant phenomics employs the word 'i-traits' to refer to a broad category of quantifiable characteristics, such as plant height and grain length, which may be quantified by image analysis techniques. In recent years, it has become more popular to use picture collections from various high-throughput plant imaging technologies to study multi-tissue phenotypes.

To collect plant image datasets from various imaging sensors, various techniques have recently been developed. Annotated Crop Image Dataset (ACID) (Pound *et al.*, 2017), X-Plant (Piovesan *et al.*, 2021), and Quantitative Plant (Lobet *et al.*, 2013 and Lobet, 2017) are three noteworthy examples. Quantitative Plant was developed in 2013; it has 31 image datasets, 13 of which include annotated examples. The computed tomography database X-Plant is used to gather three-dimensional architectural images of plants, particularly their roots and leaves. ACID mainly targets curation of annotated plant datasets for computer vision, availing insufficient metadata for four plant image datasets spanning four species. However, there are two primary constraints that are frequently noted: first, none of these resources offers detailed annotations for every image or a comprehensive integration of metadata, which is essential for improving the reusability of plant image data; second, they hinder the thorough examination of plant growth structures by not providing i-traits for the exploration of novel traits.

An extensive collection of plant photos and i-traits from cutting-edge phenotyping tools can be found in the openly accessible Open Plant Image Archive (OPIA). It is a comprehensive repository of image datasets and i-traits for model plants as well as staple crops. In contrast to other crop-related databases already in existence, OPIA is primarily concerned with compiling picture datasets that address the two core goals of plant phenotypic analysis: object recognition and image classification. OPIA is distinguished by its rigorous hand curation and image analysis, which guarantee the gathering and arrangement of extremely useful meta-information via the application of controlled vocabularies. Additionally, it provides easily navigable web interfaces and online tools that make retrieving, viewing, downloading and preparing picture data easier. Currently, OPIA contains 56 datasets totalling 5,66,225 photos with 2,417,186 annotated instances, covering 11 different plant species. Interestingly, it includes 56 i-traits made from 105 wheat and 93 rice cultivars, based on 18,644 separate RGB photos (Cao *et al.*, 2024). With careful annotation, these i-traits are cross-referenced with the GWAS Atlas and utilized in the Plant Phenotype and Trait Ontology (PPTO).

### 6.4.3 Crop simulation modelling

Crop models are mathematical representations of the quantitative data from physiological and agronomical experiments that forecast the growth and developmental status of plants (Weih *et al.*, 2022). Based on the quantitative data from physiological and agronomical trials, it is evident that, in agronomic study, mechanistic crop growth models (CGMs) have long been used to predict the development and growth of plants (Cooper *et al.*, 2021). Although these models have been applied widely, there has been little incorporation of them into crop breeding efforts. To maximize the potential of simulation models for agricultural development, breeders employ them in crop breeding programmes (Cooper *et al.*, 2021). AI driven crop simulation models combine environmental and biological processes to model plant growth and yield under various scenarios. Breeders can use these models to forecast agricultural responses to climate change scenarios, choose suitable cultivars for certain locations and optimize breeding techniques

Gene-to-phenotype (G2P) statistical models are a tool used in quantitative genetics to investigate the impact of genes on phenotypic trait variation and to make clear how crop traits are inherited (Walsh and Lynch, 2018). CGM–G2P models were developed as a result of research that showed the potential for merging quantitative genetics and mechanistic crop models to improve the estimate of the outcomes of crop improvement (Technow *et al.*, 2015; Cooper *et al.*, 2016, 2021). In CGM–G2P models, the CGM, which describes the traits and their contributions to an individual's performance in breeding programmes, provides the hierarchical structure of the G2P model (Cooper *et al.*, 2021). Using artificial neural networks (ANNs), ML algorithms and the multilayer perceptron (MLP) technique, early prediction and simulation of seed yield in rapeseed based on mineral fertilization and meteorological data was performed (Niedbała, 2019). Niedbała *et al.*, 2019 also performed multicriteria prediction and simulation of winter wheat seed yield using qualitative and quantitative datasets based on the ANN algorithm and MLP technique.

## 6.5 Advantages of Artificial Intelligence (AI) in Plant Breeding

Plant breeding or GAB could benefit from AI in the following ways (Khan *et al.*, 2022; Xu *et al.*, 2022; Bhat *et al.*, 2023):

1. AI can increase the precision and accuracy of plant-breeding techniques through the integration of AI based ML and DL models within GWAS and GS.
2. Phenotypes could be predicted with enhanced precision by training an appropriate model using multi-omics and big data technology.
3. By integrating speed breeding with AI-assisted GAB, breeders may accelerate the breeding cycles and reduce the time needed for traditional phenotype-based selection, resulting in efficient selection at very early stages.
4. By optimizing breeding experiments and reducing trial and error, AI conserves resources, including time, land and labour.
5. The integrated genomic selection or 'SpeedGS' could enhance genetic gain and shorten breeding cycle times.

## 6.6 Challenges of Artificial Intelligence (AI) in Plant Breeding

The challenges involved in using AI in plant breeding are as follows:

1. AI models require large, high-quality datasets for accurate predictions; limited or biased data can hinder the reliability of AI based outcomes.

2. Understanding the decisions made by AI models, particularly in complex DL algorithms, remains a challenge. Breeders need interpretable models to trust and implement AI-driven decisions.
3. Ethical concerns, such as ownership of data, privacy, and potential inequalities arising from AI-driven breeding, need careful consideration and regulation.
4. Challenges related to data standardization, computational infrastructure and suitable software and algorithms need to be addressed.

## 6.7 Conclusion

There is an immediate need to accelerate the development of higher-yielding, stress-tolerant and nutritious crop cultivars considering the difficulties brought about by the growing population, changing climates and high inflation. However, recent developments in the field of GAB and high-throughput phenomics offer an appealing framework for development of new crop varieties with required quality traits and with greater efficiency. The ability to analyse non-linear and epistatic interactions in GWAS and GS and, subsequently, make these variations useable for GAB, has been made possible through recent advancements in AI techniques and the use of ML algorithms or DL approaches. By leveraging AI techniques, such as genomic prediction, image-based phenotyping or high-throughput phenomics and crop simulation modelling, and by integrating speed breeding with AI-assisted GAB, breeders may overcome traditional breeding constraints and address emerging challenges in crop breeding more efficiently. Continued research and collaboration are essential to harness the full potential of AI for sustainable crop improvement and global food security.

## References

Ahmar, S., Gill, R.A., Jung, K.H., Faheem, A., Qasim, M.U. *et al.* (2020) Conventional and molecular techniques from simple breeding to speed breeding in crop plants: Recent advances and future outlook. *International Journal of Molecular Sciences* 7, 2590.

Ansarifar, J. and Wang, L. (2019) New algorithms for detecting multi-effect and multi-way epistatic interactions. *Bioinformatics* 35, 5078–5085.

Athar, H.R. and Ashraf, M. (2009) Strategies for crop improvement against salinity and drought stress: An overview. In: Ashraf, M., Ozturk, M. and Athar, H. (eds) *Salinity and Water Stress. Tasks for Vegetation Sciences*, Vol. 44. Springer, Dordrecht, Netherlands, pp. 1–16. DOI: 10.1007/978-1-4020-9065-3_1.

Bangert, W., Kielhorn, A., Rahe, F., Albert, A., Biber, P. *et al.* (2013) Field-robotbased agriculture: "Remotefarming. 1" and "BoniRob-Apps." *VDI-Berichte* 2193, 2.1.

Bhat, J.A. and Yu, D. (2021) High-throughput NGS-based genotyping and phenotyping: Role in genomics-assisted breeding for soybean improvement. *Legume Science* 3, 81.

Bhat, J.A., Ali, S., Salgotra, R.K., Mir, Z.A., Dutta, S. *et al.* (2016) Genomic selection in the era of next generation sequencing for complex traits in plant breeding. *Frontiers in Genetics* 7, 221.

Bhat, J.A., Feng, X., Mir, Z.A., Raina, A. and Siddique, K.H.M. (2023) Recent advances in artificial intelligence, mechanistic models, and speed breeding offer exciting opportunities for precise and accelerated genomics-assisted breeding. *Physiologia Plantarum* 175, e13969. DOI: 10.1111/ppl.13969.

Bhatt, D., Patel, C., Talsania, H., Patel, J., Vaghela, R. *et al.* (2021) CNN variants for computer vision: History, architecture, application, challenges and future scope. *Electronics* 10, 2470.

Breseghello, F. and Coelho, A.S.G. (2013) Traditional and modern plant breeding methods with examples in rice (*Oryza sativa* L.). *Journal of Agricultural and Food Chemistry* 61, 8277–8286.

Cao, Y., Tian, D., Tang, Z., Liu, X., Hu, W. *et al.* (2024) OPIA: An open archive of plant images and related phenotypic traits. *Nucleic Acids Research* 52, D1530–D1537.

Casanova, M.F., Hensley, M.K., Sokhadze, E.M., El-Baz, A.S., Wang, Y. *et al.* (2014) Effects of weekly low-frequency rTMS on autonomic measures in children with autism spectrum disorder. *Frontiers in Human Neuroscience* 8, 851.

Chandra, A.L., Desai, S.V., Guo, W. and Balasubramanian, V.N. (2020) Computer vision with deep learning for plant phenotyping in agriculture: A survey. arXiv:2006.11391.

Cooper, M., Technow, F., Messina, C., Gho, C. and Totir, L.R. (2016) Use of crop growth models with whole-genome prediction: Application to a maize multienvironment trial. *Journal of Crop Science* 56, 2141–2156.

Cooper, M., Powell, O., Voss-Fels, K.P., Messina, C.D., Gho, C. *et al.* (2021) Modelling selection response in plant breeding programs using crop models as mechanistic geneto-phenotype (CGM-G2P) multi-trait link functions. *In Silico Plants* 3, diaa016.

Danilevicz, M.F., Gill, M., Anderson, R., Batley, J., Bennamoun, M. *et al.* (2022) Plant genotype to phenotype prediction using machine learning. *Frontiers in Genetics* 13, 822173.

Derakhshan, B., Mahdiraji, A.R., Rabl, T. and Markl, V. (2019) Continuous deployment of machine learning pipelines. *Open Proceedings* 36, 397–408.

Falk, K.G., Jubery, T.Z., Mirnezami, S.V., Parmley, K.A., Sarkar, S. *et al.* (2020) Computer vision and machine learning enabled soybean root phenotyping pipeline. *Plant Methods* 16, 5.

Feng, W., Gao, F. and Wang, X. (2024) AI breeder: genomic predictions for crop breeding. *New Crops* 1, 100010.

Gao, P., Zhao, H., Luo, Z., Lin, Y., Feng, W. *et al.* (2023) SoyDNGP: A web-accessible deep learning framework for genomic prediction in soybean breeding. *Briefings in Bioinformatics* 24, bbad349.

Grimstad, L. and From, P.J. (2017) The thorvald II agricultural robotic system. *Robotics* 6, 24. DOI: 10.3390/robotics6040024.

Harfouche, A.L., Nakhle, F., Harfouche, A.H., Sardella, O.G., Dart, E. *et al.* (2022) A primer on artificial intelligence in plant digital phenomics: Embarking on the data to insights journey. *Trends in Plant Science* 28, 154–184.

Hesami, M. and Jones, A.M.P. (2020) Application of artificial intelligence models and optimization algorithms in plant cell and tissue culture. *Applied Microbiology and Biotechnology* 104, 9449–9485.

Hesami, M., Alizadeh, M., Jones, A.M.P. and Torkamaneh, D. (2022) Machine learning: Its challenges and opportunities in plant system biology. *Applied Microbiology and Biotechnology* 106, 3507–3530.

Hina, A., Cao, Y., Song, S., Li, S., Sharmin, R.A. *et al.* (2020) High-resolution mapping in two RIL populations refines major “QTL hotspot” regions for seed size and shape in soybean (*Glycine max* L.). *International Journal of Molecular Sciences* 3, 1040.

Jafari, M. and Shahsavar, A. (2020) The application of artificial neural networks in modeling and predicting the effects of melatonin on morphological responses of citrus to drought stress. *PLOS ONE* 15, e0240427.

Karikari, B., Chen, S., Xiao, Y., Chang, F., Zhou, Y. *et al.* (2019) Utilization of interspecific high-density genetic map of RIL population for the QTL detection and candidate gene mining for 100-seed weight in soybean. *Frontiers of Plant Science* 10, 1001.

Khan, M.H.U., Wang, S., Wang, J., Ahmar, S., Saeed, S. *et al.* (2022) Applications of artificial intelligence in climate-resilient smart-crop breeding. *International Journal of Molecular Sciences* 23, 11156. DOI: 10.3390/ijms231911156.

Kunze, L., Hawes, N., Duckett, T., Hanheide, M. and Krajník, T. (2018) Artificial intelligence for long-term robot autonomy: A survey. *IEEE Robotics & Automation Letters* 3, 4023–4030.

Lee, D., Seo, H. and Jung, M.W. (2012) Neural basis of reinforcement learning and decision making. *Annual Review of Neuroscience* 35, 287.

Libbrecht, M.W. and Noble, W.S. (2015) Machine learning applications in genetics and genomics. *Nature Reviews Genetics* 16, 321–332.

Lobet, G. (2017) Image analysis in plant sciences: Publish then perish. *Trends in Plant Science* 22, 559–566.

Lobet, G., Draye, X. and Perilleux, C. (2013) An online database for plant image analysis software tools. *Plant Methods* 9, 38.

Mahesh, B. (2020) Machine learning algorithms – a review. *International Journal of Scientific Research* 2020, 381–386.

Meuwissen, T.H., Hayes, B.J. and Goddard, M. (2001) Prediction of total genetic value using genome-wide dense marker maps genetics 157, 1819–1829.

Mir, R.R., Reynolds, M., Pinto, F., Khan, M.A. and Bhat, M.A. (2019) High throughput phenotyping for crop improvement in the genomics era. *Journal of Plant Sciences* 282, 60–72.

Montesinos-Lopez, O.A., Montesinos-Lopez, J.C., Salazar, E., Barron, A., Montesinos-Lopez, J.A. *et al.* (2021) Application of a Poisson deep neural network model for the prediction of count data in genome-based prediction. *Plant Genome* 14, e20118.

Murphy, R.R. (2019) *Introduction to AI Robotics*. MIT Press, Cambrige, MA, USA.

Niazian, M. and Niedbała, G. (2020) Machine learning for plant breeding and biotechnology. *Agriculture* 10, 436.

Niedbała, G. (2019) Simple model based on artificial neural network for early prediction and simulation winter rapeseed yield. *Journal of Integrative Agriculture* 18, 54–61.

Niedbała, G., Nowakowski, K., Rudowicz-Nawrocka, J., Piekutowska, M., Weres, J. *et al.* (2019) Multicriteria prediction and simulation of winter wheat yield using extended qualitative and quantitative data based on artificial neural networks. *Applied Science Letters* 9, 2773.

Pepe, M., Hesami, M. and Jones, A.M.P. (2021a) Machine learning-mediated development and optimization of disinfection protocol and scarification method for improved *in vitro* germination of cannabis seeds. *Plants* 10, 2397.

Pepe, M., Hesami, M., Small, F. and Jones, A.M.P. (2021b) Comparative analysis of machine learning and evolutionary optimization algorithms for precision micropropagation of *Cannabis sativa*: Prediction and validation of *in vitro* shoot growth and development based on the optimization of light and carbohydrate sources. *Frontiers of Plant Science* 12, 757869.

Piovesan, A., Vancauwenberghe, V., Van de Looverbosch, T., Verboven, P. and Nicolai, B. (2021) X-ray computed tomography for 3D plant imaging. *Trends in Plant Science* 26, 1171–1185.

Pound, M.P., Atkinson, J.A., Wells, D.M., Pridmore, T.P. and French, A.P. (2017) Deep learning for multi-task plant phenotyping. In: *Proceedings of the IEEE International Conference on Computer Vision Workshops, 22–29 October 2017*, pp. 2055–2063.

Sakib, S., Fouda, M.M., Fadlullah, Z.M. and Nasser, N. (2020) Migrating intelligence from cloud to ultra-edge smart IoT sensor based on deep learning: an arrhythmia monitoring use-case. In: *Proceedings of the 2020 International Wireless Communications and Mobile Computing*, Limassol, Cyprus, pp. 595–600.

Shivakumar, M., Nataraj, V., Kumawat, G., Rajesh, V., Chandra, S. *et al.* (2018) Speed breeding for Indian agriculture: A rapid method for development of new crop varieties. *Current Science* 7, 1819.

Spindel, J., Begum, H., Akdemir, D., Virk, P., Collard, B. *et al.* (2015) Correction: genomic selection and association mapping in rice (*Oryza sativa*): Effect of trait genetic architecture, training population composition, marker number and statistical model on accuracy of rice genomic selection in elite, tropical rice breeding lines. *PLOS Genetics* 6, e1005350.

Technow, F., Messina, C.D., Totir, L.R. and Cooper, M. (2015) Integrating crop growth models with whole genome prediction through approximate Bayesian computation. *PLOS ONE* 10, e0130855.

UN (2022) Report from UN Population Division. Available at: https://population.un.org/wpp/.

Vabalas, A., Gowen, E., Poliakoff, E. and Casson, A.J. (2019) Machine learning algorithm validation with a limited sample size. *PLOS ONE* 14, e0224365.

van Bezouw, R.F., Keurentjes, J.J., Harbinson, J. and Aarts, M.G. (2019) Converging phenomics and genomics to study natural variation in plant photosynthetic efficiency. *Plant Journal* 97, 112–133.

Voulodimos, A., Doulamis, N., Doulamis, A. and Protopapadakis, E. (2018) Deep learning for computer vision: A brief review. *Computational Intelligence and Neuroscience* 2018, 7068349.

Walsh, B. and Lynch, M. (2018) *Evolution and Selection of Quantitative Traits*. Oxford Academic, Oxford, UK.

Wang, K., Abid, M.A., Rasheed, A., Crossa, J., Hearne, S. *et al.* (2023) DNNGP, a deep neural network-based method for genomic prediction using multi-omics data in plants. *Molecular Plant Breeding* 16, 279–293.

Warwick, K. (2013) Artificial Intelligence: The Basics. Routledge, Abingdon, UK.

Watson, A., Ghosh, S., Williams, M.J., Cuddy, W.S., Simmonds, J. *et al.* (2018) Speed breeding is a powerful tool to accelerate crop research and breeding. *Nature Plants* 4, 23–29.

Weih, M., Adam, E., Vico, G. and Rubiales, D. (2022) Application of crop growth models to assist breeding for intercropping: Opportunities and challenges. *Frontiers of Plant Science* 13, 720486.

Xu, Y., Zhang, X., Li, H., Zheng, H., Zhang, J. *et al.* (2022) Smart breeding driven by big data, artificial intelligence and integrated genomic–enviromic prediction. *Molecular Plant Breeding* 15, 1664–1695.

Yan, H., Guo, H., Li, T., Zhang, H., Xu, W. *et al.* (2023) High-precision early warning system for rice cadmium accumulation risk assessment. *Science of the Total Environment* 859, 160135.

Yoosefzadeh-Najafabadi, M., Earl, H.J., Tulpan, D., Sulik, J., Eskandari, M. (2021) Application of machine learning algorithms in plant breeding: Predicting yield from hyperspectral reflectance in soybean. *Frontiers of Plant Science*. 11, 624273.

Yoosefzadeh-Najafabadi, M., Eskandari, M., Torabi, S., Torkamaneh, D., Tulpan, D. *et al.* (2022a) Machine-learning-based genomewide association studies for uncovering QTL underlying soybean yield and its components. International Journal of Molecular Sciences 23, 5538.

Yoosefzadeh-Najafabadi, M., Rajcan, I. and Vazin, M. (2022b) High-throughput plant breeding approaches: Moving along with plant-based food demands for pet food industries. *Frontiers in Veterinary Science* 9, 991844.

Yoosefzadeh-Najafabadi, M., Rajcan, I. and Eskandari, M. (2022c) Optimizing genomic selection in soybean: An important improvement in agricultural genomics. *Heliyon* 8, e11873.

Yu, X., Wu, Z., Zheng, H., Li, M. and Tan, T. (2020) How urban agglomeration improve the emission efficiency? A spatial econometric analysis of the Yangtze River Delta urban agglomeration in China. *Journal of Environmental Management* 260, 110061.

Zhang, N., Gupta, A., Chen, Z. and Ong, Y.S. (2021) Evolutionary machine learning with minions: A case study in feature selection. *IEEE Transactions on Evolutionary Computation* 26, 130–144.

# 7 Genomics-Assisted Breeding (GAB) for Trait Improvement: Unveiling Genomic Strategies for Accelerated Crop Enhancement

**Mohd Hadi Yunus[1], Aliya Firdaus[2], Zeba Khan[3]* and Mohd Yunus Khalil Ansari[4]**

*[1]Department of Horticultural Bioscience, Pusan National University, Miryang 50463, Republic of Korea; [2]Biocontrol Research Laboratory, Integral Information and Research Centre-4 (IIRC-4), Integral University, Lucknow, UP – 226026, India; [3]Department of Botany, Raja Mahendra Pratap Singh State University, Aligarh – 202002, India; [4]Department of Botany, Aligarh Muslim University, Aligarh – 202001, India*

## Abstract

This chapter provides an in-depth exploration of Genomics-Assisted Breeding (GAB) for trait improvement, an advanced marker-assisted breeding paradigm integrating genome-wide genetic selection and high-density genotyping. GAB's distinctive features include high accuracy, direct improvement, shortened breeding cycles, and elevated selection efficiency, making it particularly advantageous for enhancing complex traits. The chapter comprehensively examines cutting-edge high-throughput genomics strategies, encompassing genetic association mapping, speed breeding, and genomic selection. These methodologies significantly contribute to expediting crop improvement processes. The chapter systematically analyses diverse genomics approaches used for trait discovery and improvement across major species, offering a roadmap for implementation in other crops. Detailed discussions on the application, advantages, disadvantages, and potential improvements of each approach are supplemented with case examples to provide valuable insights for researchers and practitioners. This chapter provides a comprehensive guide for navigating the rapidly evolving field of agricultural genomics, aiding in informed decision-making and the development of sustainable crop improvement strategies. It aims to address the increasing demands for agricultural produce by offering insights into the latest advancements and methodologies in genomics. Through this knowledge, researchers and breeders can effectively enhance crop productivity and resilience, ensuring food security in the face of growing global challenges.

**Keywords:** genomics-assisted breeding, trait improvement, genetic association mapping, genomic selection, speed breeding, crop enhancement

## Abbreviations

AB-QTL – Advanced backcross quantitative trait locus/loci, AFLP – Amplified fragment length polymorphism, Bt – *Bacillus thuringiensis*, CRISPR-Cas9 – Clustered regularly interspaced short palindromic repeats/

*Corresponding author: khanzeba02@gmail.com

 *Plant Breeding Technology: Future Trends and Challenges* (ed. Zeba Khan)
DOI: 10.1079/9781800626638.0007

CRISPR-associated protein 9, GM – genetically modified, GAB – Genomic-assisted breeding, GBS – Genotyping-by-sequencing, GEBV – Genomic estimated breeding value, GS – Genomic selection, GWAS – Genome-wide association studies, HTP – High-throughput phenotyping, indel – Insertion/deletion, MAB – Marker-assisted breeding, MABC – Marker-assisted backcrossing, MAGIC – Multi-parent advanced generation inter-cross, MARS – Marker-assisted recurrent selection, MAS – Marker-assisted selection, NAM – Nested association mapping, NGS – Next-generation sequencing, NIL – near isogenic line, PCR – Polymerase chain reaction, QTL – Quantitative trait locus, RAPD – Random amplified polymorphic DNA, RFLP – Restriction fragment length polymorphism, SNP – Single nucleotide polymorphism, SSR – Simple sequence repeat, TaqMan – A technology used for PCR assays, TILLING – Targeting induced local lesions in genomes.

## 7.1 Introduction

The continuous growth of the global population, combined with the challenges posed by climate change, necessitates innovative approaches to enhance crop productivity and resilience. Traditional breeding methods often fall short in meeting the accelerated demands for improved crop varieties. Genomics-assisted breeding (GAB) represents a paradigm shift in agricultural science, harnessing advances in genomic technologies to expedite the breeding process and achieve precise trait improvements.

GAB integrates comprehensive genomic data with traditional breeding techniques, enabling the identification and selection of desirable traits at the DNA level. This approach enhances the efficiency and accuracy of breeding programmes, leading to the development of crops with superior qualities, such as higher yield, disease resistance and stress tolerance. The use of molecular markers, high-throughput genotyping, and genome-wide association studies (GWAS) are central to GAB, providing breeders with detailed genetic insights that inform selection decisions (Varshney *et al.*, 2014; Xu *et al.*, 2017).

One of the significant advancements in GAB is the deployment of marker-assisted selection (MAS). This technique allows breeders to screen large populations for specific genetic traits using molecular markers, thereby significantly reducing the time and resources required when compared to conventional phenotypic selection methods (Collard and Mackill, 2008). This technique is particularly beneficial for traits controlled by multiple genes, which are challenging to improve through traditional breeding alone. Moreover, the advent of next-generation sequencing (NGS) technologies has revolutionized the field, enabling the rapid sequencing of entire genomes and the identification of single-nucleotide polymorphisms (SNPs) that are associated with key agricultural traits (Kersey *et al.*, 2016). These genomic tools not only facilitate the discovery of new genes and alleles but also enhance our understanding of the genetic architecture underlying complex traits.

In addition to MAS, genomic selection (GS) has emerged as a powerful tool in GAB. GS uses genome-wide markers to predict the breeding value of individuals, allowing for the selection of superior genotypes early in the breeding cycle (Meuwissen *et al.*, 2001). This approach has been successfully applied in various crops, leading to significant genetic gains and accelerating the breeding process (Crossa *et al.*, 2017). The integration of genomics with advanced computational tools, such as machine learning and artificial intelligence, further augments the potential of GAB. These technologies enable the analysis of vast genomic datasets, uncovering patterns and predicting trait performance with unprecedented accuracy (Washburn *et al.*, 2020).

In this chapter, we will explore the revolutionary field of GAB and its pivotal role in accelerating crop enhancement. By integrating genomic insights with traditional breeding methods and advanced computational techniques, GAB promises accelerated development of crops with superior traits. This approach not only addresses current agricultural challenges, such as climate change and population growth, but also lays the groundwork for future innovations in food security and sustainable farming practices.

## 7.2 The Role of Precision Plant Breeding and Novel Techniques

Global food security hinges on adequate agricultural productivity and equitable access to food (Ziervogel and Ericksen, 2010). Plant breeding is primarily dedicated to developing enhanced crop varieties that offer economic benefits to farmers and nutritional advantages to consumers. The primary challenges confronting agricultural production are global climate change and its related issues (Ajeesh Krishna *et al.*, 2021). The resilience of crops to abiotic and biotic stressors is critical when selecting breeding materials. Thus, crop breeding provides a chance to include desired qualities and create new crop types. Developing more adaptable and higher performing crop varieties is critical to increasing agricultural output. To discover dependable performers, new cultivars are routinely evaluated across diverse conditions before their introduction (Tajibayev *et al.*, 2021).

Currently, plant breeding appears more promising than ever. Numerous novel breeding techniques are being employed alongside advanced knowledge and tools to drive progress in this field. Varshney *et al.* (2020) proposed five breeding methodologies (known as the 5Gs) for future crop improvement: (i) genes; (ii) genome; (iii) germplasm; (iv) genomic editing; and (v) genome breeding. The significance of this 5G approach has been underscored by many researchers (Ndlovu, 2020). It is expected that these 5Gs will play a significant role in precision breeding in the future. Neglecting any of these components may reduce the efficacy of breeding attempts. Precision breeding, which incorporates 5G methods, has helped researchers to create superior crop types. These 5G components are critical and will form the foundation of future crop development initiatives; they require superior technological facilities and cutting-edge technology.

Plant breeders and agricultural experts worldwide are adopting innovative tools such as genome editing, speed breeding and high-throughput phenotyping facilities in plant development efforts (Atefi *et al.*, 2021). Genome-assisted technologies improve crop breeding efficiency. Furthermore, high throughput phenomics improve the efficiency of crop breeding. Today, there are various promising techniques accessible in phenomics. Both genomic and phenomic techniques are essential for precision breeding and have made significant contributions to the field of crop breeding. The scope and aims of plant breeding are extensive, with the goal being crop enhancement. Thus, the right implementation of plant breeding techniques, aided by modern instruments, can contribute to increased food production and meeting future food demands.

## 7.3 Advancements in Technologies for Precision Crop Breeding

Evaluating trait-specific germplasm is critical for determining the optimum breeding materials for desirable characteristics (Tiwari *et al.*, 2018). Over the past century, agricultural genetic development has been based on phenotypic selection, which has been shown to be less successful for breeding. In crops, quantitative trait locis (QTLs) control quantitative characteristics (Dai *et al.*, 2021). Genetic mapping and molecular characterization of important QTLs enable marker-assisted breeding (MAB) in enhancing crops. Genetic mapping, whether through association or linkage, is crucial for uncovering the genetic basis of quantitative traits in crops. Numerous molecular markers have been derived from the genomes of several crop species (Kage *et al.*, 2016). These DNA-based molecular markers can identify allelic variations in genes underlying significant agronomic traits, thereby enhancing the effectiveness and precision of breeding.

Efficient GS provides the potential to enrich the genetic diversity and boost crop yields. The accessibility of genome-wide molecular markers has facilitated the implementation of GS in crops, a technique that forecasts the genomic value of breeding materials using the genomic estimated breeding value (GEBV) (Robertsen *et al.*, 2019). First-generation molecular markers, such as random amplified polymorphic DNA (RAPD) (Shamsuzzaman *et al.*, 2021), interspersed simple sequence repeats, sequence characterized amplified regions (Gonias *et al.*, 2019), single primer amplification reaction (Pandian *et al.*, 2018), and simple sequence repeats (SSRs) (Kaur *et al.*, 2018) have been effectively employed to elucidate the genetic basis of phenotypic variation in crops. These markers are useful in determining the genetic basis of GS for crop development. DNA based molecular markers can considerably enhance the efficiency and accuracy of traditional breeding through MAS (Johnson *et al.*, 2023). They are useful for identifying alleles, QTLs, and genes associated with traits of interest, allowing for the selection of crops with desirable traits.

Several QTLs related to agronomically essential features have been found in several crops, including rice (Lei *et al.*, 2018), maize (Choi *et al.*, 2019), barley (Wang *et al.*, 2014), wheat (Gahlaut *et al.*, 2017), sorghum (Rajkumar *et al.*, 2013), and small millets (Chelpuri *et al.*, 2019) under various biotic and abiotic stress conditions. These identified markers linked to significant QTLs/genes assist in enhancing agricultural yields via precision breeding. First-generation molecular markers have facilitated the rapid identification of beneficial QTLs associated with stress tolerance, providing a foundation for GAB programmes. However, first-generation molecular markers have limitations affecting the efficiency of GS for crop breeding. Locus-specific and co-dominant markers are preferable over dominant markers for MAS because they are more reproducible and reliable in genetic selection processes.

Next-generation sequencing based high-throughput molecular markers can address these challenges. The technology enables genotyping-by-sequencing (GBS), which can detect numerous molecular markers and expedite MAS approaches. Researchers have successfully utilized GWAS using GBS, which proves advantageous for GS in crop breeding. Advanced genotyping, particularly those identifying SNPs, has been utilized to uncover important QTLs and genes across various crops. Numerous GWAS have been performed in key crops like maize, rice and wheat. For example, employing GBS, Spindel *et al.* (2015) discovered QTLs associated with grain yield and flowering time in elite rice breeding lines, identifying 73,147 novel SNP markers. Huang *et al.* (2010) assessed agro-morphological characteristics in a cohort of 517 rice landraces using GBS, pinpointing significant QTLs associated with traits such as tiller count, grain dimensions (length and width), gelatinization temperature, spikelet number, amylose content, and resistance to drought. Genome-wide association studies for agro-morphological attributes and other beneficial traits have also been documented in various crops including maize, rice, barley, wheat, pearl millet, sorghum and minor millets (Huang *et al.*, 2010). The GWAS approach proves beneficial for GS in enhancing crops. Advanced GS techniques aid in creating new varieties via marker-assisted breeding (MAB), presenting avenues for precise breeding. Advancements in phenomic and genomic technologies bolster effective crop enhancement, enabling the evolution of sophisticated breeding methodologies. High-capacity genomic and phenomic facilities constitute the cornerstone of precise breeding, backing initiatives to improve crops efficiently.

## 7.4 Advancing Crop Cultivars: From Breeding Innovations to Genomic Tools

The advancement of improved crop cultivars for food, feed and industrial applications primarily relies on plant breeding techniques (Ganie *et al.*, 2021). Plant breeding has progressed from traditional methods to modern GAB (Varshney *et al.*, 2021). Over the past century, significant efforts by plant breeders worldwide have focused on improving various crop species through conventional breeding approaches (Qaim, 2020; Labroo *et al.*, 2021). These efforts have contributed to a remarkable decrease in global undernourishment, from 40% in the 1960s to less than 11% today, primarily owing to the creation of high-productivity and resilient crop varieties (Qaim, 2020).

Conventional plant breeding has steadily enhanced yields, predominantly through the adoption of high-yielding cultivars and hybrids, underscoring the pivotal role of breeding innovations. Recently, there has been a recognized need within the plant-breeding community to introduce genetic variability into breeding to enrich the genetic diversity of elite gene pools. This approach aims to enhance the precision and effectiveness of selection procedures while reducing the duration of breeding cycles (Varshney *et al.*, 2021). Varshney *et al.* (2021) proposed the GAB approach, advocating the application of genomic tools and technologies to pinpoint markers and candidate genes linked to desired traits, thereby integrating genomics into breeding methodologies. Various GAB techniques, such as marker-assisted recurrent selection (MARS), MAS, marker-assisted backcrossing (MABC) and advanced backcross QTLs (AB-QTLs), have been suggested and deployed for enhancing crops. Recently, GS has also been integrated into the GAB toolkit (Crossa *et al.*, 2017).

Marker-assisted selection involves the identification of molecular markers closely associated with genomic regions or Quantitative Trait Loci (QTLs) that control desirable traits. These QTLs can subsequently be incorporated into superior breeding materials using methods such as MABC. Successful instances of MABC include bolstering drought resistance in elite chickpea cultivars and enhancing yield and stress resilience in the widely cultivated rice variety IR64 (Kumar *et al.*, 2014; Henry *et al.*, 2015). However, while MABC is efficient for major-effect QTLs, many crop traits are influenced by numerous minor QTLs. In contrast, MARS has demonstrated effectiveness in improving traits like drought tolerance across various crop species by targeting QTLs with smaller impacts (Ali *et al.*, 2020). To address minor-effect QTLs scattered across the genome, the field of plant breeding has progressively adopted GS. Genomic selection evaluates an individual's genetic capacity using a broad range of markers spread throughout the entire genome, in contrast to MAS, which relies on a restricted number of markers (Crossa *et al.*, 2017).

Genomic selection utilizes predictive models derived from genotypic and phenotypic information of a training population (TP) to calculate GEBVs for individuals within the breeding population. These GEBVs aid in forecasting high quality individuals suitable as parents for future breeding cycles or immediate incorporation into a variety of development pipelines (Crossa *et al.*, 2017). In contrast to MAS, GS can operate without the need for predefined knowledge of prominent marker–trait associations, although integrating markers identified via GWAS can improve predictive precision (Y. Li *et al.*, 2018). The growing enthusiasm for GS among plant breeders is reinforced by recent research showcasing its effectiveness compared to conventional selection methods (Z. Cui *et al.*, 2020; Pandey *et al.*, 2020). With improvements in genotyping technologies and predictive modelling, GS is poised to establish itself as a mainstream breeding strategy alongside MABC and MAS in crop enhancement initiatives.

## 7.5 Genomics and Innovative Techniques for Sustainable Food Production

For the last 10 millennia, human beings have utilized selective breeding to cultivate crop plants with specific characteristics, aiming to enhance yield and nutritional composition through conventional agricultural methods. However, recent advancements in breeding techniques offer a diverse array of innovative options to address both biotic and abiotic stresses in crops, crucial for meeting rising food demands. The advent of Mendelian genetics marked a turning point in crop breeding, and the introduction of cutting-edge genomics tools has further revolutionized crop development. Recently, innovative approaches such as GS, advanced speed breeding, and high-throughput crop phenotyping have notably hastened the breeding timeline. Biotechnological methods, including genetic modification tools like gene transformation, have also been crucial in creating crops with sought-after characteristics. Furthermore, methodologies such as genomics, whole genome sequencing (WGS), gene discovery, and the application of molecular markers have demonstrated efficacy in tools such as mutation breeding, polyploidy breeding, cisgenesis and intragenesis (Muth *et al.*, 2008; Murovec *et al.*, 2017).

Crop breeding entails a series of decision points throughout, from assessing and mapping traits to integrating into breeding programmes, where breeders identify superior individuals from extensive segregating populations (Kaiser *et al.*, 2020). However, traditional breeding is frequently slow and inadequate for swift crop variety advancement. Typically, following initial crosses, subsequent generations necessitate multi-year assessment through replicated field trials across diverse environments to gauge genetic capability (Voss-Fels *et al.*, 2019).

Since the 1990s, molecular markers have enabled the pinpointing and choice of hybrid lines, assisting in artificial selection to enhance particular traits (Dreher *et al.*, 2002). Breeders focus on achieving multiple crop generations within a single year, accelerating the attainment of desired phenotypes (Van Bueren *et al.*, 2011; Kandemir and Saygili, 2015). Various molecular markers, including AFLP, RFLPs), SNPs and SSRs, have been instrumental in identifying QTLs in crops like rice (Oladosu *et al.*, 2019). SNP markers, in particular, are cost-effective and widely present in crop genomes (Liakos *et al.*, 2018). The current pace of yearly yield enhancement for key crops must increase twofold to satisfy forthcoming global food requirements (H. Li *et al.*, 2018). Innovative approaches in crop breeding are essential to boost staple food production and ensure global food security in the coming decades. Molecular breeding and advanced genomic studies enhance breeding efficiency and reduce development time for improved crop varieties.

Plants' shorter generation times and larger population sizes make them particularly amenable to manipulation through crossbreeding and selfing, facilitating straightforward trait selection (Stetter *et al.*, 2016). Recent innovations like 'speed breeding', as described by Lee Hickey and colleagues, utilize controlled environments and extended photoperiods to accelerate plant generations, allowing for rapid selection of desirable traits across thousands of plant variations (Watson *et al.*, 2018; Voss-Fels *et al.*, 2019). Progress in DNA sequencing technologies and automated phenotyping systems continues to propel the genomic revolution in enhancing crops.

## 7.6 Genomic Markers Revolutionizing Crop Breeding

Molecular markers are essential in traditional genetics and agricultural breeding, enabling the identification and improvement of preferred characteristics in crops. In the past, the establishment of markers, construction of genetic linkage maps, discovery of QTLs and accurate gene mapping demanded significant labour and effort. However, the emergence of NGS has transformed this field by enabling the creation of large-scale molecular markers like microsatellites (SSRs), SNPs and insertions/deletions (indels). Indels and SNPs have facilitated the establishment of high-density genetic maps, crucial for precise gene mapping and their application in crop enhancement. These markers are essential for pinpointing genetic variations associated with important agronomic traits across cultivars within a species, assisting in the choice of suitable parental lines for molecular breeding objectives. For instance, in Chinese spring wheat, genome-wide SSR analysis identified 364,347 SSR markers, with dinucleotide repeats being the most abundant (Han *et al.*, 2015). Similarly, in foxtail millet, 28,324 microsatellite-repeat motifs covering 405.3 Mb of the genome were identified, with trinucleotide repeats prevailing over dinucleotide repeats (Pandey *et al.*, 2013). In barley, 436,640 indels were discovered across the genome, integrated with SSRs, DarT markers, and gene-based SNPs into a comprehensive genetic map (G. Zhou *et al.*, 2015).

Additional marker types, such as those based on transposable elements and microRNAs (miRNAs), have proven effective in diverse crops like foxtail millet, enabling extensive genotyping across millets, cereals and bioenergy grass species. These markers have expanded their utility beyond enhancing crops to encompass evolutionary research, analysis of phylogenetic relationships, assessment of genetic diversity, map-based gene identification, and DNA profiling (Muthamilarasan and Prasad, 2015). Single nucleotide polymorphisms widely spread across genomes are notably abundant and are identified through comparative analyses of whole-genome sequences or transcriptome data from various accessions or genotypes. In rice, around 20 million SNPs were identified through aligning

3000 genome sequences against the Nipponbare reference genome (Alexandrov *et al.*, 2015). Similarly, in wheat, 46,977 SNPs linked with genes were genetically mapped using multiple mapping populations (Wang *et al.*, 2014). In maize, 6,385,011 SNPs were identified from 15 inbred lines responsive to drought, underscoring their potential in pinpointing candidate genes linked to drought tolerance (Xu *et al.*, 2014).

Recent progress in resequencing 588 *Brassica napus* accessions has generated millions of SNPs and indels, identifying 60 loci strongly linked to agronomic traits through GWAS (Lu *et al.*, 2019). These genomic resources are invaluable for exploring diversity and comprehending trait variability within populations. These markers are essential for leveraging genomics in crop improvement, enabling efficient large-scale genotyping of natural germplasm accessions and bi-parental mapping populations to uncover traits in crops (Singh *et al.*, 2020).

## 7.7 Marker Trait Association

Both the integration of both phenotypic and genotypic data is crucial for accurately establishing associations between markers and traits. Linkage mapping, a conventional technique, has been extensively utilized to unravel the genetic foundations of QTLs associated with specific traits. Furthermore, with advancements in bioinformatics and the availability of extensive genetic information, meta-QTL analysis has emerged. This method integrates published QTL data to pinpoint consistent and highly influential QTLs linked to traits of interest (Van and McHale, 2017). Genome-wide mapping using large populations has shifted towards targeted gene identification and cloning in crops (Cooper *et al.*, 2014). The advent of high-density SNP markers has facilitated GWAS that utilize natural populations. GWAS complements traditional linkage mapping by overcoming its limitations and providing a robust approach to dissecting complex traits.

Genomic selection represents a pioneering breeding approach that utilizes high-density markers and trait data to predict genomic breeding estimated values across a synthetically generated population, encompassing the entire genome (Crossa *et al.*, 2017). Although currently expensive, GS has demonstrated its superiority over marker-assisted recurrent selection in enhancing complex traits in crops. It efficiently addresses challenges related to the multitude of QTLs influencing a trait. While traditional breeding methods have markedly augmented crop production and yield, there remains a continuous need for novel approaches to meet growing global food demands. The CRISPR-Cas9 genome editing technology holds promise for swiftly addressing emerging agricultural challenges and has already been pivotal in advancing endeavours to enhance crop yield and resistance traits. Ongoing advancements in genetic and genomic tools are poised to enhance traditional breeding and evaluation processes for crop improvement (Crossa *et al.*, 2017).

## 7.8 Advancements in QTL Mapping Strategies

QTL analysis combines evaluations of observable traits with genetic information to reveal the genetic foundations of intricate characteristics in crops. The process begins with establishing a mapping population that segregates for the traits of interest, followed by precise phenotyping and the development of high-throughput genomic markers. Generating a genetic map via genotyping allows for the mapping of QTLs by integrating both phenotypic and genotypic data (Mir *et al.*, 2012). Traditionally, various biparental mapping populations such as F2, double haploids, and recombinant inbred lines have been extensively employed for QTL exploration across different crops. However, these methods have inherent limitations (Myles *et al.*, 2009). To overcome these constraints, linkage disequilibrium based association mapping has been introduced. This approach utilizes natural germplasm collections and leverages historical recombination events, offering enhanced mapping resolution and cost effectiveness compared to traditional linkage mapping (Singh and Prasad, 2017).

Genome-wide association studies, association mapping, and map-based cloning are increasingly utilized to uncover new QTLs, genes, and alleles associated with important agronomic traits in major food crops. Advanced population structures like multi-parent advanced generation inter-cross (MAGIC) and nested association mapping (NAM) are developed specifically for multi-parent association studies (Ladejobi *et al.*, 2016). State-of-the-art genotyping techniques, such as restriction-site associated sequencing (RAD-Seq), GBS and whole-genome resequencing, empower precise trait mapping through QTL-based analyses (Roorkiwal *et al.*, 2020).

These methods have effectively pinpointed molecular markers linked to diverse agronomic traits across different crops, including brassica (Qu *et al.*, 2017), soybean (Hu *et al.*, 2014), foxtail millet (Jaiswal *et al.*, 2019), chickpea (Bajaj *et al.*, 2016; Y. Li *et al.*, 2018) and rice (Feng *et al.*, 2016; Rao *et al.*, 2018). Moreover, different types of QTLs, such as proteomic (pQTLs), expression (eQTLs), phenomic (phQTLs) and metabolic (mQTLs), are increasingly employed to elucidate the genetic foundations of trait diversity (Nica and Dermitzakis, 2013; Rodziewicz *et al.*, 2019). Recent developments in NGS based strategies for mapping quantitative trait loci (QTLs), such as QTL-sequencing and MutMap, have accelerated the discovery of significant QTLs that influence agronomic traits in crops. These methodologies are instrumental in deciphering the genetic framework underlying intricate traits and offer substantial potential for augmenting crop enhancement efforts (Mir *et al.*, 2012).

## 7.9 Advancing Crop Breeding Through QTL Mapping and Genomic Tools

Critical agronomic and economically valuable traits in crops are predominantly quantitative in nature, shaped by multiple QTLs. Therefore, the identification of QTLs or causal genes/alleles is crucial for MAB. Linkage mapping, a traditional approach for investigating the genetic basis of quantitative traits, has successfully uncovered numerous QTLs across various crops (Emebiri *et al.*, 2017; Zhang *et al.*, 2018). Furthermore, advancements in bioinformatics and genetic data have facilitated meta-QTL analysis, which integrates QTL information from multiple studies to identify consistent loci beneficial for crop improvement (Van and McHale, 2017). The introduction of high-density SNP markers has transformed GWAS, which utilizes natural populations to address the limitations of traditional linkage mapping. The GWAS approach has effectively pinpointed loci associated with diverse traits in maize, such as flowering time (Buckler *et al.*, 2009), oil concentration (Li *et al.*, 2013) and disease resistance (Kump *et al.*, 2011). Similarly, GWAS has proven effective in soybeans, for understanding domestication and traits like plant height and oil content (Z. Zhou *et al.*, 2015), and rice, for traits like chilling tolerance (Schläppi *et al.*, 2017).

Genomic selection, another promising strategy, predicts breeding values by analysing high density marker data across the genome. This method has demonstrated high accuracy even for polygenic traits with low heritability, making it advantageous for complex trait improvement in crops like maize, barley, wheat and rice (Meuwissen *et al.*, 2001; Jia and Jannink, 2012). It accelerates breeding cycles and enhances phenotypic prediction, surpassing traditional marker-assisted methods by effectively handling polygenic trait complexities (Saint Pierre *et al.*, 2016). With advancements in genetic mapping and marker technology, GWAS and GS have become indispensable tools in crop breeding. They offer robust strategies for identifying and utilizing complex trait loci, thereby accelerating crop improvement efforts globally.

## 7.10 Advanced Backcross QTLs (AB-QTLs)

Advanced backcross QTL mapping is recognized as an effective strategy for simultaneously identifying QTLs and integrating desired genomic loci into elite recipient parents. This approach has been successfully applied in various crops: blast resistance in rice (Jiang *et al.*, 2020); fine mapping of seed dormancy in rice (Nguyen *et al.*, 2020); wheat breeding programmes (Huang *et al.*, 2004, 2015); enhancing disease resistance in peanuts (Kumari *et al.*, 2020); improving yield traits in pigeon pea (Saxena *et al.*, 2020); and identifying kernel width QTLs in maize (Tang *et al.*, 2020).

Exotic and wild genotypes often harbour genes that can significantly improve target traits and yields, albeit accompanied by undesirable traits that reduce progeny performance. To tackle this issue, Tanksley and Nelson (1996) introduced the Advanced Backcross-QTL (AB-QTL) method as an innovative molecular breeding strategy. This integrates the identification of QTLs with the concurrent development of improved varieties. It involves establishing a mapping population by crossing with wild or exotic germplasm, followed by the mapping and transfer of desirable traits from the wild genotype into the cultivated genetic background.

Traditional QTL mapping separates the processes of QTL detection and variety development, leading to prolonged development times and inefficient utilization of QTL information. In contrast, AB-QTL overcomes these challenges by enabling simultaneous QTL identification and utilization during variety development. This method involves analysing QTLs and utilizing molecular markers in advanced backcross generations like BC2 or BC3. The criteria for selection ensure that chosen plants resemble the recurrent parent as they advance generations and identify QTL-near isogenic lines (QTL-NILs). These QTL-NILs, obtained from BC3–BC5 selections, can directly serve as enhanced varieties or as parents in breeding programmes (Aglawe *et al.*, 2021).

## 7.11 Advancing Crop Genetics Through Association Mapping

Association mapping leverages historical recombination events, unlike linkage mapping that relies on populations with known pedigrees. This method explores the linkage between markers and traits across a diverse set of genotypes, utilizing linkage disequilibrium – the non-random association of alleles at different loci. Association mapping offers higher resolution and facilitates the discovery of a greater number of alleles compared to traditional biparental QTL mapping. Moreover, it eliminates the need for developing specific mapping populations, a time-consuming and resource-intensive step. The process of association mapping begins with selecting a diverse population, which may include natural populations or collections of breeding lines and cultivars. This population is extensively phenotyped across multiple locations and years through replicated trials to assess trait variability. Subsequently, the population undergoes genotyping using molecular markers. Analysis involves determining population structure, kinship among individuals, and linkage disequilibrium between markers and trait loci, employing various software tools for computational analysis. Association mapping finds application in a variety of genomic studies, including MAS. For further insights, readers are encouraged to explore comprehensive reviews such as the one by Myles *et al.* (2009) on association mapping.

Advancements in genomic techniques, such as genome sequencing, facilitate the development of molecular markers (Jia *et al.*, 2013; Li *et al.*, 2014; C. Sun *et al.*, 2020). Alongside these innovations, GWAS aim to uncover relationships between quantitative traits and phenotypes, offering insights into genetic diversity across genotypes. Recent reviews underscore the utility of GWAS in enhancing strategies for breeding cereal crops, including efforts in biofortification (Ashokkumar *et al.*, 2020), as well as its potential in wheat (Battenfield *et al.*, 2018) and pearl millet (Srivastava *et al.*, 2020) breeding. Moreover, association mapping leverages existing genetic diversity to select optimal parental combinations from natural populations to breed desirable traits (Ibrahim *et al.*, 2020). Recent research has effectively pinpointed novel genetic regions linked to characteristics like resistance to stripe rust in ancient wheat varieties (Wang *et al.*, 2021) and nutritional attributes in finger millet (Puranik *et al.*, 2020) using population datasets analysed through GWAS.

## 7.12 Genomics-Assisted Breeding

Genomic-assisted breeding includes two main strategies: MAS and GS. Marker-assisted selection, encompassing methods like MABC and MARS, utilizes molecular markers linked to specific traits to identify individuals with beneficial alleles (Jiang, 2013). However, many successful MAS studies in soybean have focused on major-effect QTLs or genes that substantially affect phenotypic variation (Kim *et al.*, 2020). In contrast, MAS has shown limited success with minor-effect genes that have a minor impact on phenotypic variation in complex traits such as yield, oil and protein content, and stress resistance in soybeans (Bhat *et al.*, 2016; Hina *et al.*, 2020).

Genomic selection presents a promising alternative for these complex quantitative traits. It integrates marker profiles and phenotypic data from a training population to develop predictive models, which are then used to estimate the GEBVs of breeding populations (Yin *et al.*, 2020). Before applying the model for selection, cross-validation is performed on training subsets to ensure accuracy (Bhat *et al.*, 2016). Once validated, the models enable the selection of individuals based solely on genomic data, effectively predicting their phenotypic performance (Stewart-Brown *et al.*, 2019). Recent GS research utilizing high-throughput SNP genotyping in soybeans has demonstrated encouraging outcomes for various complex traits, achieving moderate to high prediction accuracies.

A key advantage of GAB is that genotypic data collected at early growth stages can predict mature plant performance, significantly saving time, money and labour otherwise spent on extensive multi-environment and multi-year phenotypic evaluations (Bohra *et al.*, 2020). The effectiveness of breeding cycles and genetic improvements per time unit has been enhanced by this efficiency. The transition from phenotype-driven to genotype-driven selection has been accelerated by advancements in NGS technologies, which provide enhanced capacity, cost efficiency and rapidity in genome-wide genotyping (Getachew, 2019). Prior to the advent of NGS, the development of markers was costly and labour intensive, constraining GAB, particularly GS, due to limitations in marker availability. This limitation restricted the use of markers to key genomic regions essential for predicting agriculturally significant traits.

Both MAS and GS offer unique strengths and limitations, and their application in soybean breeding depends on the genetic architecture and heritability of the trait (Mir *et al.*, 2019). GAB, encompassing techniques like MABC for integrating favourable alleles into elite germplasm, MARS and GS, has significantly hastened breeding advancements across diverse crop species, resulting in the development of more than 130 publicly released cultivars in the past 15 years (Vogel, 2014). The elucidation of genetic mapping and QTL analysis using bi-parental or association mapping populations has advanced the understanding of genetic regulation in agricultural traits, facilitating MAS, QTL mapping and association studies, along with direct GS of valuable genotypes (Henry, 2015). Next-generation sequencing combined with GWAS enhances mapping resolution for accurate gene/allele/QTL localization (Liu *et al.*, 2013; Varshney *et al.*, 2014; Henry, 2015). The integration of advanced genomic technologies and comprehensive phenotyping drives next-generation breeding (Varshney *et al.*, 2005), with GS expediting the selection of superior genotypes, accelerating breeding cycles and reducing costs (Crossa *et al.*, 2017).

## 7.13 GAB for Trait Improvement

GAB starts with the identification of genetic markers correlated with QTLs or genes linked to targeted agronomic traits, which are subsequently integrated into the breeding process. (R.K. Singh and Prasad, 2017). Molecular markers aid in identifying preferred progeny at an early developmental stage through high-throughput genotyping platforms based on NGS. Different GAB methods, including GS, MABC, and MARS, have been employed for enhancing crop traits. Speed breeding has recently been introduced to expedite breeding procedures. In nations such as India, where agricultural advancements and variety release are largely reliant on breeding, updating and fortifying, breeding strategies are critical to promptly tackle agricultural challenges.

## 7.14 Marker-Assisted Selection (MAS)

Marker-assisted selection represents a pioneering approach that harnesses molecular tools in breeding. This technology allows the addition and removal of genetic material or altering the genetic materials at specific locations in the genome by replacing an existing sequence with a customized DNA segment (Shahwar *et al.*, 2024). It addresses several limitations of traditional plant breeding and expedites the process. Various markers have been developed for different crop species. Initial markers include RFLPs, while subsequent markers encompass PCR-based techniques like RAPDs, AFLPs, and SSRs, microsatellites, or sequence tagged sites (STS). Next-generation markers are sequencing-based, such as SNPs. The availability of diverse molecular markers, comprehensive genetic maps of various crops, and advanced high-throughput techniques for genotyping and phenotyping have facilitated the widespread adoption of MAS in crop improvement. Three key phases are involved in MAS: (i) establishing relationships between markers and traits; (ii) confirming the associations of these markers; and (iii) incorporating them into the selection criteria. The efficacy of MAS depends heavily on the genetic proximity between the marker and the specific gene of interest (Perumalsamy *et al.*, 2009).

## 7.15 Applications of MAS

Marker-assisted selection is widely employed across various crops to incorporate desired genomic loci to enhance traits (Boopathi, 2020). For example, in rice breeding it is utilized to improve resistance to both abiotic and biotic stresses (Hasan *et al.*, 2015; Oladosu *et al.*, 2019; Chukwu *et al.*, 2019). In wheat, MAS is applied to pyramid genes or QTLs that contribute to quality and disease resistance (Gupta *et al.*, 2010b; Gautam *et al.*, 2020). Similarly, in pea breeding, MAS aids in trait enhancement. Additionally, MAS facilitates the introgression of the maize opaque-2 allele to enhance lysine and tryptophan content (Pukalenthy *et al.*, 2020), as well as for improving drought adaptation (Ribaut and Ragot, 2007).

## 7.16 Marker-Assisted Breeding

Molecular markers are pivotal in MAS for introgressing disease resistance genes into back-crossing. Often, multiple genes are combined to achieve enduring resistance against specific pathogens. A significant early MAS success was the commercial release of a maize hybrid by Monsanto in the USA in 2006. Since then, MAS has facilitated the introduction of improved varieties via MABC for widespread cultivation. In India, MAS has been instrumental in releasing disease-resistant rice varieties like improved Samba Mahsuri, Pusa Basmati 1, Swarna Sub-1, and the maize hybrid Vivek QPM 9. Molecular markers have revolutionized the detection and mapping of QTLs, surpassing the capabilities of morphological or protein-based markers alone. They ensure the purity testing of inbred parents and enable selection based on genotype rather than phenotype, identifying desirable plants at the seedling stage for hybridization within the same season. Marker-assisted selection also facilitates integration of oligogenic and horizontal resistance with vertical resistance, which is challenging through disease tests alone. Additionally, closely linked molecular markers have been crucial in the positional cloning of numerous plant genes. These markers have driven the development of innovative breeding strategies like GS and GAB, aiming to enhance crop adaptability (Tiwari *et al.*, 2019).

## 7.17 Advancing Crop Breeding with MABC

Marker-assisted backcrossing entails transferring a specific genomic region (QTL, locus or gene) from a donor genotype into a recipient breeding line or elite cultivar. This iterative process spans several generations of backcrossing, aimed at integrating the desired trait while minimizing linkage drag. The resulting progeny from MABC retain the entire genetic background of the elite parent, augmented with genetic loci responsible for the desired phenotype from the donor parent (Gupta *et al.*, 2010a). The effectiveness of MABC hinges on the number and reliability of the molecular markers employed, the strength of marker – trait associations, the degree of undesired genetic linkage, and the size of the population across each backcrossing generation. This approach has been extensively utilized to develop advanced crop lines with enhanced tolerance to both biotic and abiotic stresses.

For instance, NILs of drought-tolerant rice were generated by transferring root QTLs from a drought-resistant japonica upland variety CT9993 into the lowland susceptible indica variety IR20. Another study, conducted by Kang *et al.* (2019) utilized an accelerated generation advancement system to expedite the development of NILs from rice stripe virus-resistant and susceptible parents, followed by MABC to produce a virus-resistant line in a short period. Similarly, Yang *et al.* (2019) integrated MABC with background selection based on genomics to introduce the blast resistance gene *Pi2* into an elite rice line. In groundnuts, MABC has been instrumental in breeding improved lines with a desirable oleic/linoleic fatty acid ratio (Bera *et al.*, 2019).

Efforts have also concentrated on integrating genes and major QTLs linked to tolerance against drought, submergence and salinity, as well as resistance to sheath blight, bacterial blight, and blast, to create stress-tolerant rice varieties with increased productivity and yield (Singh *et al.*, 2019). To address the difficulties associated with MABC in transferring numerous QTLs responsible for complex traits like yield or broad-spectrum disease resistance, MARS was developed. The MARS approach identifies and prioritizes significant QTLs or multiple genomic regions that control complex agronomic traits within or across populations, integrating them into a single genotype (Ribaut *et al.*, 2010). This approach is particularly effective in cross-pollinating species and utilizes the F2 population, enabling beneficial alleles from both parents to contribute, resulting in a genotype that represents a blend of the parental characteristics. The process entails selecting for desired traits based on phenotype and markers in the F2 population, followed by several rounds of MAS (Eathington *et al.*, 2007).

## 7.18 Advancing Crop Breeding with MARS

Selection plays a pivotal role in augmenting the genetic progress of economically valuable traits. Mass and pure line selection are standard practices in self-pollinated crops, while recurrent selection is favoured for enhancing the prevalence of advantageous alleles in cross-pollinated crops. Recurrent selection entails successive rounds of interbreeding among chosen individuals to establish a population for subsequent selection cycles. This process encompasses three primary stages: selection, evaluation of selected lines, and recombination, to increase the frequency of favourable alleles. It proves particularly effective for enhancing polygenic traits and integrating diverse genetic resources. None the less, as phenotypes are influenced by both genotype and environmental factors, the efficacy of recurrent selection may be affected. Moreover, multiple cycles of selection and interbreeding across different growing seasons prolong the selection process. Molecular marker technology can be integrated into the selection process to mitigate environmental influences. MARS is one such method that enables the incorporation of multiple QTLs governing complex traits. It facilitates the selection and interbreeding of individuals within the same crop season, thereby expediting the selection cycle and enhancing efficiency (Jiang *et al.*, 2007; Gokidi *et al.*, 2016).

## 7.19 Genetic Engineering in Agricultural Advancement

Genetically modified (GM) or transgenic crops undergo genome modification at the genetic level using diverse genetic engineering techniques. Traditional breeding is time intensive and restricted to transferring genetic material from closely related species, whereas genetic engineering allows genes to be transferred from any source into plants. However, achieving stable transgenic lines requires a successful protocol for gene insertion and rigorous selection. *Agrobacterium tumefaciens*-mediated genetic transformation is a dependable approach for establishing stable transgenic plants. Alternative methods such as electroporation, particle bombardment and sonication are employed for transient expression of foreign DNA.

The main hurdles in this method include the absence of standardized procedures for numerous crucial species and the time-consuming process of optimizing them. Achieving higher transformation efficiency necessitates meticulous adjustment of experimental conditions. Once these protocols are refined, they enable the introduction of diverse genes into target genomes to enhance characteristics and traits. Currently, 525 GM varieties across 32 crops have been brought to market, with *Zea mays* having the most extensive representation. Cultivating transgenic crops has boosted agricultural yield by approximately 22% and profitability by 68% (Kumar *et al.*, 2020).

Various genetic engineering tools have been utilized to enhance crops. For example, Bacillus thuringiensis (Bt) cotton has significantly bolstered the Indian economy, while Bt brinjal has improved the livelihoods of farmers in Bangladesh. Despite these advantages, public acceptance of GM crops remains limited due to concerns regarding environmental risks and safety issues related to human consumption. Nevertheless, scientific research has not provided evidence that GM crops pose health risks (de Vos and Swanenburg, 2018). Biotechnology plays a critical role in the future of agriculture, especially given the mounting challenges faced by the farming sector. Presently, intensive agriculture supports farmer incomes and contributes to global food supplies. However, challenges such as emerging diseases, pest outbreaks, unpredictable rainfall patterns, soil degradation from synthetic fertilizers and reliance on monoculture pose significant threats to agriculture. Plants must withstand various stresses to their environments, and biotechnological interventions aimed at modifying genes that regulate responses to multiple stresses can lead to the development of more resilient plants (Pandey *et al.*, 2020). Mitigating climate change and improving agricultural practices are crucial steps, but it is equally essential to introduce crop varieties that are resilient to climate fluctuations and can thrive under adverse conditions to ensure global food security.

## 7.20 Genome Editing

Genome editing is revolutionizing plant breeding, offering precise and efficient ways to enhance crop yields and resilience against diseases, pests and abiotic stresses. CRISPR technology permits the insertion and deletion of genetic material or the modification of genetic sequences at precise locations within the genome by substituting an existing sequence with a tailored DNA segment (Khan and Shahwar, 2024). CRISPR facilitates rapid and cost-effective crop improvements. It holds immense promise for engineering disease-resistant traits and tackling key challenges in crop improvement through precise genetic alterations (Pickar-Oliver and Gersbach, 2019) and engineering disease-resistant traits through precise genetic alterations. Genome editing employs site-specific nucleases to induce double-stranded breaks at targeted DNA sequences. These breaks are repaired using mechanisms like homologous recombination or non-homologous end joining, enabling precise genetic modifications such as insertions or deletions (Miglani, 2017; Wright *et al.*, 2018; Jun *et al.*, 2019). This technique holds substantial promise in plant breeding by enabling targeted mutagenesis, gene knockouts, gene stacking and modulation of gene expression levels.

Recent examples of CRISPR-edited crops include mushrooms resistant to browning, waxy corn with elevated amylopectin content, and false flax enriched with omega-3 oil. These products swiftly obtained approval from the US Department of Agriculture (Waltz, 2016a,b). CRISPR-Cas9 technology is also being applied to develop cereal crops resilient to climate variations and enriched with essential nutrients (Raza *et al.*, 2021). Potatoes, a crucial food crop, confront challenges such as drought, heat, nutrient deficiencies, bacterial infections, pests and viral diseases. Exploring the genetic diversity within the *Solanum* genus through traditional breeding methods has provided valuable insights, which can be further enhanced with transcriptomics using NGS. This approach identifies specific gene targets suitable for CRISPR-mediated genome editing, thereby advancing strategies for crop enhancement (Tiwari *et al.*, 2020, 2022).

## 7.21 Crop Pan-Genomes

Crop pan-genomics focuses on crucial genetic components such as SNPs, mutations and structural variants that dictate vital traits within a population. Pan-genome studies enable the discovery of genes that are missing in reference genomes due to domestication. By exploring a crop's pan-genome, which encompasses its wild and cultivated varieties and landraces, researchers can comprehensively document genetic and phenotypic diversities. This approach aids in identifying genes that are absent, common and unique in standard crop genomes (Danilevicz *et al.*, 2020). Insights from the dispensable genome help in selecting superior cultivars with gene regulation responsive to stress, thereby enhancing their resilience against various challenges (Bayer *et al.*, 2019).

## 7.22 Crop Breeding High-Throughput Phenotyping (HTP)

Improving crop production to ensure food security hinges on efficiently and economically identifying valuable agricultural traits. Traditional methods of phenotyping are often costly and time consuming, prompting the adoption of high-throughput phenotyping (HTP). Machine learning, an emerging field in artificial intelligence, involves advanced computer systems that enable machines to learn and improve autonomously without extensive programming (Singh *et al.*, 2016). Genomic selection accelerates the screening of superior germplasm, thereby expediting crop breeding cycles (Crossa *et al.*, 2017). Currently, GS benefits from advancements in machine learning and the analysis of extensive genotyping datasets linked to agriculturally significant phenotypic traits (Tong and Nikoloski, 2021)

High-throughput phenotyping technologies have transformed GAB by providing rapid, precise and large-scale measurement of plant traits. These technologies, including imaging systems (RGB, thermal, hyperspectral and fluorescence imaging), LiDAR, drones, ground-based platforms, and sensor networks, allow for comprehensive trait analysis and integration with genomic data. This integration enhances the accuracy of GS and MAS, facilitating the identification of QTLs associated with key agronomic traits such as yield, drought tolerance and disease resistance. Moreover, HTP accelerates breeding cycles by enabling the evaluation of large populations efficiently, thus speeding up the development of new crop varieties (Araus and Cairns, 2014).

Despite the significant advantages, challenges such as data management, cost and standardization persist. The vast amount of data generated requires robust storage, management, and analysis tools; and the high initial investment in HTP equipment can be prohibitive. Developing cost-effective solutions and standardizing phenotyping protocols are essential for broader adoption. Nevertheless, as HTP technologies continue to evolve and become more accessible, they are poised to play a crucial role in future crop breeding efforts, ensuring enhanced crop performance and resilience to environmental stresses (Furbank and Tester, 2011; Montes *et al.*, 2011).

## 7.23 Crop Breeding with Genomic Selection (GS)

Genomic selection, or genome-wide selection (GWS), utilizes a comprehensive array of DNA markers distributed across the genome to improve germplasm lines. This approach has the capability to capture multiple QTLs/genes with minor additive effects. Detailed phenotyping of a breeding population is not obligatory, as subsequent selection of progeny primarily depends on genotypic predictions. These predictions integrate genomic and pedigree data across multiple generations of the breeding cycle (Nakaya and Isobe, 2012). These studies highlight the effectiveness and feasibility of GS in swiftly enhancing traits, potentially leading to the accelerated release of improved genotypes. In traditional breeding, the timeline for crossing experiments, yield testing, disease resistance assessment, quality evaluation, and variety release typically spans 3–7 years, followed by an additional 4–5 years, and then another 1–3 years. Given this prolonged process, 'speed breeding' has been introduced to expedite the breeding cycle. This method involves manipulating light, photoperiod, humidity, temperature and other environmental factors to create accelerated growth conditions in controlled environments such as glasshouses and growth chambers (Ghosh *et al.*, 2018; Watson *et al.*, 2018). Speed breeding enables the generation of up to six cycles per year for crops such as barley, wheat, canola and chickpea (Hickey *et al.*, 2019).

### 7.23.1 Genomic selection (GS)

Genomic selection was introduced by Meuwissen *et al.* (2001) to predict complex traits using an adapted version of MAS. It employs molecular markers distributed throughout the genome to ensure that all QTLs are linked with at least one marker, thereby eliminating the necessity for a distinct QTL detection procedure. The approach is effective in devising innovative breeding methods based on molecular genetic markers, enhancing genetic improvement of complex traits per unit time, and cutting down on expenses. Advanced NGS and genotyping technologies significantly augment the precision of GEBVs in diverse crop plants. However, integrating HTP remains crucial for fully amalgamating the genetic advancements of complex traits through GS (Meuwissen *et al.*, 2001).

Unlike MAS, which focuses on traits controlled by a limited number of genes, GS evaluates an individual's genetic potential using a comprehensive array of markers spread throughout the genome. A predictive model for breeding value is constructed using genotyping and phenotyping data from a chosen population. The GEBVs assist in identifying superior individuals suitable as parents in hybridization programmes or for advancing successive generations in ongoing breeding efforts. Genomic selection has the capacity to enhance genetic progress per annum compared to phenotypic selection (Zhong *et al.*, 2009). This is especially beneficial when phenotypes are challenging or costly to assess, enabling early selection, economic advantages and accelerated genetic improvement.

For GS to be effective, the reference population must be sufficiently large to accurately estimate genotype–phenotype associations. This population needs regular updates with new individuals to maintain accurate associations, which may be affected by recombination, mutation or other genetic mechanisms. Genomic selection plays a pivotal role in advancing traditional crop enhancement and is indispensable for integrating GAB into commercially significant crops characterized by extensive and intricate genomes (Poland and Rife, 2012). Employing markers across the entire genome represents a robust approach to enhancing crop yield, providing high selection intensity and accuracy in choosing superior lines for preliminary breeding initiatives. Genomic selection has been extensively embraced in diverse crops, including the enhancement of hybrid rice (Y. Cui *et al.*, 2020), rice breeding programmes (Prakash *et al.*, 2020), tomato breeding, maize husk trait improvement (Z. Cui *et al.*, 2020) and wheat breeding programmes (Charmet *et al.*, 2020; Kehel *et al.*, 2020; J. Sun *et al.*, 2020; Verges and Sanford, 2020). It has also contributed to cereal crop improvement (Rahim *et al.*, 2020).

### 7.23.2 Speed breeding

Speed breeding, an innovative technology, has garnered global attention for its potential to revolutionize agriculture. Initially inspired by NASA's efforts to cultivate wheat in space, this method was pioneered by a researcher at the University of Queensland. Speed breeding significantly reduces the time required for crop generations and accelerates breeding programmes, proving invaluable for rapid gene discovery, crossbreeding, backcrossing, population mapping and trait stacking (Bhatta *et al.*, 2021). While traditional methods, such as off-season nursery or shuttle breeding, *in vitro* or embryo culture, and double haploid technology have aimed to shorten generational intervals, speed breeding surpasses these by manipulating key environmental factors such as light intensity, day length and temperature. This adjustment reduces generation time by 2.5 to 5 times compared to standard greenhouse and field conditions. In controlled greenhouse settings, crops like barley, chickpea, canola, and wheat typically achieve 2–3 generations per year. However, speed breeding allows for 4–6 generations annually, significantly expediting the development of new crop varieties. This technology has also successfully enhanced productivity in other crops, including soybean, rice, millets, sorghum, rapeseed, potato, tomato and sugarcane, presenting a promising solution to address global food security challenges (Hickey *et al.*, 2019).

### 7.23.3 Crop genetics with MAGIC populations

The concept of the MAGIC population was initially proposed by Mott *et al.* (2000) for mice, later termed by Mackay and Powell (2007). It extends the advanced intercross, also known as heterogeneous stock. Traditional mapping populations combine genomes from two contrasting parents to capture small genomic regions influencing traits. In contrast, MAGIC populations involve crossing multiple parents in various combinations, facilitating the identification of genes controlling quantitative traits. This approach accommodates complex pedigree structures and offers deeper insights into genomic architectures. Multiple inbred founder lines are intercrossed over generations, yielding inbreds with genomes contributed by multiple parents. Developing MAGIC populations requires substantial upfront investment in time and meticulous maintenance but yields populations rich in diversity, polymorphism, and allele frequency (Huang *et al.*, 2015). Generating MAGIC populations involves accumulating numerous genetic markers across founders through recombination, crucial for achieving dense, high-resolution genome mapping. Heterogeneous stocks enhance QTL detection and localization compared to F2 crosses. For comprehensive insights into the MAGIC technique, refer to Huang *et al.* (2015).

Advantages of MAGIC populations include creating extensive genetic variations through recombination among diverse founder lines, optimizing gene combinations for target traits, directly utilizing superior combinations as new varieties, identifying allelic variability for complex traits, and precisely pinpointing genetic markers linked to QTLs of interest. However, challenges include significant costs and time investments, potential relatedness among progeny due to shared recombination events, the need for large-scale phenotyping, potential limitations in progeny numbers due to parental incompatibility, and managing extensive segregation. MAGIC populations serve as diverse, permanent mapping resources developed using genomic strategies by the International Rice Research Institute (IRRI) for rice breeding programmes. These lines provide tailored, diverse germplasm, enabling breeders to map quantitative traits effectively in rice (Bandillo *et al.*, 2013). Descalsota *et al.* (2018) utilized GWAS with MAGIC populations in rice for enhancing biofortification and disease resistance. Similarly, MAGIC populations have been established in wheat, pearl millet, maize, and commercial crops like strawberry (Dell'Acqua *et al.*, 2015; Wada *et al.*, 2017). Riaz *et al.* (2020) identified genomic loci associated with blotch resistance in wheat using MAGIC populations.

### 7.23.4 TILLING and Eco-TILLING

An effective way of annotating gene function is by identifying mutations in specific genes and linking these mutations to observable phenotypes in mutant organisms or plants. The forward genetic approach begins with identifying a mutant phenotype and subsequently determining the responsible gene. This approach requires creating and screening large populations, which is time and labour intensive. In contrast, reverse genetics approaches start with identifying gene sequences and then linking these sequences to observed phenotypes resulting from gene alterations. This method is widely used for gene function annotation. Targeting induced local lesions in genomes (TILLING) exemplifies a reverse genetics technique that integrates chemical mutagenesis with PCR based screening to detect point mutations in targeted genomic regions (McCallum *et al.*, 2000; Kurowska *et al.*, 2011). The technique identifies heteroduplexes in individuals carrying mutations in specific genes within a population. It aims to discover induced single nucleotide variations or point mutations, in contrast to Eco-TILLING, which focuses on identifying natural allelic variations. Advantages offered by TILLING include facilitating the functional characterization of numerous newly identified genes, enabling the discovery of novel mutations or indels within genes, employing high-throughput screening for nucleotide polymorphisms in targeted sequences, and being applicable to any species regardless of genome size or ploidy level. It allows for the rapid and cost-effective discovery of induced alleles in plants. Coupled with high-throughput sequencing techniques, TILLING has proven efficient in the identification of the allelic mutations in target genes, making it valuable for breeding programmes. On the other hand, Eco-TILLING identifies SNPs within natural populations and is useful for establishing marker – trait associations in breeding. These methods have found wide application across various crops, including cereals, to enhance breeding programmes and dissect agronomically important traits (Irshad *et al.*, 2020). Despite its potential, TILLING requires standardization for genes of interest and adaptation for polyploid genomes such as wheat.

## 7.24 Genome Sequencing and Sequence-Based Markers

### 7.24.1 Genome sequencing

Recent advancements in sequencing technologies have revolutionized whole-genome sequencing, leading to significant strides in crop improvement programmes in recent years. These technological advances have greatly enhanced our understanding of genome organization across various crops (Varshney *et al.*, 2009). The ability to identify genetic variations underlying phenotypic traits is crucial for deciphering diverse biological processes. Next-generation sequencing technologies have, in particular, enabled researchers to unravel the genetic basis of phenotypic variation with high-resolution genomic data. Various DNA sequencing technologies have emerged, expanding the scope of genetic research in plant breeding (J. He *et al.*, 2014).

These advancements have facilitated the identification of QTLs and genes responsible for multiple traits, facilitating their precise integration into elite breeding lines through methods like GBS and by using a GAB approach (Varshney *et al.*, 2012). Many research endeavours have showcased the creation of novel cultivars and strains with improved tolerance or resistance to pests and environmental challenges through these methodologies. Furthermore, advancements in genome sequencing have spurred innovations in agricultural breeding methods like mutation breeding, MAB and genome editing, thereby greatly hastening efforts to enhance crop performance (Krishna *et al.*, 2023).

The efficiency of sequencing technologies in detecting mutations or variations in targeted crops at reduced time and cost has expedited the development of new varieties compared to traditional breeding methods. The availability of whole-genome sequences for many cereal and non-cereal crops has facilitated the discovery of novel genes and molecular markers associated with critical agronomic traits, offering immense potential for further crop enhancement. SNP markers, in particular, have been pivotal in constructing genetic maps essential for efficient crop improvement strategies. Sequencing technologies form the cornerstone of modern breeding approaches, transforming conventional breeding practices into GAB. Researchers have successfully leveraged these platforms to develop new crop varieties that meet evolving agricultural challenges and contribute to sustainable food production (Sandhu *et al.*, 2021).

### 7.24.2 Plant breeding with SNP technology

Over the past three decades, a range of DNA profiling techniques including RFLPs, RAPD and SSRs have been pivotal in molecular plant breeding. However, these methods are often laborious, time intensive and lack scalability for modern breeding programmes. Moreover, their inconsistency in accurately pinpointing markers within specific gene regions has limited their impact on plant breeding (Collard and Mackill, 2008; Xu and Crouch, 2008). In contrast, SNPs have emerged as the preferred markers in breeding due to their prevalence and compatibility with high-throughput detection methods (Mammadov *et al.*, 2012). Advances in NGS and whole-genome re-sequencing have significantly improved the identification and validation of diagnostic markers in plants. Techniques such as genotype-by-sequencing have also reduced costs and facilitated extensive genotyping of recombinant progenies (Elshire *et al.*, 2011).

For instance, the re-sequencing of diverse maize lines identified over 27 million SNPs, underscoring the utility of NGS in capturing genetic diversity (Jiao *et al.*, 2012). Similar initiatives have been undertaken in crops such as barley, soybean and lupin, resulting in the identification of sequence-based markers associated with beneficial alleles for enhancing crop performance (Yang *et al.*, 2015; Takahagi *et al.*, 2016; Valliyodan *et al.*, 2017). High density SNP genotyping arrays, such as those from Affymetrix and Illumina, have further revolutionized plant breeding by enabling genome wide SNP detection (Ganal *et al.*, 2011). These platforms, along with PCR-based assays like KASP and TaqMan, offer flexibility and accuracy in SNP genotyping across various plant species (C. He *et al.*, 2014; Long *et al.*, 2017). Moreover, diagnostic markers derived from genome or transcriptome sequencing are increasingly used to streamline MAS in breeding. These markers facilitate precise trait engineering and have been successfully applied in enhancing yield and stress tolerance in crops like lupin and soybean (Yang *et al.*, 2015; Patil *et al.*, 2016). The integration of advanced sequencing technologies with genotyping platforms has opened new avenues for developing tailored cultivars with desired traits, thus advancing crop breeding efforts to meet future food demands effectively.

## 7.25 Perspectives

Despite the promising potential of GAB, several obstacles hinder its widespread application. One major challenge is the need for advanced high-throughput phenotyping platforms in field conditions. Current phenome projects, while pioneering, fall short of fully realizing the capabilities of GAB due to high costs and limited phenotyping capacities. Moreover, effectively managing and analysing the vast amount of phenotypic data generated remains a significant bottleneck, requiring robust data management and bioinformatics tools. Another critical area of concern is understanding inheritance mechanisms associated with important agronomic traits, including epigenetic processes like DNA methylation, genomic imprinting and RNA editing. Although research in epigenetics has made significant strides, much remains to be understood about these mechanisms and their potential contributions to crop improvement. Insights gained from unravelling epigenetic processes could potentially drive molecular genetic innovations in crop breeding. Progress in plant genome sequencing has enabled the evolution of next-generation sequencing techniques for discovering alleles and pinpointing potential genes. Combining markers associated with high-throughput traits, economical genotyping techniques and accurate phenotyping platforms will be crucial in expediting the acceptance of GAB. These newly developed genetic and genomic tools are expected to complement, rather than replace, traditional breeding methods. Moving forward, the extensive adoption of MAS and GS, either independently or in synergy, holds promise for enhancing plant breeding outcomes at the genomic level.

## 7.26 Future Prospects

Although GAB holds significant promise for sustainable agriculture, several barriers currently hinder its widespread adoption. The cost of genotyping has significantly decreased with advancements in next-generation sequencing technologies. However, further reductions in genotyping and resequencing costs are necessary to facilitate the broader application of GAB, particularly in public plant breeding programmes. While genome sequences for many major crop plants are freely available, orphan crops and wild relatives still lack comprehensive genome data. Access to genome information for these species is crucial as a foundational step towards implementing GAB effectively. One of the critical prerequisites for advancing GAB is the establishment of high-throughput phenotyping platforms. Currently, there are limited publicly accessible phenomics platforms, which poses a major obstacle to leveraging GAB in crop improvement initiatives. Initiating phenome projects and developing affordable high-throughput phenotyping facilities for major crops are essential steps to fully harnessing the potential of GAB.

The field of plant genomics generates extensive data, necessitating efficient data management strategies to facilitate easy access and utilization by researchers. Bioinformaticians play a vital role in addressing these challenges, ensuring that crop-specific and trait-specific information is readily accessible in user friendly formats, thereby promoting the adoption of GAB. Additionally, efforts should focus on educating and training plant breeders and researchers in utilizing bioinformatic tools and databases effectively. While considerable progress has been made in understanding genomic-level inheritance mechanisms, further insights into epigenetic regulation would enhance the efficacy of GAB in crop improvement efforts.

## References

Aglawe, S.B., Singh, M., Rama Devi, S.J.S., Deshmukh, D.B. and Verma, A.K. (2021) Genomics assisted breeding for sustainable agriculture: Meeting the challenge of global food security. In: Upadhyay, A.K., Sowdhamini, R. and Patil, V.U. (eds) *Bioinformatics for Agriculture: High-Throughput Approaches*. Springer, Singapore, pp. 23–51. DOI: 10.1007/978-981-33-4791-5_2.

Ajeesh Krishna, T.P., Maharajan, T., Ignacimuthu, S. and Ceasar, S.A. (2021) Genomic-assisted breeding in finger millet (*Eleusine coracana* (L.) Gaertn.) for abiotic stress tolerance. In: Kole, C. (ed.) *Genomic Designing for Abiotic Stress Resistant Cereal Crops*. Springer International Publishing, pp. Cham, Switzerland, 291–317. DOI: 10.1007/978-3-030-75875-2_8.

Alexandrov, N., Tai, S., Wang, W., Mansueto, L., Palis, K. *et al.* (2015) SNP-seek database of SNPs derived from 3000 rice genomes. *Nucleic Acids Research* 43(D1), D1023–D1027. DOI: 10.1093/nar/gku1039.

Ali, M., Zhang, L., DeLacy, I., Arief, V., Dieters, M. *et al.* (2020) Modeling and simulation of recurrent phenotypic and genomic selections in plant breeding under the presence of epistasis. *The Crop Journal* 8(5), 866–877. DOI: 10.1016/j.cj.2020.04.002.

Araus, J.L. and Cairns, J.E. (2014) Field high-throughput phenotyping: The new crop breeding frontier. *Trends in Plant Science* 19(1), 52–61. DOI: 10.1016/j.tplants.2013.09.008.

Ashokkumar, K., Govindaraj, M., Karthikeyan, A., Shobhana, V.G. and Warkentin, T.D. (2020) Genomics-integrated breeding for carotenoids and folates in staple cereal grains to reduce malnutrition. *Frontiers in Genetics* 11, 497320. DOI: 10.3389/fgene.2020.00414.

Atefi, A., Ge, Y., Pitla, S. and Schnable, J. (2021) Robotic technologies for high-throughput plant phenotyping: Contemporary reviews and future perspectives. *Frontiers in Plant Science* 12, 611940. DOI: 10.3389/fpls.2021.611940.

Bajaj, D., Srivastava, R., Nath, M., Tripathi, S., Bharadwaj, C. *et al.* (2016) EcoTILLING-based association mapping efficiently delineates functionally relevant natural allelic variants of candidate genes governing agronomic traits in chickpea. *Frontiers in Plant Science* 7, 450. DOI: 10.3389/fpls.2016.00450.

Bandillo, N., Raghavan, C., Muyco, P.A., Sevilla, M.A.L., Lobina, I.T. *et al.* (2013) Multi-parent advanced generation inter-cross (MAGIC) populations in rice: progress and potential for genetics research and breeding. *Rice* 6(1), 11. DOI: 10.1186/1939-8433-6-11.

Battenfield, S.D., Sheridan, J.L., Silva, L.D., Miclaus, K.J., Dreisigacker, S. *et al.* (2018) Breeding-assisted genomics: Applying meta-GWAS for milling and baking quality in CIMMYT wheat breeding program. *PLOS ONE* 13(11), e0204757. DOI: 10.1371/journal.pone.0204757.

Bayer, P.E., Golicz, A.A., Tirnaz, S., Chan, C.K., Edwards, D. *et al.* (2019) Variation in abundance of predicted resistance genes in the *Brassica oleracea* pangenome. *Plant Biotechnology Journal* 17(4), 789–800. DOI: 10.1111/pbi.13015.

Bera, S.K., Kamdar, J.H., Kasundra, S.V., Patel, S.V., Jasani, M.D. *et al.* (2019) Steady expression of high oleic acid in peanut bred by marker-assisted backcrossing for fatty acid desaturase mutant alleles and its effect on seed germination along with other seedling traits. *PLOS ONE* 14(12), e0226252. DOI: 10.1371/journal.pone.0226252.

Bhat, J.A., Ali, S., Salgotra, R.K., Mir, Z.A., Dutta, S. *et al.* (2016) Genomic selection in the era of next generation sequencing for complex traits in plant breeding. *Frontiers in Genetics* 7. DOI: 10.3389/fgene.2016.00221.

Bhatta, M., Sandro, P., Smith, M.R., Delaney, O., Voss-Fels, K.P. *et al.* (2021) Need for speed: Manipulating plant growth to accelerate breeding cycles. *Current Opinion in Plant Biology* 60, 101986. DOI: 10.1016/j.pbi.2020.101986.

Bohra, A., Saxena, K.B., Varshney, R.K. and Saxena, R.K. (2020) Genomics-assisted breeding for pigeonpea improvement. *Theoretical and Applied Genetics* 133(5), 1721–1737. DOI: 10.1007/s00122-020-03563-7.

Boopathi, N.M. (2020) Marker-assisted selection (MAS). In: Boopathi, N.M. (ed.) *Genetic Mapping and Marker Assisted Selection*. Springer, Singapore, pp. 343–388. DOI: 10.1007/978-981-15-2949-8_9.

Buckler, E.S., Holland, J.B., Bradbury, P.J., Acharya, C.B., Brown, P.J. *et al.* (2009) The genetic architecture of maize flowering time. *Science* 325(5941), 714–718. DOI: 10.1126/science.1174276.

Charmet, G., Tran, L.-G., Auzanneau, J., Rincent, R. and Bouchet, S. (2020) BWGS: AR package for genomic selection and its application to a wheat breeding programme. *PLOS ONE* 15(4), e0222733. DOI: 10.1371/journal.pone.0222733.

Chelpuri, D., Sharma, R., Durga, K.K., Katiyar, P., Mahendrakar, M.D. *et al.* (2019) Mapping quantitative trait loci (QTLs) associated with resistance to major pathotype-isolates of pearl millet downy mildew pathogen. *European Journal of Plant Pathology* 154(4), 983–994. DOI: 10.1007/s10658-019-01718-x.

Choi, J.-K., Sa, K.J., Park, D.H., Lim, S.E., Ryu, S.-H. *et al.* (2019) Construction of genetic linkage map and identification of QTLs related to agronomic traits in DH population of maize (*Zea mays* L.) using SSR markers. *Genes & Genomics* 41(6), 667–678. DOI: 10.1007/s13258-019-00813-x.

Chukwu, S.C., Rafii, M.Y., Ramlee, S.I., Ismail, S.I., Oladosu, Y. *et al.* (2019) Marker-assisted selection and gene pyramiding for resistance to bacterial leaf blight disease of rice (*Oryza sativa* L.). *Biotechnology & Biotechnological Equipment* 33(1), 440–455. DOI: 10.1080/13102818.2019.1584054.

Collard, B.C.Y. and Mackill, D.J. (2008) Marker-assisted selection: An approach for precision plant breeding in the twenty-first century. *Philosophical Transactions of the Royal Society B: Biological Sciences* 363(1491), 557–572. DOI: 10.1098/rstb.2007.2170.

Cooper, M., Messina, C.D., Podlich, D., Totir, L.R., Baumgarten, A. *et al.* (2014) Predicting the future of plant breeding: Complementing empirical evaluation with genetic prediction. *Crop and Pasture Science* 65(4), 311–336. DOI: 10.1071/CP14007.

Crossa, J., Pérez-Rodríguez, P., Cuevas, J., Montesinos-López, O., Jarquín, D. *et al.* (2017) Genomic selection in plant breeding: Methods, models, and perspectives. *Trends in Plant Science* 22(11), 961–975. DOI: 10.1016/j.tplants.2017.08.011.

Cui, Y., Li, R., Li, G., Zhang, F., Zhu, T. *et al.* (2020) Hybrid breeding of rice via genomic selection. *Plant Biotechnology Journal* 18(1), 57–67. DOI: 10.1111/pbi.13170.

Cui, Z., Dong, H., Zhang, A., Ruan, Y., He, Y. *et al.* (2020) Assessment of the potential for genomic selection to improve husk traits in maize. *G3: Genes, Genomes, Genetics* 10(10), 3741–3749. DOI: 10.1534/g3.120.401600.

Dai, D., Ma, Z. and Song, R. (2021) Maize kernel development. *Molecular Breeding* 41(1), 2. DOI: 10.1007/s11032-020-01195-9.

Danilevicz, M.F., Fernandez, C.G.T., Marsh, J.I., Bayer, P.E. and Edwards, D. (2020) Plant pangenomics: approaches, applications and advancements. *Current Opinion in Plant Biology* 54, 18–25. DOI: 10.1016/j.pbi.2019.12.005.

de Vos, C.J. and Swanenburg, M. (2018) Health effects of feeding genetically modified (GM) crops to livestock animals: A review. *Food and Chemical Toxicology* 117, 3–12. DOI: 10.1016/j.fct.2017.08.031.

Dell'Acqua, M., Gatti, D.M., Pea, G., Cattonaro, F., Coppens, F. *et al.* (2015) Genetic properties of the MAGIC maize population: A new platform for high definition QTL mapping in *Zea mays*. *Genome Biology* 16(1), 167. DOI: 10.1186/s13059-015-0716-z.

Descalsota, G.I.L., Swamy, B.M., Zaw, H., Inabangan-Asilo, M.A., Amparado, A. *et al.* (2018) Genome-wide association mapping in a rice MAGIC plus population detects QTLs and genes useful for biofortification. *Frontiers in Plant Science* 9, 1347. DOI: 10.3389/fpls.2018.01347.

Dreher, K., Morris, M., Khairallah, M., Ribaut, J.M., Shivaji Pandey, S.P. *et al.* (2002) Is marker-assisted selection cost-effective compared with conventional plant breeding methods? The case of quality proteinmaize. In: Evenson, R.E., Santaniello, V. and Zilberman, D. (eds) *Economic and Social Issues in Agricultural*, 1st edn. CABI Publishing, Wallingford, UK, pp. 203–236. DOI: 10.1079/9780851996189.0203.

Eathington, S.R., Crosbie, T.M., Edwards, M.D., Reiter, R.S. and Bull, J.K. (2007) Molecular markers in a commercial breeding program. *Crop Science* 47, S3. DOI: 10.2135/cropsci2007.04.0015IPBS.

Elshire, R.J., Glaubitz, J.C., Sun, Q., Poland, J.A., Kawamoto, K. *et al.* (2011) A robust, simple genotyping-by-sequencing (GBS) approach for high diversity species. *PLOS ONE* 6(5), e19379. DOI: 10.1371/journal.pone.0019379.

Emebiri, L.C., Tan, M.-K., El-Bouhssini, M., Wildman, O., Jighly, A. *et al.* (2017) QTL mapping identifies a major locus for resistance in wheat to sunn pest (*Eurygaster integriceps*) feeding at the vegetative growth stage. *Theoretical and Applied Genetics* 130(2), 309–318. DOI: 10.1007/s00122-016-2812-1.

Feng, Y., Lu, Q., Zhai, R., Zhang, M., Xu, Q. *et al.* (2016) Genome wide association mapping for grain shape traits in indica rice. *Planta* 244(4), 819–830. DOI: 10.1007/s00425-016-2548-9.

Furbank, R.T. and Tester, M. (2011) Phenomics – technologies to relieve the phenotyping bottleneck. *Trends in Plant Science* 16(12), 635–644. DOI: 10.1016/j.tplants.2011.09.005.

Gahlaut, V., Jaiswal, V., Tyagi, B.S., Singh, G., Sareen, S. *et al.* (2017) QTL mapping for nine drought-responsive agronomic traits in bread wheat under irrigated and rain-fed environments. *PLOS ONE* 12(8), e0182857. DOI: 10.1371/journal.pone.0182857.

Ganal, M.W., Durstewitz, G., Polley, A., Bérard, A., Buckler, E.S. *et al.* (2011) A large maize (*Zea mays* L.) SNP genotyping array: Development and germplasm genotyping, and genetic mapping to compare with the B73 reference genome. *PLOS ONE* 6(12), e28334. DOI: 10.1371/journal.pone.0028334.

Ganie, S.A., Wani, S.H., Henry, R. and Hensel, G. (2021) Improving rice salt tolerance by precision breeding in a new era. *Current Opinion in Plant Biology* 60, 101996. DOI: 10.1016/j.pbi.2020.101996.

Gautam, T., Dhillon, G.S., Saripalli, G., Singh, V.P., Prasad, P. *et al.* (2020) Marker-assisted pyramiding of genes/QTL for grain quality and rust resistance in wheat (*Triticum aestivum* L.). *Molecular Breeding* 40(5), 49. DOI: 10.1007/s11032-020-01125-9.

Getachew, T. (2019) Pulse crops production opportunities, challenges and its value chain in Ethiopia: A review article. *Journal of Environment and Earth Science* 9(1). DOI: 10.7176/JEES/9-1-03.

Ghosh, S., Watson, A., Gonzalez-Navarro, O.E., Ramirez-Gonzalez, R.H., Yanes, L. *et al.* (2018) Speed breeding in growth chambers and glasshouses for crop breeding and model plant research. *Nature Protocols* 13(12), 2944–2963. DOI: 10.1038/s41596-018-0072-z.

Gokidi, Y., Bhanu, A.N. and Singh, M.N. (2016) Marker assisted recurrent selection: An overview. *Advances in Life Sciences* 5(17), 6493–6499.

Gonias, E.D., Ganopoulos, I., Mellidou, I., Bibi, A.C., Kalivas, A. *et al.* (2019) Exploring genetic diversity of tomato (*Solanum lycopersicum* L.) germplasm of genebank collection employing SSR and SCAR markers. *Genetic Resources and Crop Evolution* 66(6), 1295–1309. DOI: 10.1007/s10722-019-00786-6.

Gupta, P.K., Kumar, J., Mir, R.R. and Kumar, A. (2010a) Marker-assisted selection as a component of conventional plant breeding. In: Janick, J. (ed.) *Plant Breeding Reviews*, Vol. 33. Wiley, pp. 145–217. DOI: 10.1002/9780470535486.ch4.

Gupta, P.K., Langridge, P. and Mir, R.R. (2010b) Marker-assisted wheat breeding: Present status and future possibilities. *Molecular Breeding* 26(2), 145–161. DOI: 10.1007/s11032-009-9359-7.

Han, B., Wang, C., Tang, Z., Ren, Y., Li, Y. *et al.* (2015) Genome-wide analysis of microsatellite markers based on sequenced database in chinese spring wheat (*Triticum aestivum* L.). *PLOS ONE* 10(11), e0141540. DOI: 10.1371/journal.pone.0141540.

Hasan, M.M., Rafii, M.Y., Ismail, M.R., Mahmood, M., Rahim, H.A. *et al.* (2015) Marker-assisted backcrossing: A useful method for rice improvement. *Biotechnology & Biotechnological Equipment* 29(2), 237–254. DOI: 10.1080/13102818.2014.995920.

He, C., Holme, J. and Anthony, J. (2014) SNP genotyping: The KASP assay. *Crop Breeding: Methods and Protocols* 1145, 75–86. DOI: 10.1007/978-1-4939-0446-4_7.

He, J., Zhao, X., Laroche, A., Lu, Z.-X., Liu, H. *et al.* (2014) Genotyping-by-sequencing (GBS), an ultimate marker-assisted selection (MAS) tool to accelerate plant breeding. *Frontiers in Plant Science* 5, 484. DOI: 10.3389/fpls.2014.00484.

Henry, R. (2015) *Application of genomics-assisted breeding for generation of climate resilient crops: Progress and prospects*. *Frontiers in Plant Science* 6, 563. DOI: 10.3389/fpls.2015.00563.

Henry, A., Swamy, B.M., Dixit, S., Torres, R.D., Batoto, T.C. *et al.* (2015) Physiological mechanisms contributing to the QTL-combination effects on improved performance of IR64 rice NILs under drought. *Journal of Experimental Botany* 66(7), 1787–1799. DOI: 10.1093/jxb/eru506.

Hickey, L.T., Hafeez, A.N., Robinson, H., Jackson, S.A., Leal-Bertioli, S.C. *et al.* (2019) Breeding crops to feed 10 billion. *Nature Biotechnology* 37(7), 744–754. DOI: 10.1038/s41587-019-0152-9.

Hina, A., Cao, Y., Song, S., Li, S., Sharmin, R.A. *et al.* (2020) High-resolution mapping in two RIL populations refines major "QTL hotspot" regions for seed size and shape in soybean (*Glycine max* L.). *International Journal of Molecular Sciences* 21(3), 1040. DOI: 10.3390/ijms21031040.

Hu, Z., Zhang, D., Zhang, G., Kan, G., Hong, D. *et al.* (2014) Association mapping of yield-related traits and SSR markers in wild soybean (*Glycine soja* Sieb. and Zucc.). *Breeding Science* 63(5), 441–449. DOI: 10.1270/jsbbs.63.441.

Huang, B.E., Verbyla, K.L., Verbyla, A.P., Raghavan, C., Singh, V.K. *et al.* (2015) MAGIC populations in crops: Current status and future prospects. *Theoretical and Applied Genetics* 128(6), 999–1017. DOI: 10.1007/s00122-015-2506-0.

Huang, X., Wei, X., Sang, T., Zhao, Q., Feng, Q. *et al.* (2010) Genome-wide association studies of 14 agronomic traits in rice landraces. *Nature Genetics* 42(11), 961–967. DOI: 10.1038/ng.695.

Huang, X.Q., Kempf, H., Ganal, M.W. and Röder, M.S. (2004) Advanced backcross QTL analysis in progenies derived from a cross between a German elite winter wheat variety and a synthetic wheat (*Triticum aestivuml* L.). *Theoretical and Applied Genetics* 109(5), 933–943. DOI: 10.1007/s00122-004-1708-7.

Ibrahim, A.K., Zhang, L., Niyitanga, S., Afzal, M.Z., Xu, Y. *et al.* (2020) Principles and approaches of association mapping in plant breeding. *Tropical Plant Biology* 13(3), 212–224. DOI: 10.1007/s12042-020-09261-4.

Irshad, A., Guo, H., Zhang, S. and Liu, L. (2020) TILLING in cereal crops for allele expansion and mutation detection by using modern sequencing technologies. *Agronomy* 10(3), 405. DOI: 10.3390/agronomy10030405.

Jaiswal, V., Gupta, S., Gahlaut, V., Muthamilarasan, M., Bandyopadhyay, T. *et al.* (2019) Genome-wide association study of major agronomic traits in foxtail millet (*Setaria italica* L.) using ddRAD sequencing. *Scientific Reports* 9(1), 5020. DOI: 10.1038/s41598-019-41602-6.

Jia, Y. and Jannink, J.-L. (2012) Multiple-trait genomic selection methods increase genetic value prediction accuracy. *Genetics* 192(4), 1513–1522. DOI: 10.1534/genetics.112.144246.

Jia, J., Zhao, S., Kong, X., Li, Y., Zhao, G. *et al.* (2013) *Aegilops tauschii* draft genome sequence reveals a gene repertoire for wheat adaptation. Nature 496(7443),91–95. DOI: 10.1038/nature12028.

Jiang, G.-L. (2013) Molecular markers and marker-assisted breeding in plants. In: Andersen, S.B. (ed.) *Plant Breeding from Laboratories to Fields*. InTech Open, London, pp. 45–83.

Jiang, G.-L., Shi, J. and Ward, R.W. (2007) QTL analysis of resistance to Fusarium head blight in the novel wheat germplasm CJ 9306. I. Resistance to fungal spread. *Theoretical and Applied Genetics* 116(1), 3–13. DOI: 10.1007/s00122-007-0641-y.

Jiang, H., Feng, Y., Qiu, L., Gao, G., Zhang, Q. *et al.* (2020) Identification of blast resistance QTLs based on two advanced backcross populations in rice. *Rice* 13(1), 31. DOI: 10.1186/s12284-020-00392-6.

Jiao, Y., Zhao, H., Ren, L., Song, W., Zeng, B. *et al.* (2012) Genome-wide genetic changes during modern breeding of maize. *Nature Genetics* 44(7), 812–815. DOI: 10.1038/ng.2312.

Johnson, K.V.H., Krishna, T.P.A., Dash, M., Thiyageshwari, S., Ceasar, S.A. *et al.* (2023) Food and nutritional security: Innovative approaches for improving micronutrient use efficiency in soybean (*Glycine max* (L.) Merrill) under hostile soils. *Journal of Soil Science and Plant Nutrition* 23(1), 56–70. DOI: 10.1007/s42729-022-01025-1.

Jun, R., Hu, X., KeJian, W. and Wang, C. (2019) Development and application of CRISPR/Cas system in rice. *Rice Science* 26, 69–76. DOI: 10.1016/j.rsci.2019.01.001.

Kage, U., Kumar, A., Dhokane, D., Karre, S. and Kushalappa, A.C. (2016) Functional molecular markers for crop improvement. *Critical Reviews in Biotechnology* 36(5), 917–930. DOI: 10.3109/07388551.2015.1062743.

Kaiser, N., Douches, D., Dhingra, A., Glenn, K.C., Herzig, P.R. *et al.* (2020) The role of conventional plant breeding in ensuring safe levels of naturally occurring toxins in food crops. *Trends in Food Science & Technology* 100, 51–66. DOI: 10.1016/j.tifs.2020.03.042.

Kandemir, N. and Saygili, I. (2015) Apomixis: New horizons in plant breeding. *Turkish Journal of Agriculture and Forestry* 39(4), 549–556. DOI: 10.3906/tar-1409-74.

Kang, J.-W., Shin, D., Cho, J.-H., Lee, J.-Y., Kwon, Y. *et al.* (2019) Accelerated development of rice stripe virus-resistant, near-isogenic rice lines through marker-assisted backcrossing. *PLOS ONE* 14(12), e0225974. DOI: 10.1371/journal.pone.0225974.

Kaur, G., Joshi, A. and Jain, D. (2018) SSR-marker assisted evaluation of genetic diversity in mungbean (*Vigna radiata* (L.) Wilcezk) genotypes. *Brazilian Archives of Biology and Technology* 61. Available at: https://www.scielo.br/j/babt/a/WpYwnVFHFzhntxvC4RvTGLw/?format=html&lang=en (accessed 19 November 2024).

Kehel, Z., Sanchez-Garcia, M., El Baouchi, A., Aberkane, H., Tsivelikas, A. *et al.* (2020) Predictive characterization for seed morphometric traits for genebank accessions using genomic selection. *Frontiers in Ecology and Evolution* 8, 32. DOI: 10.3389/fevo.2020.00032.

Kersey, P.J., Allen, J.E., Christensen, M., Davis, P., Falin, L.J. *et al.* (2016) Ensembl Genomes 2013: Scaling up access to genome-wide data. *Nucleic Acids Research* 44(D1), D574–D580. DOI: 10.1093/nar/gkt979.

Khan, Z. and Shahwar, D. (2024) Genome editing in plants via CRISPR/Cas9: A genomic scissor borrowed from bacterial immune system. In: *Genome Editing and Global Food Security*. Routledge, pp. 1–16.

Kim, J.-M., Kim, K.-H., Jung, J., Kang, B.K., Lee, J. *et al.* (2020) Validation of marker-assisted selection in soybean breeding program for pod shattering resistance. *Euphytica* 216(11), 166. DOI: 10.1007/s10681-020-02703-w.

Krishna, T.P.A., Maharajan, T. and Ceasar, S.A. (2023) The role of membrane transporters in the biofortification of zinc and iron in plants. *Biological Trace Element Research* 201(1), 464–478. DOI: 10.1007/s12011-022-03159-w.

Kumar, A., Dixit, S., Ram, T., Yadaw, R.B., Mishra, K.K. *et al.* (2014) Breeding high-yielding drought-tolerant rice: Genetic variations and conventional and molecular approaches. *Journal of Experimental Botany* 65(21), 6265–6278.

Kumar, K., Gambhir, G., Dass, A., Tripathi, A.K., Singh, A. *et al.* (2020) Genetically modified crops: Current status and future prospects. *Planta* 251(4), 91. DOI: 10.1007/s00425-020-03372-8.

Kumari, V., Gowda, M.V.C. and Yeri, S.B. (2020) Utilization of advanced backcross population derived from synthetic amphidiploid for dissecting resistance to late leaf spot in peanut (*Arachis hypogaea* L.). *Tropical Plant Biology* 13(1), 50–61. DOI: 10.1007/s12042-019-09246-y.

Kump, K.L., Bradbury, P.J., Wisser, R.J., Buckler, E.S., Belcher, A.R. *et al.* (2011) Genome-wide association study of quantitative resistance to southern leaf blight in the maize nested association mapping population. *Nature Genetics* 43(2), 163–168. DOI: 10.1038/ng.747.

Kurowska, M., Daszkowska-Golec, A., Gruszka, D., Marzec, M., Szurman, M. *et al.* (2011) TILLING – a shortcut in functional genomics. *Journal of Applied Genetics* 52(4), 371–390. DOI: 10.1007/s13353-011-0061-1.

Labroo, M.R., Studer, A.J. and Rutkoski, J.E. (2021) Heterosis and hybrid crop breeding: A multidisciplinary review. *Frontiers in Genetics* 12, 643761. DOI: 10.3389/fgene.2021.643761.

Ladejobi, O., Elderfield, J., Gardner, K.A., Gaynor, R.C., Hickey, J. *et al.* (2016) Maximizing the potential of multi-parental crop populations. *Applied & Translational Genomics* 11, 9–17. DOI: 10.1016/j.atg.2016.10.002.

Lei, L., Zheng, H.L., Wang, J.G., Liu, H.L., Sun, J. *et al.* (2018) Genetic dissection of rice (*Oryza sativa* L.) tiller, plant height, and grain yield based on QTL mapping and metaanalysis. *Euphytica* 214(7), 109. DOI: 10.1007/s10681-018-2187-2.

Li, H., Peng, Z., Yang, X., Wang, W., Fu, J. *et al.* (2013) Genome-wide association study dissects the genetic architecture of oil biosynthesis in maize kernels. *Nature Genetics* 45(1), 43–50. DOI: 10.1038/ng.2484.

Li, H., Rasheed, A., Hickey, L.T. and He, Z. (2018) Fast-forwarding genetic gain. *Trends in Plant Science* 23(3), 184–186. DOI: 10.1016/j.tplants.2018.01.007.

Li, J.-Y., Wang, J. and Zeigler, R.S. (2014) The 3,000 rice genomes project: New opportunities and challenges for future rice research. *Gigascience* 3(1), 2047-217X-3-8. DOI: 10.1186/2047-217X-3-8.

Li, Y., Ruperao, P., Batley, J., Edwards, D., Khan, T. *et al.* (2018) Investigating drought tolerance in chickpea using genome-wide association mapping and genomic selection based on whole-genome resequencing data. *Frontiers in Plant Science* 9, 190. DOI: 10.3389/fpls.2018.00190.

Liakos, K.G., Busato, P., Moshou, D., Pearson, S. and Bochtis, D. (2018) Machine learning in agriculture: A review. *Sensors* 18(8), 2674. DOI: 10.3390/s18082674.

Liu, S., Wang, X., Wang, H., Xin, H., Yang, X. *et al.* (2013) Genome-wide analysis of *ZmDREB* genes and their association with natural variation in drought tolerance at seedling stage of *Zea mays* L. *PLOS Genetics* 9(9), e1003790. DOI: 10.1371/journal.pgen.1003790.

Long, Y.M., Chao, W.S., Ma, G.J., Xu, S.S. and Qi, L.L. (2017) An innovative SNP genotyping method adapting to multiple platforms and throughputs. *Theoretical and Applied Genetics* 130(3), 597–607. DOI: 10.1007/s00122-016-2838-4.

Lu, K., Wei, L., Li, X., Wang, Y., Wu, J. *et al.* (2019) Whole-genome resequencing reveals *Brassica napus* origin and genetic loci involved in its improvement. *Nature Communications* 10(1), 1154. DOI: 10.1038/s41467-019-09134-9.

Mackay, I. and Powell, W. (2007) Methods for linkage disequilibrium mapping in crops. *Trends in Plant Science* 12, 57–63. DOI: 10.1016/j.tplants.2006.12.001.

Mammadov, J., Aggarwal, R., Buyyarapu, R. and Kumpatla, S. (2012) SNP markers and their impact on plant breeding. *International Journal of Plant Genomics* 2012, 728398. DOI: 10.1155/2012/728398.

McCallum, C.M., Comai, L., Greene, E.A. and Henikoff, S. (2000) Targeting induced local lesions in genomes (TILLING) for plant functional genomics. *Plant Physiology* 123(2), 439–442. DOI: 10.1104/pp.123.2.439.

Meuwissen, T.H., Hayes, B.J. and Goddard, M. (2001) Prediction of total genetic value using genome-wide dense marker maps. *Genetics* 157(4), 1819–1829. DOI: 10.1093/genetics/157.4.1819.

Miglani, G.S. (2017) Genome editing in crop improvement: Present scenario and future prospects. *Journal of Crop Improvement* 31(4), 453–559. DOI: 10.1080/15427528.2017.1333192.

Mir, R.R., Zaman-Allah, M., Sreenivasulu, N., Trethowan, R. and Varshney, R.K. (2012) Integrated genomics, physiology and breeding approaches for improving drought tolerance in crops. *Theoretical and Applied Genetics* 125(4), 625–645. DOI: 10.1007/s00122-012-1904-9.

Mir, R.R., Reynolds, M., Pinto, F., Khan, M.A. and Bhat, M.A. (2019) High-throughput phenotyping for crop improvement in the genomics era. *Plant Science* 282, 60–72. DOI: 10.1016/j.plantsci.2019.01.007.

Montes, J.M., Technow, F., Dhillon, B.S., Mauch, F. and Melchinger, A.E. (2011) High-throughput non-destructive biomass determination during early plant development in maize under field conditions. *Field Crops Research* 121(2), 268–273. DOI: 10.1016/j.fcr.2010.12.017.

Mott, R., Talbot, C.J., Turri, M.G., Collins, A.C. and Flint, J. (2000) A method for fine mapping quantitative trait loci in outbred animal stocks. *Proceedings of the National Academy of Sciences USA* 97, 12649–12654. DOI: 10.1073/pnas.230304397.

Murovec, J., Pirc, Ž. and Yang, B. (2017) New variants of CRISPR RNA-guided genome editing enzymes. *Plant Biotechnology Journal* 15(8), 917–926. DOI: 10.1111/pbi.12736.

Muth, J., Hartje, S., Twyman, R.M., Hofferbert, H., Tacke, E. *et al.* (2008) Precision breeding for novel starch variants in potato. *Plant Biotechnology Journal* 6(6), 576–584. DOI: 10.1111/j.1467-7652.2008.00340.x.

Muthamilarasan, M. and Prasad, M. (2015) Advances in Setaria genomics for genetic improvement of cereals and bioenergy grasses. *Theoretical and Applied Genetics* 128(1), 1–14. DOI: 10.1007/s00122-014-2399-3.

Myles, S., Peiffer, J., Brown, P.J., Ersoz, E.S., Zhang, Z. *et al.* (2009) Association mapping: Critical considerations shift from genotyping to experimental design. *Plant Cell* 21(8), 2194–2202. DOI: 10.1105/tpc.109.068437.

Nakaya, A. and Isobe, S.N. (2012) Will genomic selection be a practical method for plant breeding. *Annals of Botany* 110(6), 1303–1316. DOI: 10.1093/aob/mcs109.

Ndlovu, N. (2020) Application of genomics and phenomics in plant breeding for climate resilience. *Asian Plant Research Journal* 6(4), 53–66. DOI: 10.9734/aprj/2020/v6i430137.

Nguyen, T., Fu, K., Mou, C., Yu, J., Zhu, X. *et al.* (2020) Fine mapping of *qSdr9*, a novel locus for seed dormancy (SD) in weedy rice, and development of NILs with a strong SD allele. *Molecular Breeding* 40(8), 81. DOI: 10.1007/s11032-020-01148-2.

Nica, A.C. and Dermitzakis, E.T. (2013) Expression quantitative trait loci: Present and future. *Philosophical Transactions of the Royal Society B: Biological Sciences* 368(1620), 20120362. DOI: 10.1098/rstb.2012.0362.

Oladosu, Y., Rafii, M.Y., Samuel, C., Fatai, A., Magaji, U. *et al.* (2019) Drought resistance in rice from conventional to molecular breeding: A review. *International Journal of Molecular Sciences* 20(14), 3519. DOI: 10.3390/ijms20143519.

Pandey, G., Misra, G., Kumari, K., Gupta, S., Parida, S.K. *et al.* (2013) Genome-wide development and use of microsatellite markers for large-scale genotyping applications in foxtail millet (*Setaria italica* L.). *DNA Research* 20(2), 197–207. DOI: 10.1093/dnares/dst002.

Pandey, M.K., Chaudhari, S., Jarquin, D., Janila, P., Crossa, J. *et al.* (2020) Genome-based trait prediction in multi-environment breeding trials in groundnut. *Theoretical and Applied Genetics* 133(11), 3101–3117. DOI: 10.1007/s00122-020-03658-1.

Pandian, S., Marichelvam, K., Satish, L., Ceasar, S.A., Pandian, S.K. *et al.* (2018) SPAR markers-assisted assessment of genetic diversity and population structure in finger millet (*Eleusine coracana* (L.) Gaertn) mini-core collection. *Journal of Crop Science and Biotechnology* 21(5), 469–481. DOI: 10.1007/s12892-018-0034-0.

Patil, G., Do, T., Vuong, T.D., Valliyodan, B., Lee, J.D. *et al.* (2016) Genomic-assisted haplotype analysis and the development of high-throughput SNP markers for salinity tolerance in soybean. Science Report 6, 19199. DOI: 10.1038/srep19199.

Perumalsamy, S., Bharani, M., Sudha, M., Nagarajan, P., Arul, L. *et al.* (2009) Functional marker-assisted selection for bacterial leaf blight resistance genes in rice (*Oryza sativa* L.). *Plant Breeding* 129, 400–406. DOI: 10.1111/j.1439-0523.2009.01705.x.

Pickar-Oliver, A. and Gersbach, C.A. (2019) The next generation of CRISPR-Cas technologies and applications. *Nature Reviews Molecular Cell Biology* 20(8), 490–507. DOI: 10.1038/s41580-019-0131-5.

Poland, J.A. and Rife, T.W. (2012) Genotyping-by-sequencing for plant breeding and genetics. *Plant Genome* 5(3), plantgenome2012.05.0005. DOI: 10.3835/plantgenome2012.05.0005.

Prakash, P.T., Arbelaez Velez, J.D., Verdeprado, H., Lopena, V., Imee Zhella, R. *et al.* (2020) Empowering global rice breeding programs using genomic selection [W291]. Available at: https://agritrop.cirad.fr/594832 (accessed 19 November 2024).

Pukalenthy, B., Manickam, D., Adhimoolam, K., Mahesh, S.G., Ramanathan, N. *et al.* (2020) Marker aided introgression of *opaque 2* (*o2*) allele improving lysine and tryptophan in maize (*Zea mays* L.). *Physiology and Molecular Biology of Plants* 26(9), 1925–1930. DOI: 10.1007/s12298-020-00857-4.

Puranik, S., Sahu, P.P., Beynon, S., Srivastava, R.K., Sehgal, D. *et al.* (2020) Genome-wide association mapping and comparative genomics identifies genomic regions governing grain nutritional traits in finger millet (*Eleusine coracana* L.Gaertn.). *PLANTS, PEOPLE, PLANET* 2(6), 649–662. DOI: 10.1002/ppp3.10120.

Qaim, M. (2020) Role of new plant breeding technologies for food security and sustainable agricultural development. *Applied Economic Perspectives and Policy* 42(2), 129–150. DOI: 10.1002/aepp.13044.

Qu, C., Jia, L., Fu, F., Zhao, H., Lu, K. *et al.* (2017) Genome-wide association mapping and identification of candidate genes for fatty acid composition in *Brassica napus* L. using SNP markers. *BMC Genomics* 18(1), 232. DOI: 10.1186/s12864-017-3607-8.

Rahim, M.S., Bhandawat, A., Rana, N., Sharma, H., Parveen, A. *et al.* (2020) Genomic selection in cereal crops: Methods and applications. In: Gosal, S.S. and Wani, S.H. (eds) *Accelerated Plant Breeding, Vol. 1.* Springer International Publishing, Cham, Switzerland, pp. 51–88. DOI: 10.1007/978-3-030-41866-3_3.

Rajkumar, F., Fakrudin, B., Kavil, S.P., Girma, Y., Arun, S.S. *et al.* (2013) Molecular mapping of genomic regions harbouring QTLs for root and yield traits in sorghum (*Sorghum bicolor* L. Moench). *Physiology and Molecular Biology of Plants* 19(3), 409–419. DOI: 10.1007/s12298-013-0188-0.

Rao, I.S., Neeraja, C.N., Srikanth, B., Subrahmanyam, D., Swamy, K.N. *et al.* (2018) Identification of rice landraces with promising yield and the associated genomic regions under low nitrogen. *Scientific Reports* 8(1), 9200. DOI: 10.1038/s41598-018-27484-0.

Raza, A., Tabassum, J., Kudapa, H. and Varshney, R.K. (2021) Can omics deliver temperature resilient ready-to-grow crops. *Critical Reviews in Biotechnology* 41(8), 1209–1232. DOI: 10.1080/07388551.2021.1898332.

Riaz, A., KockAppelgren, P., Hehir, J.G., Kang, J., Meade, F. *et al.* (2020) Genetic analysis using a multi-parent wheat population identifies novel sources of septoria tritici blotch resistance. *Genes* 11(8), 887. DOI: 10.3390/genes11080887.

Ribaut, J.-M. and Ragot, M. (2007) Marker-assisted selection to improve drought adaptation in maize: The backcross approach, perspectives, limitations, and alternatives. *Journal of Experimental Botany* 58(2), 351–360. DOI: 10.1093/jxb/erl214.

Ribaut, J.M., Vicente, M.C. and Delannay, X. (2010) Molecular breeding in developing countries: Challenges and perspectives. *Current Opinion in Plant Biology* 13(2), 213–218. DOI: 10.1016/j.pbi.2009.12.011.

Robertsen, C.D., Hjortshøj, R.L. and Janss, L.L. (2019) Genomic selection in cereal breeding. *Agronomy* 9(2), 95. DOI: 10.3390/agronomy9020095.

Rodziewicz, P., Chmielewska, K., Sawikowska, A., Marczak, Ł., Łuczak, M. *et al.* (2019) Identification of drought responsive proteins and related proteomic QTLs in barley. *Journal of Experimental Botany* 70(10), 2823–2837. DOI: 10.1093/jxb/erz075.

Roorkiwal, M., Bharadwaj, C., Barmukh, R., Dixit, G.P., Thudi, M. *et al.* (2020) Integrating genomics for chickpea improvement: Achievements and opportunities. *Theoretical and Applied Genetics* 133(5), 1703–1720. DOI: 10.1007/s00122-020-03584-2.

Saint Pierre, C., Burgueño, J., Crossa, J., Fuentes Dávila, G., Figueroa López, P. *et al.* (2016) Genomic prediction models for grain yield of spring bread wheat in diverse agro-ecological zones. *Scientific Reports* 6(1), 27312. DOI: 10.1038/srep27312.

Sandhu, N., Yadav, S., Catolos, M., Cruz, M.T.S. and Kumar, A. (2021) Developing climate-resilient, direct-seeded, adapted multiple-stress-tolerant rice applying genomics-assisted breeding. *Frontiers in Plant Science* 12, 637488. DOI: 10.3389/fpls.2021.637488.

Saxena, R.K., Kale, S., Mir, R.R., Mallikarjuna, N., Yadav, P. *et al.* (2020) Genotyping-by-sequencing and multilocation evaluation of two interspecific backcross populations identify QTLs for yield-related traits in pigeonpea. *Theoretical and Applied Genetics* 133(3), 737–749. DOI: 10.1007/s00122-019-03504-z.

Schläppi, M.R., Jackson, A.K., Eizenga, G.C., Wang, A., Chu, C. *et al.* (2017) Assessment of five chilling tolerance traits and GWAS mapping in rice using the USDA mini-core collection. *Frontiers in Plant Science* 8, 957. DOI: 10.3389/fpls.2017.00957.

Shahwar, D., Khan, Z. and Park, Y. (2024) Molecular markers for marker-assisted breeding for biotic and abiotic stress in melon (*Cucumis melo* L.): A review. *International Journal of Molecular Sciences* 25(12), 6307. DOI: 10.3390/ijms25126307.

Shamsuzzaman, M., Bhattacharjya, D.K., Islam, M.S. and Hoque, M.E. (2021) Molecular diversity analysis of somaclonal variants of potato (*Solanum tuberosum* L.) by random amplified polymorphic DNA markers. *Annual Research & Review in Biology* 36, 63–76. DOI: 10.9734/arrb/2021/v36i330353.

Singh, A., Ganapathysubramanian, B., Singh, A.K. and Sarkar, S. (2016) Machine learning for high-throughput stress phenotyping in plants. *Trends in Plant Science* 21(2), 110–124. DOI: 10.1016/j.tplants.2015.10.015.

Singh, R.K. and Prasad, M. (2017) Genome-wide association studies for improving agronomic traits in foxtail millet. In: Prasad, M. (ed.) *The Foxtail Millet Genome*. Springer International Publishing, Cham, Switzerland, pp. 63–75. DOI: 10.1007/978-3-319-65617-5_6.

Singh, R.K., Shweta, S., Muthamilarasan, M., Rani, R. and Prasad, M. (2019) Study on aquaporins of *Setaria italica* suggests the involvement of *SiPIP3;1* and *SiSIP1;1* in abiotic stress response. *Functional & Integrative Genomics* 19(4), 587–596. DOI: 10.1007/s10142-018-00653-0.

Singh, R.K., Prasad, A., Muthamilarasan, M., Parida, S.K. and Prasad, M. (2020) Breeding and biotechnological interventions for trait improvement: status and prospects. *Planta* 252(4), 54. DOI: 10.1007/s00425-020-03465-4.

Spindel, J., Begum, H., Akdemir, D., Virk, P., Collard, B. *et al.* (2015) Genomic selection and association mapping in rice (*Oryza sativa*): Effect of trait genetic architecture, training population composition, marker number and statistical model on accuracy of rice genomic selection in elite, tropical rice breeding lines. *PLOS Genetics* 11(2), e1004982. DOI: 10.1371/journal.pgen.1004982.

Srivastava, R.K., Singh, R.B., Pujarula, V.L., Bollam, S., Pusuluri, M. *et al.* (2020) Genome-wide association studies and genomic selection in pearl millet: Advances and prospects. *Frontiers in Genetics* 10, 1389. DOI: 10.3389/fgene.2019.01389.

Stetter, M.G., Zeitler, L., Steinhaus, A., Kroener, K., Biljecki, M. *et al.* (2016) Crossing methods and cultivation conditions for rapid production of segregating populations in three grain amaranth species. *Frontiers in Plant Science* 7, 816. DOI: 10.3389/fpls.2016.00816.

Stewart-Brown, B.B., Song, Q., Vaughn, J.N. and Li, Z. (2019) Genomic selection for yield and seed composition traits within an applied soybean breeding program. *G3: Genes, Genomes, Genetics* 9(7), 2253–2265. DOI: 10.1534/g3.118.200917.

Sun, C., Dong, Z., Zhao, L., Ren, Y., Zhang, N. *et al.* (2020) The wheat 660K SNP array demonstrates great potential for marker-assisted selection in polyploid wheat. *Plant Biotechnology Journal* 18(6), 1354–1360. DOI: 10.1111/pbi.13361.

Sun, J., Khan, M., Amir, R. and Gul, A. (2020) Genomic selection in wheat breeding. In: Ozturk, M. and Gul, A. (eds) *Climate Change and Food Security with Emphasis on Wheat*. Elsevier, London, pp. 321–330. Available at: www.sciencedirect.com/science/article/pii/B9780128195277000236

Tajibayev, D., Yusov, V.S., Chudinov, V.A., Mal'chikov, P.N., Rozova, M.A. *et al.* (2021) Genotype by environment interactions for spring durum wheat in Kazakhstan and Russia. *Ecological Genetics and Genomics* 21, 100099. DOI: 10.1016/j.egg.2021.100099.

Takahagi, K., Uehara-Yamaguchi, Y., Yoshida, T., Sakurai, T., Shinozaki, K. *et al.* (2016) Analysis of single nucleotide polymorphisms based on RNA sequencing data of diverse bio-geographical accessions in barley. *Scientific Reports* 6(1), 33199. DOI: 10.1038/srep33199.

Tang, B., Li, Y., Mu, Z., Chen, L., Guo, H. *et al.* (2020) Fine mapping and candidate gene analysis of qKW7b, a major QTL for kernel width in maize. *Molecular Breeding* 40(7), 67. DOI: 10.1007/s11032-020-01146-4.

Tanksley, S.D. and Nelson, J.C. (1996) Advanced backcross QTL analysis: A method for the simultaneous discovery and transfer of valuable QTLs from unadapted germplasm into elite breeding lines. *Theoretical and Applied Genetics* 92(2), 191–203. DOI: 10.1007/BF00223376.

Tiwari, J.K., Buckseth, T., Singh, R.K., Kumar, M. and Kant, S. (2020) Prospects of improving nitrogen use efficiency in potato: Lessons from transgenics to genome editing strategies in plants. *Frontiers in Plant Science* 11, 597481. DOI: 10.3389/fpls.2020.597481.

Tiwari, J.K., Buckseth, T., Zinta, R., Bhatia, N., Dalamu, D. *et al.* (2022) Germplasm, breeding, and genomics in potato improvement of biotic and abiotic stresses tolerance. *Frontiers in Plant Science* 13, 805671. DOI: 10.3389/fpls.2022.805671.

Tiwari, S., Yadav, S.K., Sahu, V.K. and Tripathi, M.K. (2018) Current status and future prospects of marker assisted breeding for genetic improvement of minor millets. *International Journal of Current Microbiology and Applied Sciences* 7(12), 2587–2590. DOI: 10.20546/ijcmas.2018.712.293.

Tiwari, S., Gupta, N. and Tripathi, M.K. (2019) Genomics assisted molecular breeding for crop improvement: Status and prospects. *Advance and Innovative Research* 6(1), 27.

Tong, H. and Nikoloski, Z. (2021) Machine learning approaches for crop improvement: Leveraging phenotypic and genotypic big data. *Journal of Plant Physiology* 257, 153354. DOI: 10.1016/j.jplph.2020.153354.

Valliyodan, B., Ye, H., Song, L., Murphy, M., Shannon, J.G. *et al.* (2017) Genetic diversity and genomic strategies for improving drought and waterlogging tolerance in soybeans. *Journal of Experimental Botany* 68(8), 1835–1849. DOI: 10.1093/jxb/erw433.

Van Bueren, E.L., Jones, S.S., Tamm, L., Murphy, K.M., Myers, J.R. *et al.* (2011) The need to breed crop varieties suitable for organic farming, using wheat, tomato and broccoli as examples: A review. *NJAS-Wageningen Journal of Life Sciences* 58(3–4), 193–205. DOI: 10.1016/j.njas.2010.04.001.

Van, K. and McHale, L.K. (2017) Meta-analyses of QTLs associated with proteinand oil contents and compositions in soybean [*Glycine max* (L.) Merr.] seed. *International Journal of Molecular Sciences* 18(6), 1180. DOI: 10.3390/ijms18061180.

Varshney, R.K., Graner, A. and Sorrells, M.E. (2005) Genomics-assisted breeding for crop improvement. *Trends in Plant Science* 10(12), 621–630. DOI: 10.1016/j.tplants.2005.10.004.

Varshney, R.K., Nayak, S.N., May, G.D. and Jackson, S.A. (2009) Next-generation sequencing technologies and their implications for crop genetics and breeding. *Trends in Biotechnology* 27(9), 522–530. DOI: 10.1016/j.tibtech.2009.05.006.

Varshney, R.K., Ribaut, J.-M., Buckler, E.S., Tuberosa, R., Rafalski, J.A. *et al.* (2012) Can genomics boost productivity of orphan crops? *Nature Biotechnology* 30(12), 1172–1176. DOI: 10.1038/nbt.2440.

Varshney, R.K., Terauchi, R. and McCouch, S.R. (2014) Harvesting the promising fruits of genomics: Applying genome sequencing technologies to crop breeding. *PLOS Biology* 12(6), e1001883. DOI: 10.1371/journal.pbio.1001883.

Varshney, R.K., Sinha, P., Singh, V.K., Kumar, A., Zhang, Q. *et al.* (2020) 5Gs for crop genetic improvement. *Current Opinion in Plant Biology* 56, 190–196. DOI: 10.1016/j.pbi.2019.12.004.

Varshney, R.K., Bohra, A., Yu, J., Graner, A., Zhang, Q. *et al.* (2021) Designing future crops: Genomics-assisted breeding comes of age. *Trends in Plant Science* 26(6), 631–649. DOI: 10.1016/j.tplants.2021.03.010.

Verges, V.L. and Sanford, D.A. (2020) Genomic selection at preliminary yield trial stage: Training population design to predict untested lines. *Agronomy* 10(1), 60. DOI: 10.3390/agronomy10010060.

Vogel, B. (2014) *Smart Breeding: The Next Generation. Marker Assisted Selection: A Biotechnology for Plant Breeding Without Genetic Engineering*. Greenpeace International.

Voss-Fels, K.P., Stahl, A. and Hickey, L.T. (2019) Q&A: modern crop breeding for future food security. *BMC Biology* 17(1), 18. DOI: 10.1186/s12915-019-0638-4.

Wada, T., Oku, K., Nagano, S., Isobe, S., Suzuki, H. *et al.* (2017) Development and characterization of a strawberry MAGIC population derived from crosses with six strawberry cultivars. *Breeding Science* 67(4), 370–381. DOI: 10.1270/jsbbs.17009.

Waltz, E. (2016a) CRISPR-edited crops free to enter market, skip regulation. *Nature Biotechnology* 34(6), 582. DOI: 10.1038/nbt0616-582.

Waltz, E. (2016b) Gene-edited CRISPR mushroom escapes US regulation. *Nature* 532(7599), 293. DOI: 10.1038/nature.2016.19754.

Wang, J., Yang, J., Jia, Q., Zhu, J., Shang, Y. *et al.* (2014) A new QTL for plant height in barley (*Hordeum vulgare* L.) showing no negative effects on grain yield. *PLOS ONE* 9(2), e90144. DOI: 10.1371/journal.pone.0090144.

Wang, Y., Yu, C., Cheng, Y., Yao, F., Long, L. *et al.* (2021) Genome-wide association mapping reveals potential novel loci controlling stripe rust resistance in a Chinese wheat landrace diversity panel from the southern autumn-sown spring wheat zone. *BMC Genomics* 22(1), 34. DOI: 10.1186/s12864-020-07331-1.

Washburn, J.D., Mejia-Guerra, M.K., Ramstein, G., Kremling, K.A., Valluru, R. et al. (2019) Evolutionarily informed deep learning methods for predicting relative transcript abundance from DNA sequence. *Proceedings of the National Academy of Sciences USA* 116, 5542–5549. DOI: 10.1073/pnas.1814551116.

Watson, A., Ghosh, S., Williams, M.J., Cuddy, W.S., Simmonds, J. *et al.* (2018) Speed breeding is a powerful tool to accelerate crop research and breeding. *Nature Plants* 4(1), 23–29. DOI: 10.1038/s41477-017-0083-8.

Wright, W.D., Shah, S.S. and Heyer, W.-D. (2018) Homologous recombination and the repair of DNA double-strand breaks. *Journal of Biological Chemistry* 293(27), 10524–10535. DOI: 10.1074/jbc.TM118.000372.

Xu, Y. and Crouch, J.H. (2008) Marker-assisted selection in plant breeding: From publications to practice. *Crop Science* 48(2), 391–407. DOI: 10.2135/cropsci2007.04.0191.

Xu, J., Yuan, Y., Xu, Y., Zhang, G., Guo, X. *et al.* (2014) Identification of candidate genes for drought tolerance by whole-genome resequencing in maize. *BMC Plant Biology* 14(1), 83. DOI: 10.1186/1471-2229-14-83.

Xu, Y., Li, P., Zou, C., Lu, Y., Xie, C. *et al.* (2017) Enhancing genetic gain in the era of molecular breeding. *Journal of Experimental Botany* 68(11), 2641–2666. DOI: 10.1093/jxb/erx135.

Yang, D., Tang, J., Yang, D., Chen, Y., Ali, J. *et al.* (2019) Improving rice blast resistance of Feng39S through molecular marker-assisted backcrossing. *Rice* 12(1), 70. DOI: 10.1186/s12284-019-0329-3.

Yang, H., Jian, J., Li, X., Renshaw, D., Clements, J. *et al.* (2015) Application of whole genome re-sequencing data in the development of diagnostic DNA markers tightly linked to a disease-resistance locus for marker-assisted selection in lupin (*Lupinus angustifolius*). *BMC Genomics* 16(1), 660. doi: 10.1186/s12864-015-1878-5.

Yin, L., Zhang, H., Zhou, X., Yuan, X., Zhao, S. *et al.* (2020) KAML: Improving genomic prediction accuracy of complex traits using machine learning determined parameters. *Genome Biology* 21(1), 146. DOI: 10.1186/s13059-020-02052-w.

Zhang, D., Wu, S., An, X., Xie, K., Dong, Z. *et al.* (2018) Construction of a multicontrol sterility system for a maize male-sterile line and hybrid seed production based on the *ZmMs7* gene encoding a PHD-finger transcription factor. *Plant Biotechnology Journal* 16(2), 459–471. DOI: 10.1111/pbi.12786.

Zhong, S., Dekkers, J.C., Fernando, R.L. and Jannink, J.-L. (2009) Factors affecting accuracy from genomic selection in populations derived from multiple inbred lines: A barley case study. *Genetics* 182(1), 355–364. DOI: 10.1534/genetics.108.098277.

Zhou, G., Zhang, Q., Tan, C., Zhang, X. and Li, C. (2015) Development of genome-wide indel markers and their integration with SSR, dart and SNP markers in single barley map. *BMC Genomics* 16(1), 804. DOI: 10.1186/s12864-015-2027-x.

Zhou, Z., Jiang, Y., Wang, Z., Gou, Z., Lyu, J. *et al.* (2015) Resequencing 302 wild and cultivated accessions identifies genes related to domestication and improvement in soybean. *Nature Biotechnology* 33(4), 408–414. DOI: 10.1038/nbt.3096.

Ziervogel, G. and Ericksen, P.J. (2010) Adapting to climate change to sustain food security. *WIREs Climate Change* 1(4), 525–540. DOI: 10.1002/wcc.56.

# 8 Marker-assisted Recurrent Selection in Plant Breeding

**Parisa Bolouri[1,2]* and Sevval Dikmen[1]**

*[1]Department of Genetics and Bioengineering, Yeditepe University, Istanbul, Turkey;*
*[2]Department of Life Science, Khazar University, Baku, Azerbaijan*

**Abstract**

Marker-assisted recurrent selection (MARS) is a breeding strategy that combines traditional recurrent selection with the use of molecular markers. This approach allows breeders to select plants with desirable traits more efficiently and accurately, leading to faster and more targeted genetic improvement. There are a lot of molecular markers available that can be used in MARS. Techniques such as amplified fragment length polymorphism (AFLP) and random amplified polymorphic DNA (RAPD) identify genetic variations by examining random regions of DNA. The versatility of genetic markers extends to various plant species and specific breeding goals, making them indispensable for precision in crop improvement. These markers can be used to identify and select for specific genes or DNA sequences that are associated with desirable traits such as yield, disease resistance or drought tolerance. By using these markers, breeders can make more informed decisions about which plants to select for further breeding, saving time and resources. However, the use of molecular markers in recurrent selection also presents challenges, including the need for specialized equipment and expertise, as well as the potential for marker–trait associations to vary in different environments or populations. Overall, MARS offers a promising approach to improve plant breeding efficiency and effectiveness. In this chapter we will explore the principles, benefits and challenges of MARS in plant breeding.

**Keywords:** marker-assisted recurrent selection (MARS), plant breeding, molecular markers

**Abbreviations**

AFLP – Amplified fragment length polymorphism, CAPS – Cleaved amplified polymorphic sequences, CRISPR – Clustered regularly interspaced short palindromic repeats, dPAC – derived polymorphic amplified cleaved sequence, GMOs – Genetically modified organisms, GS – Genomic selection, HSTs – High-throughput sequencing technologies, MAB – Marker-assisted breeding, MABB – Marker-assisted backcross breeding, MABC – Marker-assisted backcrossing, MARS – Marker-assisted recurrent selection, MAS – Marker-assisted selection, NGS – next-generation sequencing, PCR – polymerase chain reaction, QTLs – Quantitative trait loci, RAPD – random amplified polymorphic DNA, RFLP – Restriction fragment length polymorphism, SNPs – Single nucleotide polymorphisms, SSR – Simple sequence repeat

*Corresponding author: bolouri88@gmail.com

 *Plant Breeding Technology: Future Trends and Challenges* (ed. Zeba Khan)
DOI: 10.1079/9781800626638.0008

## 8.1 Introduction

Plant breeding is a method of creating new plant species with desired markers using existing ones. This is accomplished through a variety of breeding methods, such as plant insertion, selection and hybridization, aimed at improving economically desirable traits. Selection, in particular, is a very important breeding strategy, which focuses on improving the economic characteristics of multiple plant varieties. Depending on the source, it can be divided into two main categories: natural and artificial (Wang *et al.*, 2010).

The natural selection process involves the selection of superior plants without human intervention, under the influence of natural factors such as temperature, rainfall, and biotic and abiotic factors. This method ensures the preservation of genetic diversity by allowing all plants in the population to reproduce (Goldman, 2013).

In contrast, artificial selection involves human intervention in selecting superior plants within a population. Then, cultivation methods are applied to these selected plants depending on their reproduction style. For autogamous or self-pollinated yields, refined line and mass selection are used to create homozygous pure line varieties. In cross-pollinated crops, recurrent selection is used for population improvement, including simple recurrent selection, recurrent selection for general combining ability, reciprocal recurrent selection, and recurrent selection for specific combining ability. Clonal selection is used to isolate the best clones from new generations in vegetatively propagated plants. As a result, currently most are plants grown by artificial selection (Charmet *et al.*, 1999; Gazal *et al.*, 2015; Şahin and Kavuncu, 2019).

The phenotypic iterative selection process involves repeatedly selecting plants and combining them to form a population for the subsequent selection process. This process involves cycles of selection, evaluation and recombination, to increase the frequency of suitable alleles in the population. Developing polygenic markers by increasing the frequency of desirable genes for different economic traits is an effective strategy. This method also helps maintain high genetic diversity through repeated crossbreeding among selected plants of the heterozygous population (Orf, 2008; Singh *et al.*, 2021a; Sinha *et al.*, 2023).

While the effectiveness and efficiency of selection can be affected by environmental factors, the use of molecular marker technology can accelerate this process. Now, let us take a closer look at the importance of integrating this technology into the selection process (Goldman, 2013).

Molecular marker technology can make selection more effective, by reducing the impact of environmental factors and speeding up the process. Methods such as marker-assisted recurrent selection (MARS), in particular, accelerate the breeding process and increase the precision and efficiency of trait selection by integrating molecular markers into recurrent selection (Goldman, 2013; Wang *et al.*, 2018).

The integration of molecular marker technology into recurrent selection has significantly increased the accuracy and efficiency of identification of desired genotypes, thereby reducing the time and resources required for traditional phenotypic selection. Moreover, MARS enables the simultaneous selection of multiple quantitative trait loci (QTLs) (Ali *et al.*, 2020; Kushanov *et al.*, 2021; Singh *et al.*, 2023) and has enabled genetic analyses and molecular marker-assisted breeding (MAB) for qualitative and quantitative traits (Shahwar *et al.*, 2023)

Compared to marker-assisted backcrossing (MABC), MARS offers a more comprehensive approach by separately addressing the limitations of marker-assisted breeding and recurrent selection. Furthermore, the practical utility of MARS extends to its compatibility with other molecular breeding strategies, such as genomic selection (GS), providing a holistic and integrated approach to plant breeding (Jiang, 2013; Ashkani *et al.*, 2015).

Overall, marker-assisted iterative selection stands as an important strategy in modern molecular breeding programmes and offers a promising route for the development of improved plant strains with desired markers (Suvarna and Yashaswini, 2023).

## 8.2 Molecular Breeding Techniques: Advancements and Applications in Plant Improvement

The term 'molecular breeding' encompasses a variety of contemporary plant breeding methods.

These include marker-assisted backcross breeding (MABB), marker-assisted selection (MAS) and GS (Bernardo, 2010). Methods such as MARS enable MAS-linked markers to be made more efficient and cost effective than traditional methods based on DNA banding design (Collard, 2008). MAS is particularly useful for monogenic traits, but MARS and GS are more suitable for polygenic markers (Bernardo, 2010). Marker-assisted backcross breeding is a simple form of MAS in which a disease-resistant gene is introduced into an elite plant line but may not be as accurate as more complex approaches such as MARS and GS (Brauner *et al.*, 2019). Genomic selection operates markers across the entire genome and is a type of MAS that ensures that all QTLs are in linkage disequilibrium with at least one trait. This approach utilizes high density markers and high-throughput genotyping to improve quantitative traits in large plant breeding populations and is the driving force behind the advancement of molecular breeding strategies (Eathington *et al.*, 2019). In addition to these techniques, molecular breeding also harnesses the power of high-throughput sequencing technologies, such as next-generation sequencing (NGS), to analyse the entire genome of plants and identify specific genes and variations associated with desired traits (Jiang, 2013). This comprehensive approach allows breeders to understand the genetic structure of plants in unprecedented detail, thus facilitating the growth of improved strains with improved traits, such as higher yields, more beneficial nutrient content and increased resistance to both biotic and abiotic stresses (Goddard and Hayes, 2007). Combining cutting-edge molecular tools with traditional breeding methodologies continues to redefine the plant development landscape, and offers a promising path to address global food security challenges and promote sustainable agricultural practices (Ahmar *et al.*, 2020; Ceccarelli, 2020).

Additionally, combining molecular breeding methodologies with advanced computational techniques and comprehensive data analytics has further accelerated the pace of variety development. Leveraging the capabilities of bioinformatics and machine learning, breeders are now equipped to predict the consequences of complex traits and fine tune their breeding approaches with unparalleled accuracy. This harmonious integration of biological insights and data-driven methodologies is leading to specialty crop development initiatives by facilitating the breeding of resilient and high-performance plant varieties that can adapt to different environmental conditions (Ahmar *et al.*, 2020).

## 8.3 MARS: Enhancing Breeding Efficiency and Trait Expression

Marker-assisted recurrent selection is a recurrent selection method that uses molecular features to identify and select various genomic regions that influence the expression of complex traits (Fig. 8.1). The aim of this method is to establish the highest performing genotype within a single population or among related populations (Mondini *et al.*, 2009; Abdulmalik *et al.*, 2017; Chugh *et al.*, 2022). The approach allows crossbreeding and genotypic selection between selected populations for a single selection cycle in the same yielding season (Kumar *et al.*, 2023). Therefore, MARS has the potential to increase the efficiency of recurrent selection and accelerate the process, especially helping to accelerate the integration of a large number of good genes or QTLs from various sources based on recurrent selection (Gazal *et al.*, 2015).

In MARS, de novo quantitative trait locus (QTL) mapping is performed for each breeding population, which helps integrate multiple QTLs with a small number of genetic consequences driving complex markers (Xu and Jonathan, 2008). Once the major QTLs of interest have been identified from the MARS breeding population, selected individuals are subsequently fertilized to create lines with optimal QTL reinforcement (Türkoğlu *et al.*, 2023a).

MARS provides a robust framework to accelerate the breeding process by effectively combining desired traits from various genetic sources, making it a powerful tool for creating improved lines with optimal genetic traits (Obermeier and Friedt, 2015). By integrating molecular markers and genotypic selection, MARS offers a powerful approach to increase the expression of complex markers and, ultimately, contribute to the development of high-performance genotypes (Bouchez *et al.*, 2002). According to Gokidi (Gokidi *et al.*, 2016), MARS is particularly valuable for enhancing breeding efficiency and integrating multiple QTLs from

a) INITIAL POPULATION: The starting plant population identified before commencing genetic improvement.

b) MARKER SELECTION: Careful selection of markers representing specific genetic traits.

c) SELECTION ITERATION: The initial selection followed by repeated recurrent selection steps
.

d) POPULATION DEVELOPMENT: Evolution of the population towards the desired genetic traits after each iteration.

e) FINAL POPULATION: The ultimate plant population obtained after completing the MARS process, possessing the desired genetic characteristics.

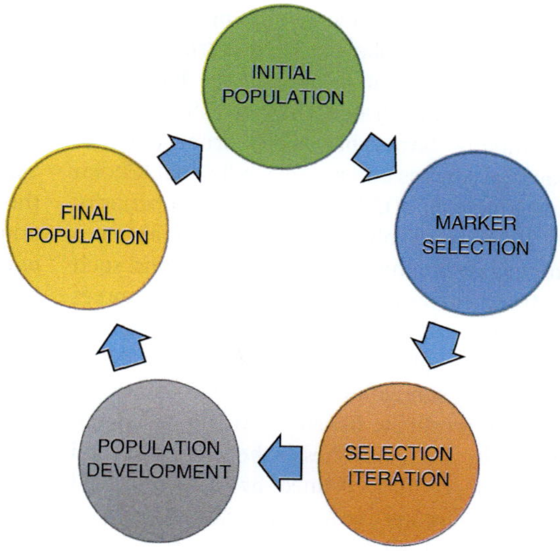


**Fig. 8.1.** Marker-assisted recurrent selection (MARS) strategy – basic steps. (Adapted from Gokidi *et al.*, 2016.)

various sources. With its ability to identify and integrate multiple QTLs, MARS stands as a useful tool for prospective breeding and creation of lines with the optimal complement of QTLs (Miedaner *et al.*, 2020; Singh *et al.*, 2021a).

## 8.4 MARS Procedure: Maximizing Trait Estimation Efficiency

MARS affects the estimation of trait effects by genotyping F3 or F2 populations and phenotyping F5 or F4 generations derived from F3 or F2, observed with two to three recombination cycles based on the fact that trait alleles for small effect QTLs are observed (Eathington *et al.*, 2007; Jiang, 2013).

In the initial phase of MARS, fingerprinting, with exact molecular markers of progeny from a typical breeding population resulting from a biparental cross, is performed. This means identifying QTLs in 200–300 or more than 300 generations from the F3 population using precise molecular markers to facilitate analysis of genotypic significance per generation. In the next phase, approximately 200–300 offspring from the F3-derived population (i.e. F3:4 or F3:5) are evaluated in multi-location trials to collect phenotyping data. Based on observed genotyping with phenotyping data, several plants are selected and allowed for managed pollination or hybridization over three or two cycles (Türkoğlu *et al.*, 2023a). By implementing a set of basic programmes and using genetic information obtained before flowering, multiple cycles (typically three to four) of molecular marker-assisted selection and controlled pollination can be completed in a single year (Türkoğlu *et al.*, 2023a). The resulting recombinant lines (either F3 or F3:4 generations) then undergo persistent self-pollination for 2–3 years, resulting from material produced by three or two recombination cycles or intermingling between selected plants. Following this process, the recombined lines are subjected to a final round of phenotypic screening at multiple locations. This screening aims to identify the most promising combinations, which are then selected for release as varieties (Nayak *et al.*, 2016; Lakhneko *et al.*, 2021) (Fig. 8.2).

## 8.5 A Comparative Analysis of MARS and Traditional Phenotypic Recurrent Selection Methods

The dominance of MARS over phenotypic recurrent selection schemes has been consistently supported by the previous findings of various researchers (Cholin and Kulkarni, 2023). Numerous computer simulations have consistently demonstrated that MARS excels in assembling favourable alleles from different parental sources into a single individual (Cholin and Kulkarni, 2023). However, the superiority of MARS over phenotypic selection becomes most evident in highly diverse populations, such as those encountered in breeding programmes involving subtropical or tropical materials, or those geographically isolated from each other. Additional simulation analyses have indicated that the proximate benefit of MARS on phenotypic selection diminishes as the fraction of total genotypic conflict, defined by QTLs included in the selection index, decreases (Kadirvel *et al.*, 2015). These insights underscore the significance of innovative approaches in plant breeding, as exemplified by recent studies. In a study conducted by Rahman *et al.* (2020), the application of MARS was investigated with a focus on enhancing plant traits, such as root rot resistance. By leveraging phenotyping and genotyping methods, Rahman *et al.* (2020) demonstrated the effectiveness of MARS in improving targeted traits compared to traditional selection methods (Swarup *et al.*, 2020).

In the light of these studies, it is better understood how important progressive approaches are in plant breeding. Similarly, the adoption of innovative technologies and approaches to challenge abiotic stresses encountered in a globally widespread agricultural crop such as maize is a significant development in the field of plant breeding. However, maize yield is greatly affected by various climate and soil factors and can face obstacles due to abiotic stresses, such as drought, salinity and extreme temperatures. To cope with these obstacles, the field of plant breeding has been transformed by the contribution of high-throughput sequencing technologies (HSTs). These advances have facilitated the identification of useful QTLs, associations between markers and traits, and genes and alleles that contribute to plant improvement (Fatiukha *et al.*, 2020). At present, well established omics techniques, such as genomics, transcriptomics, proteomics and metabolomics are being

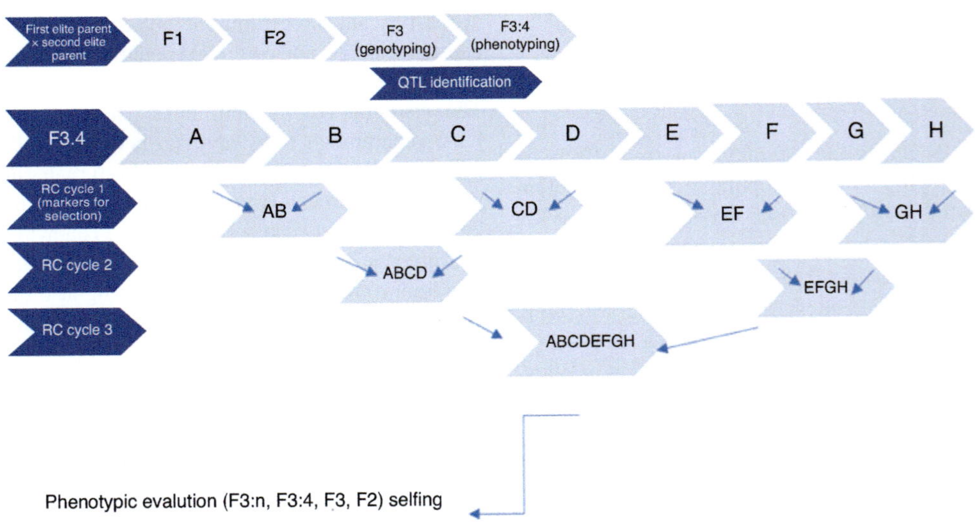


**Fig. 8.2.** Marker-assisted recurrent selection procedure. Schematic representation of procedures carried out in MARS, where the individuals A, B, C, D, E, F, G and H from F3:4 are used for intercrossing, represented by recombination cycles (RC) 1, 2 and 3; and appropriate genotypes are selected based on the segregating markers at each recombinant cycles. The genetic composition is then fixed with two, three cycles of selfing. (Adapted from Nayak *et al.*, 2016.)

integrated into maize breeding studies. These approaches have led to the emergence of new biomarkers that may help improve the ability of maize to cope with various challenges (Raj and Nadarajah, 2022).

## 8.6 A Comparative Analysis of MARS and GS in Plant Breeding Strategies

In phenotypic recurrent selection, plant genotypes are selected and constructed based on visible traits for one or more traits. In this process, selected individuals are subjected to self-pollination and simultaneously tested using a testing device. Then, the self-pollinated grains from these plants are collected one by one and stored for use in the third year (Degenkolbe *et al.*, 2013).

Test cross progenies are carefully evaluated in repeated yield trials and the best cross progeny is determined. Self-pollinated seeds of the plant producing superior test cross progeny are sown in cross blocks to create all possible crosses (Darrah *et al.*, 2019). This cycle can be repeated in subsequent years by harvesting equal amounts of seed from the crosses (Katara *et al.*, 2021).

This entire process creates a cycle that can be repeated two or three times to increase the frequency of good alleles in the population. However, despite the advantages in improving various agronomic traits, the efficiency of selection is not always satisfactory due to the influence of environmental factors on phenotypic selection and the long time required for genotypic selection. This is also related to factors such as the fact that at least two to three cropping seasons are required to complete a selection cycle (Begna, 2021).

On the other hand, in MARS, plant genotypes are selected based on precise molecular markers linked to genes or QTLs and crosses between selected individuals are also made during the exact crop season for one selection cycle. This can increase the efficiency of the selection procedure and accelerate the integration of good genes from various parents into a single genotype (Gazal *et al.*, 2015).

The dominance of MARS over phenotypic recurrent selection schemes has been consistently supported by various researchers' previous findings (Cholin and Kulkarni, 2023). Numerous computer simulations have consistently demonstrated that MARS excels in assembling favourable alleles from different parental sources into a single individual (Cholin and Kulkarni, 2023).

However, the superiority of MARS over phenotypic selection becomes most evident in highly diverse populations, such as those encountered in breeding programmes involving subtropical or tropical materials, or those geographically isolated from each other. Additional simulation analyses have indicated that the proximate benefit of MARS on phenotypic selection diminishes as the fraction of total genotypic conflict (defined by QTLs included in the selection index) decreases (Kadirvel *et al.*, 2015).

These insights underscore the significance of innovative approaches in plant breeding, as exemplified by recent studies. In a study conducted by Rahman *et al.* (2020) the application of MARS was investigated with a focus on enhancing plant traits, such as root rot resistance. By leveraging phenotyping and genotyping methods, Rahman *et al.* (2020) demonstrated the effectiveness of MARS in improving targeted traits compared to traditional selection methods (Rahman *et al.*, 2020; Swarup *et al.*, 2020)

Similarly, Cagliari *et al.* (2019) discussed the utilization of novel methods like GS and haplotype based breeding. Genomic selection allows for genetic improvement by leveraging within-family genetic variation, estimating genome-wide breeding values using SNP markers. On the other hand, haplotype based breeding identifies superior haplotypes for specific traits, offering promise in developing stress-resistant plant varieties as an alternative to conventional breeding approaches (Cagliari *et al.*, 2019; Rahman *et al.*, 2020). In a complementary study, Raj and Nadarajah (2022) explored genetic improvement through forward breeding, which involves the strategic combination of multiple favourable alleles. This approach enables the development of genotypes with a broad genetic background by transferring multiple QTLs (Raj and Nadarajah, 2022). Through the integration of these diverse breeding methods, novel strategies have emerged in plant breeding aimed at genetic enhancement and the development of stress-resistant varieties (Ramos Guimaraes *et al.*, 2021).

## 8.7 Diversity of Molecular Markers

The successful use of MARS in plant breeding depends on the utilization of various molecular features. These markers serve as useful tools for identifying and describing genetic diversity. Additionally, Kocaman *et al.* (2020) discussed the application of various molecular markers in biotechnology, emphasizing their versatility and significance in different contexts within the field (Kocaman *et al.*, 2020) (Table 8.1).

### 8.7.1 Amplified fragment length polymorphism (AFLP) in plant genetic diversity studies

Amplified Fragment Length Polymorphism (AFLP) serves as a powerful molecular marker analysis method widely used in assessing genetic diversity, as demonstrated in several studies (Babu *et al.*, 2021).

Studies in the field of plant genetic diversity generally aim to examine the agricultural morphological characteristics and genetic diversity of local populations of plant species. In this context, Zommita *et al.* (2023) conducted a study on local populations of wheat (*Triticum monococcum* ssp. *monococcum*) fields in Turkey, in which the AFLP marker method was used to evaluate genetic diversity as well as agricultural morphological characteristics. These studies highlight the importance of molecular tools for in-depth genetic analyses, which contributes to our understanding of genetic diversity in plant species. These studies, in which various molecular markers such as AFLP and inter simple sequence repeats (ISSR) were discussed, contributed to the expansion of the range of tools and scope of analyses used in genetic research (Kocaman *et al.*, 2020).

Akkale *et al.* (2010) examined the genetic diversity of potato (*Solanum tuberosum* L.) genotypes in Turkey. In this research, a total of 26 potato varieties were used, and genetic variations were determined using the AFLP marker method (Akkale *et al.*, 2010). AFLP analysis using DNA samples obtained from young and fresh leaves using the Doyle and Doyle (1990) method was carried out using the Licor AFLP kit (cat no: 829–06195) and digestion, pre-amplification and selective amplification steps were applied in accordance with the manufacturer's recommendations. The results obtained provided important information about the genetic diversity of potato genotypes in Turkey and provided an in depth understanding of the genetic diversity of this crop. This study made a valuable contribution to potato cultivation and constituted a fundamental resource for the conservation and management of potato diversity in Turkey (Akkale *et al.*, 2010).

### 8.7.2 Random amplified polymorphic DNA (RAPD)

Random amplified polymorphic DNA (RAPD) analysis is extensively utilized for genetic mapping, as well as taxonomic and phylogenetic studies of numerous organisms (Khan *et al.*, 2012). The technique involves polymerase chain reaction (PCR) amplification from random starting points to discern polymorphisms within specific DNA regions, providing a versatile tool for assessing genetic variations among distinct tomato genotypes (Cholin and Kulkarni, 2023). Despite the widespread use of RAPD, due to its simplicity and cost effectiveness, Babu *et al.* (2021) shed light on certain limitations associated with this technique, such as potential drawbacks in resolution and repeatability. However, the article emphasizes the effectiveness of RAPD in uncovering the genetic diversity within tomato varieties (Meng *et al.*, 2009). The study successfully employed RAPD markers to reveal nuanced genetic variations among the tested tomato varieties, offering valuable insights crucial for breeding and conservation programmes. It is noteworthy that, despite its utility, the authors recommend a cautious approach and suggest complementing RAPD with other molecular marker techniques to enhance results validation and reliability. This not only underscores the importance of genetic diversity analysis in tomato varieties but also positions RAPD as a valuable tool within this context, while acknowledging the need for a comprehensive approach for a more robust assessment (Babu *et al.*, 2021; El-Mansy *et al.*, 2021).

**Table 8.1.** Comparison of the different characteristics of the most frequently used molecular markers techniques (Gokidi *et al.*, 2016).

| Molecular markers | RAPD | AFLP | SSR | CAPS | SCAR | IRAP/ REMAP | RAMP | SSCP | RFLP | SNP |
|---|---|---|---|---|---|---|---|---|---|---|
| Genetic sequence necessity | No | No | No | Yes | Yes | Yes | No | Yes | Yes | Yes |
| Mechanization | Yes | Yes | Yes | Yes | Yes | Yes | Yes | No | No | Yes |
| DNA amount needed | Low | Medium | Low | Low | Low | Low | Low | Low | High | Low |
| Test cost per trial | Low | Medium | Low/medium | Medium | Low | Low | Medium | High | High | Low |
| Equipment requirement | Low | Medium | Low/medium | High | Medium | High | High | High | High | Medium |
| Replicating convenience | Low | High | Medium | High | High | High | Medium | Medium | High | High |
| Dominancy (D)/co-dominancy (C) | D | D | C | C | C | D | D | C | C | C |
| Frequency | High | High | Medium | Low | Low | High | Medium | Low | High | High |

RAPD – Random amplified polymorphic DNA, AFLP – Amplified fragment length polymorphism, SSR – Simple sequence repeat, CAPS – Cleaved amplified polymorphic sequences, SCAR – Sequence characterized amplified region, IRAP – inter-retrotransposon amplified polymorphism, REMAP – retrotransposon-microsatellite amplified polymorphism, RAMP – Random amplified microsatellite polymorphism, SSCP – Single-strand conformation polymorphism, RFLP – Restriction fragment length polymorphism.

### 8.7.3 The role of restriction fragment length polymorphism (RFLP)

Restriction fragment length polymorphism (RFLP) markers are one of the treatment genetic restrictions that have an important role in detecting differences in DNA sequences. These markers are examined by analysing specific restrictions on enzyme usage and fragment lengths formed by cleavage of different regions in DNA. The resulting DNA fragments of different sizes are then analysed using techniques such as gel electrophoresis (Livini *et al.*, 1992). These RFLP markers are considered an important research tool in studying genetic diversity in maize. The work of Livini *et al.* (1992) showed that RFLP-based genetic relatedness estimates can be used to successfully classify inbred lines belonging to heterotic groups in maize. This study revealed that RFLP markers are an effective tool in the process of correctly assigning various inbred lines of maize to heterotic groups. Similarly, Melchinger (1999) found that RFLP markers play an important role in determining genetic similarity at the DNA level in maize (Zehr *et al.*, 1992). RFLP markers are considered a valuable molecular tool used to assess genetic differences and similarities between maize populations. These studies demonstrate that RFLP markers play an important role in understanding genetic diversity in maize breeding (Livini *et al.*, 1992). However, the slow progress and high cost of this method have directed researchers to faster and more economical alternatives (Jarcho, 2001). Among these alternatives, simple sequence repeats (SSRs) and single nucleotide polymorphisms (SNPs), which are PCR based markers, occupy an important place (Livini *et al.*, 1992). These approaches enable faster results and reduced costs. Additionally, the development of sequencing technology in maize has enabled a more comprehensive examination of genetic diversity and facilitated the selection of improved maize varieties (Zommita *et al.*, 2023).

### 8.7.4 Simple sequence repeat (SSR) markers

Simple sequence repeat (SSR), a widely used method in genetic analysis, involves short DNA sequences called microsatellites. These repeated sequences are used to identify genetic differences between individuals and provide high resolution analysis of genetic diversity. Genotyping processes are typically PCR based and this method is widely used in genotyping, genetic mapping and population genetics (Chukwu *et al.*, 2020). In the broad field of agricultural research, the application of microsatellite markers, especially SSRs, has emerged as a crucial tool. Shedding light on the importance of these markers, Vieira *et al.* (2016) provides a comprehensive understanding of the use of microsatellites in genetic and molecular research in their work (Vieira *et al.*, 2016). In the context of agricultural research, this information is compatible with the application of SSR markers, a special category of microsatellites. The high resolution and versatility of SSR markers make them indispensable in genetic diversity analyses, aiding in the selection of desired traits and contributing to the development of crop breeding programmes (Jlassi *et al.*, 2021). In a study by Bolouri *et al.* (2021) on bread wheat (*Triticum aestivum*), 425 SSR markers were investigated for cold resistance properties and four effective QTLs were identified among them, one of which has a major effect and is located on chromosome A, and accounts for about 12% of the phenotypic changes (Bolouri *et al.*, 2021).

As highlighted by Vieira *et al.* (2016), microsatellite markers present a high degree of variability and play an important role in unravelling genetic diversity by revealing polymorphisms in DNA sequences. This information is valuable for understanding genetic structures, unravelling population genetics, and gaining insight into evolutionary processes. On the other hand, Wang *et al.* (2018) emphasized the practical application of SSR markers in assessing genetic diversity among elite varieties in malting barley seed (Wang *et al.*,). This study supports ongoing efforts in breeding programmes by providing valuable information on the genetic structure of malting barley varieties. Essentially, these two seminal studies highlight the important role of SSR markers in advancing agricultural research and lay a solid foundation for genetic analyses and crop-improvement strategies (Wang *et al.*, 2018).

### 8.7.5 Single-nucleotide polymorphisms (SNP)

Single nucleotide polymorphisms serve as powerful genetic markers with broad applications across diverse species. This genetic diversity at the single nucleotide level has been the focus of genetic research shedding light on the complexities of various organisms (Ahmadikhah *et al.*, 2009).

For example, the Information Commons for Rice (IC4R) database contains 18 million SNPs identified by resequencing 5152 rice accessions. However, due to the complexity of these raw SNPs, they are not suitable for general use. Therefore, these data were subjected to combined bioinformatics processing to facilitate the use of SNPs in different research fields, such as population genetics, evolutionary analysis, association studies and genomic breeding in rice. The resulting database from these efforts, called SnpReady for Rice (SR4R), provides access to comprehensive data on the genetic diversity of rice populations. SR4R provides four reference SNP panels: 2,097,405 HapMap SNPs, 156,502 tag SNPs, 1180 fixed SNPs and 38 barcode SNPs. In this way, rice researchers can perform various analyses and different population genetic studies (Xu *et al.*, 2021).

Alongside these advances, SNP markers have been developed from leaf samples collected from strawberry farmers in La Trinidad, Benguet, Philippines, and used to detect possible misidentifications among commercially grown strawberry varieties (Ledesma *et al.*, 2023). For this purpose, sequence tags expressed from the *anthocyanidin reductase* (*ANR*) gene were scanned and possible SNPs were identified. The 11 SNP markers developed were used as discriminators among the collected samples, and grouped the varieties into five genotypic clusters. These molecular markers may serve as potential tools that can be used in future variety validation efforts, germplasm management and breeding programmes (Vieira *et al.*, 2016).

SNP markers can be used as an important tool to identify and monitor genes affecting drought tolerance or disease resistance in a particular plant species. In particular, these markers are widely used to understand the genetic structure of plants and to identify specific genotypes in breeding programmes. These factors include the WRKY family of transcription factors. The aim of this study is to evaluate the polymorphism of the *TaWRKY2-D1* gene among various bread wheat varieties, interspecific hybrids and wild wheat relatives within the genus *Triticum* and to determine whether this gene may be a potential molecular target for marker-assisted selection. Initial sequencing was performed on six Ukrainian commercial varieties of the *TaWRKY2-D1* gene, and some sequence changes were detected in the, approximately, 1.8 kb promoter region of the gene and the coding region consisting of four exons and three introns. Based on the obtained sequence information, five sets of primers covering different gene regions were designed to determine *TaWRKY2-D1* genetic diversity in 202 wheat varieties, including accessions from 77 CIMMYT collections, 72 commercial varieties grown in Ukraine, and 53 hybrid and wild wheat species. The combination of DNA markers developed enabled effective and reproducible determination of the genetic diversity of the varieties. In particular, the primer set targeting introns in exon 3 of the gene emerged as the most informative for screening the heterogeneity of *TaWRKY2-D1*. The developed molecular markers serve as effective and informative tools for selecting drought tolerance genotypes (Singh *et al.*, 2021a).

As seen in the previous examples, the critical role of genetic diversity and molecular markers in agricultural practices is increasing. In particular, the widespread use of SNP markers to determine the agronomic characteristics of genotypes increases the effectiveness of agricultural breeding programmes. SNP markers are considered an important tool in genetic diversity analysis, heritability monitoring, association studies and GS. These provide valuable information to plant breeders in selecting and improving genotypes with desirable traits, such as disease resistance, drought tolerance and yield potential. Therefore, the extensive use of SNP markers and other molecular markers in agricultural practices will contribute to improving future agricultural production processes and increasing sustainability (Vieira *et al.*, 2016; Singh *et al.*, 2021; Xu *et al.*, 2021b).

### 8.7.6 Cleaved amplified polymorphic sequences (CAPS)

Cleaved amplified polymorphic sequences (CAPS) are a molecular marker analysis method utilized for analysing genetic diversity within populations. This technique involves the use of restriction enzymes to identify polymorphisms in specific DNA sequences. Subsequently, these polymorphisms are amplified through PCR amplification, making them more discernible for genetic analysis (Kaundun *et al.*, 2023).

The study conducted by El-Mansy *et al.* (2021) focused on genetic diversity research in tomato varieties under thermal stress. The study evaluated genetic diversity using a variety of markers, including CAPS, as well as morphological and cytological techniques. The use of CAPS markers, together with other methods, has provided a comprehensive understanding of genetic variability within tomato populations under thermal stress. This research highlights the flexibility of CAPS markers as well as demonstrating their value in elucidating genetic traits under harsh environmental conditions. In the study, using a total of 150 CAPS markers, significant genetic variability emerged within the tomato population under heat stress (El-Mansy *et al.*, 2021).

Additionally, Kaundun *et al.* (2023) introduce the derived polymorphic amplified cleaved sequence (dPACS) assay, an innovative method for genetic research (Kaundun *et al.*, 2023). Although not directly related to the tomato breeding study, this reference highlights the continuous innovation in molecular marker analysis methods, potentially inspiring future developments in the field. The work of Shavrukov (2016) compares the application of SNP and CAPS markers in genetic analysis in wheat and barley (Shavrukov, 2016). While not specific to tomatoes, this reference emphasizes the general applicability of CAPS markers across different plant species, providing insights into their versatility. Furthermore, Kaundun *et al.* (2023) present the dPACS as an unknown PCR-RFLP method for detecting known single nucleotide and deletion/insertion polymorphisms. Although not directly related to the tomato breeding study, this reference showcases advancements in CAPS related methodologies, potentially influencing future applications in similar studies (Kaundun *et al.*, 2023). In summary, CAPS markers, as exemplified in the referenced tomato breeding study, serve as powerful tools for elucidating genetic diversity and optimizing genetic improvement strategies in challenging environmental conditions. As indicated by recent studies, ongoing developments are contributing to the continued improvement of CAPS methodologies, increasing their effectiveness in genetic analyses of various plant species (Pozharskiy *et al.*, 2023).

## 8.8 MARS: Benefits

### 8.8.1 Advantages of MARS in plant breeding

Marker-assisted recurrent selection offers several benefits over standard breeding methods, showcasing its versatility and efficiency in genetic improvement (Anand *et al.*, 2022).

#### *8.8.1.1 Faster and more efficient selection*

MARS uses molecular markers, speeding up the breeding process by bypassing the time-consuming aspects of traditional phenotypic selection. This leads to quicker identification and advancement of desirable traits (Alwi, 2005).

#### *8.8.1.2 Enhanced precision*

By leveraging genetic information, MARS enables breeders to pinpoint specific traits with greater accuracy, resulting in more precise and targeted breeding efforts. This ultimately leads to the growth of superior crop types with desired features (Anand *et al.*, 2022).

#### *8.8.1.3 Greater accuracy and precision than MAS*

By combining both phenotypic and genotypic information, MARS provides a more accurate and precise selection compared to MAS. This dual approach enhances the strategy's capability to pinpoint plants with both desired traits and favorable marker alleles. MARS also allows for the identification of plants with potential hidden genetic potential, leading to more comprehensive and effective breeding programmes. This advanced level of accuracy and precision ultimately results in the growth of best crop types with improved traits and increased resilience to environmental stressors (Avise, 2012).

#### *8.8.1.4 Reduced linkage drag compared to MABC*

Unlike MABC, which may transfer large genomic regions along with the target gene, MARS uses recombination to minimize linkage drag. This results in a more focused and controlled integration of desired traits (Hospital, 2001). This approach allows for a more precise incorporation of the target gene into the breeding line, reducing the presence of avoided genetic material from the donor parent. By leveraging recombination, MARS enables breeders to select for and introduce specific traits without the burden of linked regions that may carry undesirable characteristics. This fine-tuned control over trait integration enhances the efficiency and precision of genetic improvement in plant breeding programmes (Babu *et al.*, 2004).

#### *8.8.1.5 More efficient than GS in certain cases*

In some cases, MARS is more efficient than GS. When a smaller set of markers is available, MARS can outperform GS, which requires extensive markers and training populations. MARS can achieve efficiency with a more modest set, making it a practical choice. It also serves as a valuable tool for addressing complex traits involving multiple genes. By incorporating both genetic and phenotypic data, MARS allows for a more comprehensive understanding of trait inheritance, enabling breeders to tackle polygenic traits more effectively. Furthermore, MARS offers a flexible framework that can adapt to different breeding goals and target traits, making it a versatile approach for various crop improvement programmes. This adaptability is particularly advantageous in the face of changing environmental conditions and evolving market demands, allowing breeders to tailor their selection processes to meet specific challenges and opportunities. As technology and our understanding of genetics continue to advance, MARS is poised to play an increasingly crucial role in the ongoing quest to enhance crop productivity, resilience and quality (Bernardo and Charcosset, 2006; Baliyan *et al.*, 2017).

## 8.9 MARS: Applications

The examples and impacts of the MARS technique in specific plant species highlight its crucial role in genetic improvement within the agricultural sector (McLennon *et al.*, 2021).

### 8.9.1 Species-specific utilizations and applications

MARS has successfully improved genetics in major crops such as wheat, rice, corn and tomatoes. The main goal is to identify and improve genetic traits in specific plant species. Research using MARS aims to increase plant productivity, strengthen disease resistance and improve adaptation abilities (Beyene *et al.*, 2016).

Studies on staple crops such as wheat, rice, maize and tomato clearly demonstrate the effectiveness of MARS in genetic improvement strategies. These projects play an important role in agricultural production by determining genetic diversity and selecting plants with desired characteristics. Application of the MARS method plays a critical role in identifying and improving plant traits as well as genetic improvement in agriculture. This approach has the potential to increase productivity, strengthen disease resistance and improve adaptation abilities by effectively using genetic resources in agricultural practices. In this context, evaluating MARS from a broad perspective in agricultural research and practices can contribute to the development of sustainable agricultural strategies and global food security (McLennon *et al.*, 2021).

Using genetic improvement techniques, particularly adopting methods such as MARS, researchers have targeted specific genetic traits to increase plant productivity. These efforts have helped plants better adapt to environmental stresses, making them more resilient to drought, high temperatures and diseases. In addition, by increasing the nutritional content, the aim is to obtain more nutritious products for human health. Application of MARS with such approaches can help to develop sustainable agricultural practices and solve food security problems. Continued analysis and growth in this field may further expand the impact of genetic-improvement techniques in plant species (Rahman *et al.*, 2020).

### 8.9.2 Agricultural sustainability and conservation

The impact of MARS on agricultural sustainability underscores the potential of genetic improvement to align with ecological balance. By aiming to bolster the natural resistance of plants to diseases, this approach contributes to reducing the reliance on pesticides in agricultural practices. For example, Türkoğlu *et al.* (2023a) investigated the genetic diversity and population structure in Turkish bread wheat genotypes using SSR markers. Their findings shed light on the genetic makeup of wheat varieties, which can inform breeding efforts aimed at enhancing disease resistance and reducing pesticide usage (Türkoğlu *et al.*, 2023a).

The decrease in pesticide usage represents a significant step towards mitigating environmental impacts (Bjørnstad *et al.*, 2002). By reducing chemical inputs, MARS helps to preserve soil, water and air quality; fosters ecosystem balance and promotes the conservation of biological diversity. Consequently, the positive effects of MARS on agricultural sustainability demonstrate its potential to mitigate environmental impacts, paving the way for a more sustainable future in agriculture (Bolouri *et al.*, 2021).

### 8.9.3 Disease resistance strategies in agriculture

MARS effectively showcases the potential of genetic improvement to bolster resistance against diseases in agriculture. This approach targets the creation of plant varieties endowed with heightened resilience to pathogens by fortifying their innate defence mechanisms (Bolouri *et al.*, 2021). For instance, Türkoğlu *et al.* (2023b) demonstrated the successful implementation of MARS in enhancing resistance against *Fusarium* wilt in pea plants, resulting in reduced susceptibility to this devastating disease (Türkoğlu *et al.*, 2023b). Through genetic modifications in plant structures, MARS can engender robust defences against disease agents, thereby enhancing both the health and productivity of cultivated plants.

The augmentation of resistance to pathogens holds promise for reducing reliance on chemical control methods in agricultural settings, thus mitigating associated environmental impacts. Additionally, the cultivation of plant varieties endowed with heightened disease resistance contributes to the promotion of sustainability and efficiency in agriculture (Zou *et al.*, 2024). By reducing crop losses due to diseases, MARS not only enhances agricultural productivity but also fosters a more sustainable and ecologically balanced farming system.

### 8.9.4 Enhancing productivity through agricultural practices

The impact of MARS on plant productivity underscores the key role of genetic improvement strategies in the agricultural sector. This strategy can yield various positive effects, tailored to specific plant types. For example, Bovill *et al.* (2010) demonstrated the potential of MARS to increase resistance to chronic root rot in wheat. This research provides a concrete example of how MARS can be used to enhance resistance to specific diseases (Bovill *et al.*, 2010). Similarly, Charmet *et al.* (1999) investigated how MARS can be used to combine cumulative additive and interactive QTLs in wheat (Charmet *et al.*, 1999).

Additionally, the capacity of MARS to enhance resistance to diseases can positively influence productivity by ensuring that plants are healthier and more resilient. For instance, a study shows that MARS can enhance resistance to leaf spot disease in cotton plants, potentially reducing harvest losses (Cholin and Kulkarni, 2023). Moreover, the effects of MARS on adaptability can increase the capacity of plants to adapt to various environmental conditions. For example, research indicates that MARS can increase tolerance to salt stress, allowing plants to thrive in saltier soils (Kaundun *et al.*, 2023). In conclusion, the effects of MARS on plant productivity are extensive, highlighting the importance of this strategy in agricultural production. However, further research is needed to fully understand these effects in different plant species and under various conditions.

### 8.9.5 Promoting biodiversity and enhancing adaptation in agricultural practices

Marker-assisted recurrent selection plays a crucial position in the future of agriculture by offering the potential to increase plant diversity and enhance their ability to adapt to various environments. The implementation of this strategy aims to increase genetic diversity and enable plants to adapt more effectively to changing environmental conditions (Charmet *et al.*, 1999).

One prominent advantage offered by MARS is the increase in diversity, enriching genetic differences among plants. Similar to natural selection processes, this allows plants to become more resilient to various stress factors. Increasing genetic diversity, especially to cope with global challenges like climate change, holds critical importance for agricultural sustainability (Cholin and Kulkarni, 2023).

Strengthening adaptation abilities is a key objective of MARS. Adapting to factors such as changing climate conditions, soil properties and disease pathogens, enables plants to grow more healthily and more efficiently. MARS accelerates this adaptation process, enhancing the capacity of agriculture to respond more effectively to new and unexpected challenges (Chukwu *et al.*, 2020).

## 8.10 MARS in Different Plant Species and Markers

### 8.10.1 Cereal crops

MARS has been extensively applied in grain crops such as grain, corn and rice. In grain breeding programmes, MARS has played a crucial role in enhancing traits related to yield, disease resistance, and adaptability to diverse environmental conditions (Feng *et al.*, 2022; Mallick *et al.*, 2022; Pozharskiy *et al.*, 2023). Similarly, in rice, MARS has been used to enhance markers like grain quality, drought resistance and pest tolerance (Gazal *et al.*, 2015). The adaptability of MARS to diverse cereal crops underscores its significance in addressing various challenges faced by these staple foods (Alexandrov *et al.*, 2015; Anand *et al.*, 2022).

### 8.10.2 Fruits and vegetables

The application of MARS extends to fruit and vegetable crops, where it has proven effective in improving key traits. For instance, in tomato breeding, MARS has been utilized to enhance characteristics such as flavour, shelf life and disease resistance. In fruit crops like apples or berries, MARS has facilitated the selection of plants with improved fruit quality, nutritional content and resistance to specific diseases. This broadens the scope of MARS, making it a valuable tool in the genetic enhancement of diverse horticultural crops (Gazal *et al.*, 2015).

### 8.10.3 Disease resistance

MARS has shown remarkable success in breeding for disease resistance across various plant species. In the context of different crops, including cereals and vegetables, MARS has been instrumental in identifying and selecting plants with heightened resistance to prevalent diseases. This includes resistance to fungal, bacterial and viral pathogens. The ability of MARS to expedite the growth of plant types with improved disease resistance is crucial for sustainable and resilient agriculture (Gazal *et al.*, 2015; Singh *et al.*, 2021a).

### 8.10.4 Multi-trait selection

MARS is not only successful in improving single features, but also shows superior performance in multiple feature selection. Breeding programmes often aim to improve multiple traits simultaneously to create balanced, high-performance plant varieties. MARS provides a comprehensive approach to genetic improvement by facilitating the identification and combination of desired traits. This is particularly valuable for addressing the complex challenges faced in modern agriculture (Kaundun *et al.*, 2023).

Kaundun *et al.* (2023) provided a detailed review of the dPACS assay. This method serves as an important tool in the investigation of genetic diversity and genomic variations. The dPACS test is designed to detect polymorphisms quickly and accurately in DNA sequences. The potential of this method can greatly benefit multiple-trait selection processes in plant breeding and genetic-improvement programmes (Kaundun *et al.*, 2023).

MARS has several important advantages over other commonly used techniques in multiple feature selection. First, MARS offers a flexible modelling approach and can handle many different types of variables simultaneously. In other words, MARS can better capture complex relationships in the dataset and, therefore, make more accurate predictions. Additionally, MARS can automatically detect interactions between variables, thus providing greater explainability of the model. On the other hand, other methods often have difficulty handling interactions between variables or may require adding them manually, which can increase the complexity of the model and increase the risk of overfitting. Finally, MARS can identify the effects of independent variables and thus remove unnecessary variables from the model, increasing the likelihood of obtaining a simpler and generalizable model. These features highlight the superiority of MARS over other techniques in multiple feature selection (Razzaq *et al.*, 2021).

## 8.11 Technological Advancements in Marker-Assisted Recurrent Selection

### 8.11.1 High-throughput genotyping

Recent technological breakthroughs have guided the growth of high-throughput genotyping platforms. These platforms allow the rapid and cost-effective analysis of a large number of molecular markers across the whole genome. This high-throughput genotyping capability accelerates the MARS process, allowing breeders to efficiently screen and select individuals with desired traits (Alexandrov *et al.*, 2015).

### 8.11.2 Next-generation sequencing (NGS)

Next-generation sequencing has revolutionized molecular marker finding and genotyping. The technology provides unparalleled accuracy and depth of sequencing, allowing for the identification of novel markers and a comprehensive understanding of genomic variations. The integration of NGS into MARS facilitates more precise trait mapping and the discovery of previously undetected genetic variations associated with desired traits (Singh *et al.*, 2020).

### 8.11.3 Bioinformatics and data analytics

Advancements in bioinformatics and data analytics play a crucial role in managing the extended amount of genomic data generated through MARS. Powerful computational tools and algorithms aid in the analysis, interpretation and visualization of complex genetic information. This contributes to more informed decision making during the selection process, ensuring that breeders capitalize on the full potential of molecular markers (Azodi *et al.*, 2020).

### 8.11.4 CRISPR-Cas9 gene editing

The source of CRISPR-Cas9 gene editing technology has provided plant breeders with unprecedented control over the plant genome (Khan and Shahwar, 2023). While not directly a part of MARS, CRISPR-Cas9 complements the MARS process by enabling targeted modifications of specific genes associated with desirable traits (Abuhammad, 2021). This precise genome editing tool allows for the development of custom-designed plants with improved markers identified through MARS (Bailey-Serres *et al.*, 2019).

### 8.11.5 SNP arrays and marker panels

Single nucleotide polymorphism arrays and marker panels have become integral components of MARS. These tools offer a focused and efficient approach to genotyping, allowing breeders to assess specific regions of interest. SNP arrays, in particular, are widely used for their reliability and scalability, contributing to the streamlined implementation of MARS in various crops (Bera *et al.*, 2019).

### 8.11.6 Automation and robotics

The automation of laboratory processes and robotic systems has expedited the genotyping workflow. Automated sample preparation, DNA extraction, and marker analysis enhance the efficiency of MARS, reducing the turnaround time for breeding programmes. This automation also minimizes human error and increases the reproducibility of results (Bayer *et al.*, 2019).

## 8.12 Challenges and Solutions in Molecular Markers

Marker-assisted recurrent selection stands as a potent tool in the realm of plant breeding, but, akin to any technological innovation, it grapples with certain challenges and limitations. Delving deeper into these issues is paramount to harnessing the full potential of MARS in genetic improvement initiatives. The key challenges and limitations associated with MARS are discussed in the next sections, along with some additional insights.

### 8.12.1 Costs, resources and financial implications

Implementing MARS presents a significant challenge due to the substantial financial investment and technical resources required. This limitation, particularly notable for breeders operating within limited budgets or resource-constrained environments, may hinder the widespread adoption of MARS. However, potential solutions exist to address this challenge. One approach involves exploring collaborative initiatives and partnerships to share costs and resources among different breeding programmes. For example, establishing regional breeding consortia where multiple programmes pool resources for genotyping can help alleviate the financial burden on individual programmes. Despite these collaborative efforts, some breeding programmes may still struggle to afford the advanced genotyping technologies needed for MARS implementation. Nonetheless, addressing these challenges through collaborative efforts can pave the way for broader adoption and utilization of MARS in agricultural breeding programmes (Bajaj *et al.*, 2016).

### 8.12.2 Strengthening technological infrastructure

A challenge for MARS is the need for access to advanced genotyping technologies and bioinformatics tools for its successful implementation. Thus, breeding programmes without access to modern genotyping facilities may struggle to fully utilize the advantages offered by MARS. To address this issue, it is important to facilitate technology transfer and capacity-building programmes aimed at broadening access to advanced genotyping technologies. Accessibility to these resources can be significantly enhanced by, for example, collaborating with research institutions and organizations to set up training sessions, and to provide support to breeders in regions with inadequate technological infrastructure. Additionally, breeders in developing regions often encounter difficulties in accessing state-of-the-art genotyping equipment; hence, establishing partnerships to address this issue is crucial for promoting the adoption of MARS (Kumar *et al.*, 2023).

### 8.12.3 Analysing complex molecular marker data

Analysing complex molecular marker data presents significant challenges in modern breeding programmes. Processing the extensive datasets generated by these markers often requires advanced computational skills and bioinformatics expertise. Breeders without a background in bioinformatics may struggle to interpret and utilize the vast amount of genomic data effectively. To address this limitation, it is essential to develop user-friendly bioinformatics tools and provide comprehensive training programmes aimed at enhancing breeders' analytical skills. By creating intuitive software interfaces and offering training sessions, breeders with varying levels of computational expertise can access and analyse molecular marker data more efficiently. However, understanding complex statistical analyses remains a potential barrier for breeders with limited computational skills. Therefore, comprehensive training programmes that cover both bioinformatics and statistical analysis are essential to ensure the successful utilization of molecular marker data in breeding programmes (Kumar *et al.*, 2023).

### 8.12.4 Environmental interaction in marker-assisted selection

The relationship between markers and traits can exhibit variability under different environmental conditions, presenting a significant challenge in MAS. Achieving consistent associations between markers and traits across diverse environments remains elusive, thereby affecting the reliability of MAS. To address this challenge, it is crucial to invest in research aimed at understanding and modelling gene–environment interactions, which can enhance the robustness of marker–trait associations. For instance, conducting multi-environment trials can help identify markers associated with traits under various agro-climatic conditions, thereby bolstering the reliability of MAS. However, it is important to note that a marker linked to drought resistance in one region may not exhibit the same performance in another region with different climate conditions (Das *et al.*, 2019; Cui *et al.*, 2020).

### 8.12.5 Linkage disequilibrium breakdown and its management

The challenge posed by the breakdown of linkage between markers and target genes over successive generations necessitates proactive strategies for effective management. As this breakdown occurs, the reliability of markers as indicators of desired alleles diminishes, leading to decreased selection accuracy. To address this limitation, it is crucial to implement strategies such as recurrent genotyping, which allows for the continuous monitoring and rejuvenation of marker–trait linkages across breeding cycles. For instance, regularly reassessing marker–trait associations and refreshing breeding populations can help sustain the efficacy of selected markers. Despite being once correlated with disease resistance, markers may lose their association over breeding cycles, highlighting the importance of ongoing vigilance and adaptation in MAS programmes (Jlassi *et al.*, 2021).

### 8.12.6 Limited allelic variation and its implications for MARS

The challenge of limited allelic variation poses significant hurdles for the effective implementation of MARS. Relying on genetic deviation within a population, MARS struggles to achieve its potential in crops with constrained genetic diversity. This limitation stems from the insufficient availability of allelic variation necessary for robust MAS. To address this challenge, it is imperative to explore avenues for introducing genetic diversity through controlled crossbreeding or gene editing techniques (Nayak *et al.*, 2016). For instance, the utilization of gene-editing technologies can introduce novel alleles into crops with limited genetic variation, thereby expanding the allelic pool suitable for MARS applications. However, it is important to note that crops with restricted genetic bases may inherently lack the diversity required for effective MAS, underscoring the need for innovative strategies to overcome this limitation (Kaundun *et al.*, 2023).

### 8.12.7 Ethical considerations and the role of public perception in MARS adoption

The use of molecular markers and advanced genetic technologies raises ethical concerns, potentially impacting the widespread adoption of MARS due to public perception and acceptance of genetically modified organisms or edited crops. Engaging in transparent communication and education campaigns to address public concerns about genetically modified organisms (GMOs) is essential in overcoming this challenge. For instance, conducting public outreach programmes to explain the safety measures and benefits of MARS-derived crops can foster public trust and acceptance. However, public resistance to genetically modified crops may hinder the acceptance of MARS-derived varieties. Therefore, public participation and understanding play a significant role in the widespread adoption of MARS technology (Elçi and Hançer, 2014).

#### 8.12.8 Lengthy breeding cycles in MARS

Marker-assisted recurrent selection often faces the challenge of requiring multiple breeding cycles to accumulate favourable alleles. This extended duration may impede the overall pace of genetic improvement compared to other quicker breeding methods. To address this limitation, optimizing breeding strategies to shorten cycles while maintaining allele accumulation efficiency is crucial. For example, implementing GS methods alongside MARS can expedite the selection of individuals with desired markers, thereby reducing the overall breeding cycle duration. However, it is important to acknowledge that developing crops with complex traits may necessitate several generations, further extending the time required for MARS implementation. Thus, balancing breeding efficiency with trait complexity is essential for maximizing the effectiveness of MARS (de C. Lara *et al.*, 2019).

#### 8.12.9 Addressing complex traits in MARS

Marker-assisted recurrent selection encounters challenges when dealing with traits controlled by numerous genes and environmental factors. This complexity limits the effectiveness of MARS in addressing traits with intricate genetic architectures. To overcome this limitation, investing in advanced genomic research is essential to unravel the genetic basis of complex traits and develop more precise markers. For instance, utilizing genome-wide association studies (GWAS) can identify additional markers associated with complex traits, thereby enhancing the efficacy of MARS. Traits such as yield, influenced by multiple genes and environmental conditions, present particular challenges for MARS due to their complexity. Therefore, advancing genomic research and marker development is vital for extending the applicability of MARS to traits with complex genetic architectures (Oh *et al.*, 2023).

#### 8.12.10 Navigating regulatory challenges in MARS

Strict regulations on GMOs present significant hurdles for the release and commercialization of crops developed through MARS (Kaundun *et al.*, 2023). These regulatory challenges may impede the real-world application of MARS derived varieties, limiting their accessibility to farmers and consumers. To address this issue, collaboration with regulatory bodies is crucial to establish clear guidelines and streamline the approval process for MARS derived crops (Elçi and Hançer, 2014). For example, engaging in proactive discussions with regulatory authorities and providing them with comprehensive data on safety and efficacy can facilitate efficient approvals. However, stringent approval processes for GMOs may still delay or restrict the market release of crops developed through MARS. Therefore, concerted efforts are needed to overcome regulatory hurdles and ensure the successful adoption of MARS derived varieties in agricultural systems (Elçi and Hançer, 2014).

### 8.13 Success Stories and Studies in MARS Application

#### 8.13.1 Increased disease resistance in wheat

There have been promising results yielded by MARS applications in enhancing disease resistance in wheat varieties. For instance, a targeted programme focused on wheat successfully bolstered its resilience against prevalent fungal infections. By employing MAS techniques, breeders pinpointed and integrated key resistance alleles into breeding programmes. This strategic approach resulted in a notable decrease in yield losses (Cagliari *et al.*, 2019). Such successes underscore the potential of MARS in fortifying crops against detrimental pathogens, thereby contributing to food security and agricultural sustainability.

### 8.13.2 Improved nutritional traits in maize

Successful outcomes in improving the nutritional content of maize have been achieved by MARS applications. For example, maize breeding using MARS focused on increasing the levels of essential nutrients. Through precise MAS methods, varieties with high levels of essential nutrients were developed, addressing nutritional deficiencies in specific regions (Crossa *et al.*, 2017; Calayugan *et al.*, 2020).

### 8.13.3 Faster maturation in rice varieties

Marker-assisted recurrent selection has yielded significant results in accelerating the maturation periods of rice varieties. By targeting specific genomic regions associated with maturation, breeders have made rapid progress in producing early-maturing rice cultivars (Frisch and Melchinger, 2005).

### 8.13.4 Drought tolerance in barley

To enhance drought tolerance in barley, MARS has played a pivotal role. Prioritizing the development of barley varieties resilient to drought conditions, breeders utilized MAS to identify genetic markers linked to drought tolerance. This strategic approach has led to the creation of barley cultivars capable of maintaining yield even under water-stressed environments (Nayak *et al.*, 2016).

### 8.13.5 Improved fruit quality in tomatoes

To enhance fruit quality traits in tomatoes, MARS has played a crucial role. Prioritizing improvements in flavour, texture and nutritional content, breeders utilized MAS to target specific genomic regions associated with these traits. This strategic approach has resulted in the successful development of tomato varieties with heightened consumer appeal (Rahman *et al.*, 2020).

## 8.14 Future Perspectives of MARS

The trajectory of MARS holds exciting possibilities for the end of plant breeding, with advancements and evolving technologies shaping its prospective role. The key considerations for the future perspectives of MARS are discussed in the next sections.

### 8.14.1 Precision breeding with CRISPR technology

To combat the escalating challenges posed by climate change and address the demand for stress-tolerant grain varieties, gene-editing technologies, particularly the CRISPR-Cas9 system, are playing a pivotal role in agricultural production. Modern cereal varieties have significantly bolstered global food security; however, they have predominantly been cultivated for optimal environmental conditions. Yet, the imperative to adapt to shifting climatic conditions underscores the necessity for developing new varieties resilient to diverse stress factors (Sanchez *et al.*, 2023).

The CRISPR-Cas9 system presents a promising avenue for achieving this objective by facilitating precise and targeted alterations in major grain species like wheat, corn, rice, oats, rye and sorghum. Despite encountering challenges, such as the prolonged duration required to produce transgene-free lines and regulatory constraints in specific regions, recent advancements in gene editing techniques have opened doors for the creation of high yielding, stress-tolerant and disease-resistant cereal varieties (Gouda *et al.*, 2021).

For instance, in one study, researchers embarked on the mission to cultivate varieties resilient to low-temperature stress in wheat plants. Wheat exhibits high susceptibility to low-temperature stress, which can result in yield reduction under such conditions. To address this challenge, scientists employed genetic modifications using the CRISPR-Cas9 system to enhance the resistance of the wheat plants to low-temperature stress. Through these genetic alterations, the wheat plants demonstrated increased resilience to low-temperature stress, leading to enhanced productivity and quality (Sami *et al.*, 2021).

This study underscores the significance and potential of the CRISPR-Cas9 system as a crucial tool in enhancing cereal crops, contributing to the development of resilient agricultural systems capable of withstanding the challenges imposed by changing environmental conditions (Sami *et al.*, 2021).

### 8.14.2 Expansion to Underutilized Crops

To broaden the scope of MARS applications, researchers have sought to expand their efforts from major crops to less used or regionally specific crops. This expansion aims to meet critical needs such as improving food security, adapting agriculture to diverse climates and promoting biodiversity. A notable example of these diversification efforts is a study focused on improving drought stress tolerance of quinoa, a crop that has a high nutritional value but is underutilized (Herzog and Frisch, 2013).

Quinoa is known for its ability to adapt to harsh growing conditions and is attracting attention as a promising alternative to traditional crops in regions at risk of drought. However, its sensitivity to long dry periods poses a significant obstacle to its widespread cultivation. In response to this problem, researchers have developed drought-tolerant quinoa varieties by combining MARS with traditional breeding methods (Türkoğlu *et al.*, 2023a)

Through integration of genomic data and phenotypic information, researchers identified genomic regions associated with drought tolerance in quinoa. They then selected and crossed varieties with the desired traits using targeted breeding approaches guided by MARS predictions. The resulting quinoa lines have increased drought tolerance without compromising yield or nutritional quality (Türkoğlu *et al.*, 2023a).

This study demonstrates that MARS can be applied to underutilized crops such as quinoa and can increase the diversity and resilience of agricultural systems on a global scale by addressing specific challenges such as drought stress (Türkoğlu *et al.*, 2023a).

### 8.14.3 Integration of Multi-Omics Approaches

The future of MARS involves integrating multi-omics approaches, including genomics, metabolomics and transcriptomics. This holistic understanding of plant genomes will supply deeper senses into the complex relations between genes and their products. The comprehensive data derived from multi-omics analyses will refine selection processes, making MARS even more effective in capturing the genetic basis of complex traits (Herzog and Frisch, 2013).

In a recent study published in *Nature Communications*, researchers aimed to understand the genetic basis of drought tolerance in rice by integrating genomics, metabolomics and transcriptomics. Using a combination of advanced sequencing techniques and bioinformatics tools, they identified key genomic regions associated with drought tolerance and monitored the metabolic and transcriptional changes underlying this trait. By integrating multi-omics data, researchers gained a holistic understanding of the molecular mechanisms involved in drought response in rice plants. This comprehensive information allowed them to develop MARS strategies to increase drought tolerance in rice varieties. The study highlights the power of integrating multi-omics approaches in unravelling complex traits and developing crop-breeding programmes for sustainable agriculture (van Etten *et al.*, 2023).

### 8.14.4 Data-Driven Decision Support Systems

The evolution of MARS includes the incorporation of advanced analytics and machine learning algorithms. Data-driven decision support systems will play a pivotal role in handling large-scale genomic datasets. These systems will aid breeders in making informed selections, predicting trait expressions and optimizing breeding strategies, thereby improving the efficiency of MARS (Nayak *et al.*, 2016).

As an example, a study titled, 'A data-driven crop model for maize yield prediction' (Chang *et al.*, 2023) stands out. This study introduces an advanced data-driven approach to estimate the drought resistance of maize crops. Using large variation phenotypic and genomic datasets, researchers conducted a summary analysis to learn about the responses of maize plants to drought stress. Machine backups and data analysis techniques we used to support the complexity of plant varieties and genetic variations and to determine genetic suitability to drought. The study emphasizes the importance of data-driven decision support results and emphasizes the role of multivariate adaptive regression splines, as discussed by Xu *et al.* (2016).

### 8.14.5 Climate-Resilient Crop Breeding

With the challenges posed by climate change in mind, the future of MARS emphasizes the development of crop varieties resilient to changing environmental conditions. Identifying and integrating markers associated with traits such as drought, salt, heat and pest resilience will be crucial (Hospital, 2009).

In this context, a study titled, 'Drought tolerance in rice: focus on recent mechanisms and approaches' (Panda *et al.*, 2021) is noteworthy. This study utilized a multi-locus approach to increase the drought tolerance of rice plants. Researchers have conducted a comprehensive analysis for the detection and integration of genetic markers to improve drought tolerance of rice plants. The study shows how MARS can play a role in adaptation to climate change and how it can be an effective tool in breeding crops to withstand environmental stressors (Zargar *et al.*, 2022).

### 8.14.6 International Collaboration and Data Sharing

To enhance the effectiveness of MARS, international collaboration and data sharing are essential. Open science initiatives will play a crucial role in facilitating the exchange of genomic information, marker data and successful breeding strategies among researchers worldwide. This collaborative approach is expected to accelerate progress in plant breeding and contribute to achieving a more sustainable and food-secure future (Jlassi *et al.*, 2021).

In this context, a study titled 'Meeting the challenges facing wheat production: the strategic research agenda of the Global Wheat Initiative' (Langridge *et al.*, 2022) is noteworthy. This study highlights the importance of global initiatives such as the Wheat Initiative, which provide a platform for wheat breeding among researchers worldwide. These initiatives facilitate the dissemination and sharing of knowledge of innovative breeding methods and successful strategies, such as MARS. International collaboration and data sharing play a critical role in contributing to accelerating progress in plant breeding (Zurita-Silva *et al.*, 2014).

### 8.14.7 Ethical and Societal Considerations

As MARS continues to advance, there is a growing recognition of the need for public engagement and awareness regarding the ethical implications of genetic technologies. In this context, transparent communication, ethical considerations and inclusive decision-making processes are paramount. Future perspectives emphasize the importance of addressing societal concerns related to genetically improved crops through open dialogue and collaborative efforts. By fostering public understanding and involvement, we can ensure that MARS development aligns with ethical principles and societal values (Kadirvel *et al.*, 2015).

A research team has conducted a comprehensive assessment of the ethical and social aspects of agricultural applications of genetic editing technologies. The study meticulously examined the potential benefits and worrisome aspects of genetically improved crops. Researchers have used methods such as public opinion surveys and focus group discussions to understand the effects of such technologies on society. An extensive literature review was also conducted to evaluate the agricultural and environmental impacts of genetic editing. The findings provided important clues for the ethical use and social acceptance of genetic editing technologies. The research highlighted that transparent communication and participatory decision-making processes can play a critical role in increasing the social and ethical compliance of these technologies (Zurita-Silva *et al.*, 2014).

## 8.15 Commercial Adoption and Regulatory Clarity

To ensure the future success of MARS, it is imperative to establish clearer regulatory frameworks and foster increased commercial adoption. This will require collaborative efforts between researchers and policymakers to navigate complex regulatory landscapes. The goal is to ensure that MARS-derived varieties not only meet rigorous safety standards but also gain acceptance in global markets (Zampieri *et al.*, 2023).

For example, researchers from the Chinese Academy of Agricultural Sciences conducted a study (Wang *et al.*, 2023) to examine the importance of MARS for commercial adoption and regulatory clarity. In this study, researchers aimed to identify strategies for commercial adoption of new rice varieties developed by MARS and overcoming regulatory hurdles. By reviewing existing regulations and collaborating with industry stakeholders, the researchers developed recommendations to support commercial adoption of MARS derived rice varieties. This work has been recognized as an important step to increase commercial use of MARS and promote regulatory clarity (Kumar Singh *et al.*, 2024).

## 8.16 Conclusion

This chapter discusses how plant breeding has evolved with innovative approaches, such as molecular marker technologies and MARS, and the important steps in developing plant varieties resistant to abiotic stresses. Research shows the superiority of molecular marker technologies such as MARS over traditional selection methods. While MARS accelerates selection by identifying plant genotypes with molecular markers, phenotypic recurrent selection is often affected by environmental factors and can take a long time.

Therefore, the chapter highlights the importance of adopting molecular markers and next-generation technologies in plant breeding and the role of integration of various breeding methods in plant genetic improvement and stress resistance development efforts. It is stated that in the future, further development of these innovative approaches may contribute to the promotion of sustainable agricultural practices and the development of stress-tolerant plant varieties in response to global food security challenges.

In summary, this chapter addresses the role of innovative tools such as MARS in plant breeding. The challenges and solutions of MARS shed an important light on the importance and future potential of molecular marker technologies, genetic improvement and the future direction of agriculture.

## References

Abdulmalik, R.O., Menkir, A., Meseka, S.K., Unachukwu, N., Ado, S.G. *et al.* (2017) Genetic gains in grain yield of a maize population improved through marker assisted recurrent selection under stress and non-stress conditions in West Africa. *Frontiers in Plant Science* 8, 841. DOI: 10.3389/fpls.2017.00841.

Abuhammad, S., Khabour, O.F. and Alzoubi, K.H. (2021) Researchers views about perceived harms and benefits of gene editing: A study from the MENA region. *Heliyon* 7(4), e06860. DOI: 10.1016/j.heliyon.2021.e06860.

Ahmadikhah, A., Mirarab, M., Pahlevani, M.H. and Nayyeripasand, L. (2009) Identification of molecular markers linked to WA-cytoplasmic male sterility in rice and development of specific SCAR markers. Plant Genome 8, 0031. DOI: 10.3835/plantgenome2014.07.0031.

Ahmar, S., Gill, R.A., Jung, K.H., Faheem, A., Qasim, M.U. *et al.* (2020) Conventional and molecular techniques from simple breeding to speed breeding in crop plants: Recent advances and future outlook. *International Journal of Molecular Sciences* 21(7), 2590. DOI: 10.3390/ijms21072590.

Akkale, C., Yildirim, Z., Yildirim, M., Kaya, C., Öztürk, G. *et al.* (2010) Assessing genetic diversity of some potato (*Solanum tuberosum* L.) genotypes grown in Turkey using the AFLP marker technique. *Turkish Journal of Field Crops* 15(1), 73–78.

Alexandrov, N., Tai, S., Wang, W., Mansueto, L., Palis, K. *et al.* (2015) SNP-seek database of SNPs derived from 3000 rice genomes. *Nucleic Acids Research* 43(D1), D1023–D1027. DOI: 10.1093/nar/gku1039.

Ali, M., Zhang, L., DeLacy, I., Arief, V., Dieters, M. *et al.* (2020) Modeling and simulation of recurrent phenotypic and genomic selections in plant breeding under the presence of epistasis. *The Crop Journal* 8(5), 866–877. DOI: 10.1016/j.cj.2020.04.002.

Alwi, Z.B. (2005) The use of SNPs in pharmacogenomics studies. *Malaysian Journal of Medical Sciences* 12(2), 4–12.

Anand, A., Subramanian, M. and Kar, D. (2022) Breeding techniques to dispense higher genetic gains. *Frontiers in Plant Science* 13, 1076094. DOI: 10.3389/fpls.2022.1076094.

Ashkani, S., Rafii, M.Y., Shabanimofrad, M., Miah, G., Sahebi, M. *et al.* (2015) Molecular breeding strategy and challenges towards improvement of blast disease resistance in rice crop. *Frontiers in Plant Science* 6, 886. DOI: 10.3389/fpls.2015.00886.

Avise, J.C. (2012) *Molecular Markers, Natural History and Evolution*. Springer Science & Business Media.

Azodi, C.B., Pardo, J., VanBuren, R., de Los Campos, G. and Shiu, S.H. (2020) Transcriptome-based prediction of complex traits in maize. *Plant Cell* 32(1), 139–151. DOI: 10.1105/tpc.19.00332.

Babu, K.N., Sheeja, T.E., Minoo, D., Rajesh, M.K., Samsudeen, K. *et al.* (2021) Random amplified polymorphic DNA (RAPD) and derived techniques. *Methods in Molecular Biology (Clifton, N.J.)* 2222, 219–247. DOI: 10.1007/978-1-0716-0997-2_13.

Babu, R., Nair, S.K., Prasanna, B.M. and Gupta, H.S. (2004) Integrating marker-assisted selection in crop breeding–prospects and challenges. *Current Science* 87(5) 607–619.

Bailey-Serres, J., Parker, J.E., Ainsworth, E.A., Oldroyd, G.E. and Schroeder, J.I. (2019) Genetic strategies for improving crop yields. *Nature* 575(7781), 109–118. DOI: 10.1038/s41586-019-1679-0.

Bajaj, D., Srivastava, R., Nath, M., Tripathi, S., Bharadwaj, C. *et al.* (2016) EcoTILLING-based association mapping efficiently delineates functionally relevant natural allelic variants of candidate genes governing agronomic traits in chickpea. *Frontiers in Plant Science* 7, 450. DOI: 10.3389/fpls.2016.00450.

Baliyan, N., Malik, R., Rani, R., Mehta, K., Vashisth, U. *et al.* (2017) Integrating marker-assisted background analysis with foreground selection for pyramiding bacterial blight resistance genes into Basmati rice. *Comptes Rendus Biologies* 341(1), 1–8. DOI: 10.1016/j.crvi.2017.11.003.

Bayer, P.E., Golicz, A.A., Tirnaz, S., Chan, C.K.K., Edwards, D. *et al.* (2019) Variation in abundance of predicted resistance genes in the *Brassica oleracea* pangenome. *Plant Biotechnology Journal* 17(4), 789–800. DOI: 10.1111/pbi.13015.

Begna, T. (2021) Conventional breeding methods widely used to improve self-pollinated crops. *International Journal of Research Studies in Agricultural Sciences* 7, 2454–6224. DOI: 10.20431/2454-6224.0701001.

Bera, S.K., Kamdar, J.H., Kasundra, S.V., Patel, S.V., Jasani, M.D. *et al.* (2019) Steady expression of high oleic acid in peanut bred by marker-assisted backcrossing for fatty acid desaturase mutant alleles and its effect on seed germination along with other seedling traits. *PLOS ONE* 14, e0226252. DOI: 10.1371/journal.pone.0226252.

Bernardo, R. (2010) Genomewide selection with minimal crossing in self-pollinated crops. *Crop Science* 50, 624–627. DOI: 10.2135/cropsci2009.05.0250.

Bernardo, R. and Charcosset, A. (2006) Usefulness of gene information in marker-assisted recurrent selection: A simulation appraisal. *Crop Science* 46(2), 614–621. DOI: 10.2135/cropsci2005.05-0088.

Beyene, Y., Semagn, K., Mugo, S., Prasanna, B.M., Tarekegne, A. *et al.* (2016) Performance and grain yield stability of maize populations developed using marker-assisted recurrent selection and pedigree selection procedures. *Euphytica* 208, 285–297. DOI: 10.1007/s10681-015-1590-1.

Bjørnstad, Å., Patil, V., Tekauz, A., Marøy, A.G., Skinnes, H. *et al.* (2002) Resistance to scald (*Rhynchosporium secalis*) in barley (*Hordeum vulgare*) studied by near-isogenic lines: I. Markers and differential isolates. *Phytopathology* 92(7), 710–720. DOI: 10.1094/PHYTO.2002.92.7.710.

Bolouri, P., Haliloğlu, K., Mohammadi, S.A., Türkoğlu, A., İlhan, E. *et al.* (2021) Identification of novel QTLs associated with frost tolerance in winter wheat (*Triticum aestivum* L.). *Plants* 12(8), 1641. DOI: 10.3390/plants12081641.

Bouchez, A., Hospital, F., Causse, M., Gallais, A. and Charcosset, A. (2002) Marker-assisted introgression of favorable alleles at quantitative trait loci between maize elite lines. *Genetics* 162(4), 1945–1959. DOI: 10.1093/genetics/162.4.1945.

Bovill, W.D., Horne, M., Herde, D., Davis, M., Wildermuth, G.B. *et al.* (2010) Pyramiding QTL increases seedling resistance to crown rot (*Fusarium pseudograminearum*) of wheat (*Triticum aestivum*). *Theoretical and Applied Genetics* 121, 127–136. DOI: 10.1007/s00122-010-1296-7.

Brauner, P.C., Schipprack, W., Utz, H.F., Bauer, E., Mayer, M. *et al.* (2019) Testcross performance of doubled haploid lines from european flint maize landraces is promising for broadening the genetic base of elite germplasm. *Theoretical and Applied Genetics* 132, 1897–1908. DOI: 10.1007/s00122-019-03325-0.

Cagliari, D., Dias, N.P., Galdeano, D.M., Dos Santos, E.Á., Smagghe, G. *et al.* (2019) Management of pest insects and plant diseases by non-transformative RNAi. *Frontiers in Plant Science* 10, 1319. DOI: 10.3389/fpls.2019.01319.

Calayugan, M.I.C., Formantes, A.K., Amparado, A., Descalsota-Empleo, G.I., Nha, C.T. *et al.* (2020) Genetic analysis of agronomic traits and grain iron and zinc concentrations in a doubled haploid population of rice (*Oryza sativa* L.). *Scientific Reports* 10(1), 2283. DOI: 10.1038/s41598-020-58524-1.

Ceccarelli, S. and Grando, S. (2020) Evolutionary plant breeding as a response to the complexity of climate change. *iScience* 23(12), 101815. DOI: 10.1016/j.isci.2020.101815.

Chang, Y., Latham, J., Licht, M. and Wang, L. (2023) A data-driven crop model for maize yield prediction. *Communications Biology* 6, 439. DOI: 10.1038/s42003-023-04833-y.

Charmet, G., Robert, N., Perretant, M.R., Gay, G., Sourdille, P. *et al.* (1999) Marker-assisted recurrent selection for cumulating additive and interactive QTLs in recombinant inbred lines. *Theoretical and Applied Genetics* 99, 1143–1148. DOI: 10.1007/s001220051318.

Cholin, S.S. and Kulkarni, C.C. (2023) Prospects of marker-assisted recurrent selection: Current insights and future implications. *Tropical Plant Biology* 16(4), 259–275. DOI: 10.1007/s12042-023-09348-8.

Chugh, V., Kaur, D., Purwar, S., Kaushik, P., Sharma, V. *et al.* (2022) Applications of molecular markers for developing abiotic-stress-resilient oilseed crops. *Life (Basel, Switzerland)* 13(1), 88. DOI: 10.3390/life13010088.

Chukwu, S.C., Rafii, M.Y., Ramlee, S.I., Ismail, S.I., Oladosu, Y. *et al.* (2020) Recovery of recurrent parent genome in a marker-assisted backcrossing against rice blast and blight infections using functional markers and SSRs. *Plants (Basel)* 9(11), 1411. DOI: 10.3390/plants9111411.

Collard, B.C. and Mackill, D.J. (2008) Marker-assisted selection: An approach for precision plant breeding in the twenty-first century. *Philosophical Transactions of the Royal Society of London. Series B, Biological Sciences* 363(1491), 557–572. DOI: 10.1098/rstb.2007.2170.

Crossa, J., Pérez-Rodríguez, P., Cuevas, J., Montesinos-López, O., Jarquín, D. *et al.* (2017) Genomic selection in plant breeding: Methods, models, and perspectives. *Trends in Plant Science* 22(11), 961–975. DOI: 10.1016/j.tplants.2017.08.011.

Cui, Y., Li, R., Li, G., Zhang, F., Zhu, T. *et al.* (2020) Hybrid breeding of rice via genomic selection. *Plant Biotechnology Journal* 18(1), 57–67. DOI: 10.1111/pbi.13170.

Darrah, L.L., McMullen, M.D. and Zuber, M.S. (2019) Breeding, genetics, and seed corn production. In: Serna-Saldivar, S.O. (ed.) *Corn*, 3rd edn. AACC International Press, pp. 19–41. DOI: 10.1016/B978-0-12-811971-6.00002-4.

Das, A., Sharma, N. and Prasad, M. (2019) CRISPR/Cas9: A novel weapon in the arsenal to combat plant diseases. *Frontiers in Plant Science* 9, 2008. DOI: 10.3389/fpls.2018.02008.

de C. Lara, L.A., Santos, M.F., Jank, L., Chiari, L., Vilela, M.D.M. *et al.* (2019) Genomic selection with allele dosage in *Panicum maximum* Jacq. *G3: Genes, Genomes, Genetics* 9(8), 2463–2475. DOI: 10.1534/g3.118.200986.

Degenkolbe, T., Do, P.T., Kopka, J., Zuther, E., Hincha, D.K. *et al.* (2013) Identification of drought tolerance markers in a diverse population of rice cultivars by expression and metabolite profiling. *PLOS ONE* 8(5), e63637. DOI: 10.1371/journal.pone.0063637.

Doyle, J.J. and Doyle, J.L. (1990) Isolation of plant DNA from fresh tissue. *Focus* 12(1), 13–15.

Eathington, S., Crosbie, T., Edwards, M., Reiter, R. and Bull, J. (2007) Molecular markers in a commercial breeding program. *Crop Science* 47 (S3), S154–S163. DOI: 10.2135/cropsci2007.04.0015.

Elçi, E. and Hançer, T. (2014) Genetic analysis of maize (*Zea mays* L.) hybrids using microsatellite markers. DOI: 10.15832/tbd.81254.

El-Mansy, A.B., Abd El-Moneim, D., Alshamrani, S.M., Safhi, F.A., Abdein, M.A. *et al.* (2021) Genetic diversity analysis of tomato (*Solanum lycopersicum* L.) with morphological, cytological, and molecular markers under heat stress. *Horticulturae* 7(4), 65. DOI: 10.3390/horticulturae7040065.

Fatiukha, A., Filler, N., Lupo, I., Lidzbarsky, G., Klymiuk, V. *et al.* (2020) Grain protein content and thousand kernel weight qtls identified in a durum × wild emmer wheat mapping population tested in five environments. *Theoretical and Applied Genetics* 133, 119–131. DOI: 10.1007/s00122-019-03444-8.

Feng, Z., Li, M., Xu, Z., Gao, P., Wu, Y. *et al.* (2022) Development of rice variety with durable and broad-spectrum resistance to blast disease through marker-assisted introduction of *Pigm* gene. *Frontiers in Plant Science* 13, 937767. DOI: 10.3389/fpls.2022.937767.

Frisch, M. and Melchinger, A.E. (2005) Selection theory for marker-assisted backcrossing. *Genetics* 170(2), 909–917. DOI: 10.1534/genetics.104.035451.

Gazal, A., Ahmed, Z., Lone, A., Abidi, I.A. and Ali, G. (2015) Molecular breeding for resilience in maize – A review. *Journal of Applied and Natural Science* 7(2), 1057–1063. DOI: 10.31018/jans.v7i2.731.

Goddard, M.E. and Hayes, B.J. (2007) Genomic selection. *Journal of Animal Breeding and Genetics = Zeitschrift Fur Tierzuchtung Und Zuchtungsbiologie* 124(6), 323–330. DOI: 10.1111/j.1439-0388.2007.00702.x.

Gokidi, Y., Bhanu, A. and Singh, M.N. (2016) Marker assisted recurrent selection: An overview. *Advances in Life Sciences* 5(17), 6493–6499.

Goldman, I.L. (2013) Biodiversity in plant breeding. In: Lvein, S.A. (ed.) *Encyclopedia of Biodiversity*, 2nd edn. Academic Press, pp. 459–469. DOI: 10.1016/B978-0-12-822562-2.00024-4.

Gouda, A.C., Warburton, M.L., Djedatin, G.L., Kpeki, S.B., Wambugu, P.W. *et al.* (2021) Development and validation of diagnostic SNP markers for quality control genotyping in a collection of four rice (*Oryza*) species. *Scientific Reports* 11(1), 18617. DOI: 10.1038/s41598-021-97689-3.

Herzog, E. and Frisch, M. (2013) Efficient marker-assisted backcross conversion of seed-parent lines to cytoplasmic male sterility. *Plant Breeding* 132(1), 33–41. DOI: 10.1111/pbr.12021.

Hospital, F. (2001). Size of donor chromosome segments around introgressed loci and reduction of linkage drag in marker-assisted backcross programs. *Genetics*, 158(3), 1363-1379. PMID: , PMCID: PMC1461714. DOI: https://doi.org/10.1093/genetics/158.3.1363.

Hospital, F. (2009) Challenges for effective marker-assisted selection in plants. *Genetica*, 136(2), 303–310. DOI: https://doi.org/10.1007/s10709-008-9307-1.

Jarcho, J. (2001) Restriction fragment length polymorphism analysis. *Current Protocols in Human Genetics* 1(1), 2.7.1–2.7.15. DOI: 10.1002/0471142905.hg0207s01.

Jiang, G.-L. (2013) Molecular markers and marker-assisted breeding in plants. In: *Andersen, S.B. (ed.) Plant Breeding from Laboratories to Fields.* InTech Open, London. DOI: 10.5772/52583.

Jlassi, I., Bnejdi, F., Saadoun, M., Hajji, A., Mansouri, D. *et al.* (2021) SSR markers and seed quality traits revealed genetic diversity in wheat (*Triticum durum* Desf.). Molecular Biology Reports 48, 3185–3193 DOI: 10.1007/s11033-021-06385-y.

Kadirvel, P., Senthilvel, S., Geethanjali, S., Sujatha, M. and Varaprasad, K.S. (2015) Genetic markers, trait mapping and marker-assisted selection in plant breeding. In: *Plant Biology and Biotechnology: Vol. II: Plant Genomics and Biotechnology*. Springer, New Delhi, pp. 65–88. DOI: 10.1007/978-81-322-2283-5_4.

Katara, J. L., Parameswaran, C., Devanna, B. N., Verma, R. L., Anilkumar, C. *et al.* (2021). Genomic supported breeding: Need and current perspective for improvement of rice in India. *ORYZA-International Rice Journal*, 58(1), 61–68. DOI: DOI: 10.35709/ory.2021.58.spl.1.

Kaundun, S.S., Hutchings, S.J., Downes, J. and Baker, K. (2023) Derived polymorphic amplified cleaved sequence (dPACS) assay. In: Shavrukov, M. (ed.) *Marker-Assisted Breeding of Aromatic Rice Cultivars*. Springer, pp. 373–385. DOI: 10.1007/978-1-0716-3024-2_27.

Khan, Z. and Shahwar, D. (2023) Genome editing in plants via CRISPR/Cas9; A genomic scissor borrowed from bacterial immune system. In: Khan, Z., Shahwar, D. and Heikal, Y. (eds). *Genome Editing and Global Food Security: Molecular Engineering Technologies for Sustainable Agriculture*. Taylor & Francis, London, pp. 1–16.

Khan, Z., Ansari, M.Y.K. and Gupta, H. (2012). Induction of mutations in *Cichorium intybus* L. by base analogue 6-aminopurine (6-AP) and their detection with random amplified polymorphic DNA (RAPD) analysis. *African Journal of Biotechnology*, 11(56), 11901–11906. DOI: 10.5897/AJB11.2158.

Kocaman, B., Toy, S., and Marakli, S. (2020). Application of different molecular markers in biotechnology. *International Journal of Science Letters*, 2, 98–113. DOI: https://doi.org/10.38058/ijsl.770081.

Kumar, M., Kumar, S., Sandhu, K.S., Kumar, N., Saripalli, G. *et al.* (2023) GWAS and genomic prediction for pre-harvest sprouting tolerance involving sprouting score and two other related traits in spring wheat. *Molecular Breeding* 43(3), 14. DOI: 10.1007/s11032-023-01357-5.

Kumar Singh, V., Pundir, S., Chaturvedi, D., Kaur, A., Pandey, A. *et al.* (2024) Enhancing maize (*Zea mays* L.) crop through advanced techniques: A comprehensive approach. In: Kaushik, P. (ed.) *New Prospects of Maize*. InTech Open, London. DOI: 10.5772/intechopen.114029.

Kushanov, F.N., Turaev, O.S., Ernazarova, D.K., Gapparov, B.M., Oripova, B.B. *et al.* (2021) Genetic diversity, QTL mapping, and marker-assisted selection technology in cotton (*Gossypium* spp). *Frontiers in Plant Science* 12, 779386. DOI: 10.3389/fpls.2021.779386.

Lakhneko, O., Stepanenko, A., Kuzminskiy, Y., Borisjuk, N. and Morgun, B. (2021) Survey of drought-associated *TAWRKY2-D1* gene diversity in bread wheat and wheat relatives. *Molecular Biotechnology* 63(10), 953–962. DOI: 10.1007/s12033-021-00350-7.

Langridge, P., Alaux, M., Almeida, N.F., Ammar, K., Baum, M. *et al.* (2022) Meeting the challenges facing wheat production: The strategic research agenda of the global wheat initiative. *Agronomy* 12(11), 2767. DOI: 10.3390/agronomy12112767.

Ledesma, N. A., Matulac, J. M., Sevilleja, J. E., & Enriquez, M. L. (2023). Detecting misidentifications of strawberry cultivars in the Philippines using single nucleotide polymorphism markers. *The Journal of Horticultural Science and Biotechnology*, 98(4), 495-507. DOI: DOI: 10.1080/14620316.2022.2162444.

Livini, C., Ajmone-Marsan, P., Melchinger, A.E., Messmer, M.M. and Motto, M. (1992) Genetic diversity of maize inbred lines within and among heterotic groups revealed by RFLPs. *Theoretical and Applied Genetics* 84(1–2), 17–25. DOI: 10.1007/BF00223976.

Mallick, N., Jha, S.K., Agarwal, P., Mall, A.M.N., Kumar, S. *et al.* (2022) Marker-assisted improvement of bread wheat variety HD2967 for leaf and stripe rust resistance. *Plants (Basel)* 11(9), 1152. DOI: 10.3390/plants11091152.

McLennon, E., Dari, B., Jha, G., Sihi, D. and Kankarla, V. (2021) Regenerative agriculture and integrative permaculture for sustainable and technology-driven global food production and security. *Agronomy Journal* 113(6), 4541–4559. DOI: 10.1002/agj2.20814.

Meng, F.-j., Xu, X.-y., Huang, F.-l. and Li, J.-f. (2009) Analysis of genetic diversity in cultivated and wild tomato varieties in the chinese market by RAPD and SSR. DOI: 10.1016/S1671-2927(09)60234-0.

Melchinger, A.E. (1999) Genetic diversity and selection in maize. *Theoretical and Applied Genetics*, 99(5), 747–753.

Miedaner, T., Boeven, A.L.G.-C., Gaikpa, D.S., Kistner, M.B. and Grote, C.P. (2020) Genomically supported breeding for quantitative disease resistance in small grains and maize. *International Journal of Molecular Sciences* 21(24), 9717. DOI: 10.3390/ijms21249717.

Mondini, L., Noorani, A. and Pagnotta, M.A. (2009) Assessing plant genetic diversity by molecular tools. *Diversity* 1(1), 19–35. DOI: 10.3390/d1010019.

Nayak, S., Singh, V. K. and Varshney, R. (2016) Marker-assisted selection. In: *Encyclopedia of Applied Plant Sciences*. Academic Press, pp. 183–197. DOI: DOI: 10.1016/B978-0-12-394807-6.00192-1.

Obermeier, C. and Friedt, W. (2015) Applied oilseed rape marker technology and genomics. In: *Applied Plant Genomics and Biotechnology*. Woodhead Publishing, pp. 253–295. DOI: 10.1016/B978-0-08-100068-7.00016-1.

Oh, J.E., Kim, J.E., Kim, J., Lee, M.H., Lee, K. *et al.* (2023) Development of an SNP marker set for marker-assisted backcrossing using genotyping-by-sequencing in tetraploid perilla. *Molecular Genetics and Genomics* 298(6), 1435–1447. DOI: 10.1007/s00438-023-02066-6.

Orf, J.H. (2008) Breeding, genetics, and production of soybeans. In: Johnson, L.A., White, P.J. and Galloway, R. (eds) *Soybeans*. AOCS Press, Urbana, Illinois, USA, pp. 33–65. DOI: 10.1016/B978-1-893997-64-6.50005-6.

Panda, D., Mishra, S.S. and Behera, P.K. (2021) Drought tolerance in rice: Focus on recent mechanisms and approaches. *Rice Science* 28(2), 119–132. DOI: 10.1016/j.rsci.2021.01.002.

Pozharskiy, A., Kostyukova, V., Khusnitdinova, M., Adilbayeva, K., Nizamdinova, G. *et al.* (2023) Genetic diversity of the breeding collection of tomato varieties in Kazakhstan assessed using SSR, SCAR and CAPS markers. *PeerJ* 11, e15683. DOI: 10.7717/peerj.15683.

Rahman, M., Davies, P., Bansal, U., Pasam, R., Hayden, M. *et al.* (2020) Marker-assisted recurrent selection improves the crown rot resistance of bread wheat. *Molecular Breeding* 40(1), 28. DOI: 10.1007/s11032-020-1105-1.

Raj, S.R.G. and Nadarajah, K. (2022) QTL and candidate genes: Techniques and advancement in abiotic stress resistance breeding of major cereals. *International Journal of Molecular Sciences* 24(1), 6. DOI: 10.3390/ijms24010006.

Ramos Guimaraes, P.H., Guimarães Santos Melo, P., Centeno Cordeiro, A.C., Pereira Torga, P., Nakano Rangel, P.H. *et al.* (2021) Index selection can improve the selection efficiency in a rice recurrent selection population. *Euphytica* 217(5), 95. DOI: 10.1007/s10681-021-02819-7.

Razzaq, M.K., Aleem, M., Mansoor, S., Khan, M.A., Rauf, S. *et al.* (2021) Omics and CRISPR-Cas9 approaches for molecular insight, functional gene analysis, and stress tolerance development in crops. *International Journal of Molecular Sciences* 22(3), 1292. DOI: 10.3390/ijms22031292.

Şahin, Ö. and Kavuncu, O. (2019) QTL analysis methods. *Kastamonu University Journal of Engineering and Sciences* 5(1), 51–57.

Sami, A., Xue, Z., Tazein, S., Arshad, A., Zhu, H. Z. *et al.* (2021) CRISPR-Cas9-based genetic engineering for crop improvement under drought stress. *Bioengineered* 12(1), 5814–5829. DOI: 10.1080/21655979.2021.1969831.

Sanchez, D., Sadoun, S.B., Mary-Huard, T., Allier, A., Moreau, L. *et al.* (2023) Improving the use of plant genetic resources to sustain breeding programs' efficiency. *Proceedings of the National Academy of Sciences USA* 120(14), e2205780119. DOI: 10.1073/pnas.2205780119.

Shahwar, D., Khan, Z. and Park, Y. (2023) Molecular marker-assisted mapping, candidate gene identification, and breeding in melon (*Cucumis melo* L.): A review. *International Journal of Molecular Sciences* 24(20), 15490. DOI: 10.3390/ijms242015490.

Shavrukov, Y. (2016) Comparison of SNP and CAPS markers application in genetic research in wheat and barley. *BMC Plant Biology* 16(Suppl. 1), 11. DOI: 10.1186/s12870-015-0689-9.

Singh, D.P., Singh, A.K. and Singh, A. (2021a) Molecular tools in crop improvement and cultivar development. In: Singh, D.P., Singh, A.K. and Singh, A. (eds) *Plant Breeding and Cultivar Development*. Academic Press, pp. 535–562. DOI: 10.1016/B978-0-12-817563-7.00014-3.

Singh, D.P., Singh, A.K. and Singh, A. (2021b) Recurrent selection in self-pollinated crops. In: *Plant Breeding and Cultivar Development*. Academic Press, pp. 331–343. DOI: 10.1016/B978-0-12-817563-7.00007-6.

Singh, M., Nara, U., Kumar, A., Thapa, S., Jaswal, C. *et al.* (2021c) Enhancing genetic gains through marker-assisted recurrent selection: From phenotyping to genotyping. *Cereal Research Communications*. DOI: 10.1007/s42976-021-00207-4.

Singh, N.K., Joshi, A., Sahoo, S., Tufchi, M. and Rakshit, S. (2023) *Molecular Breeding for Improving Yield in Maize: Recent Advances and Future Perspectives*. Elsevier EBooks. DOI: 10.1016/b978-0-323-85243-2.00010-6.

Singh, R.K., Prasad, A., Muthamilarasan, M., Parida, S.K. and Prasad, M. (2020) Breeding and biotechnological interventions for trait improvement: Status and prospects *Planta* 252(4). DOI: 10.1007/s00425-020-03465-4.

Sinha, D., Mauyra, A.K., Abdi, G., Majeed, M., Agarwal, R. *et al.* (2023) Integrated genomic selection for accelerating breeding programs of climate-smart cereals. *Genes* 14(7), 1484. DOI: 10.3390/genes14071484.

Suvarna, A.K. and Yashaswini, R. (2023) Marker assisted recurrent selection for crop improvement. In: Singh, D.P., Singh, A.K. and Singh, A. (eds) Suvarna, A.K. and Yashaswini, R. (eds) *Molecular Marker Techniques: A Potential Approach of Crop Improvement*. Springer Nature, Singapore, pp. 55–67.

Swarup, S., Cargill, E., Crosby, K., Flagel, L., Kniskern, J. *et al.* (2020) Genetic diversity is indispensable for plant breeding to improve crops. *Crop Science* 61(2), 839–852. DOI: 10.1002/csc2.20377.

Türkoğlu, A, Bolouri, P., Haliloğlu, K., Eren, B., Demirel, F. *et al.* (2023a) Modeling callus induction and regeneration in hypocotyl explant of fodder pea (*Pisum sativum* var. *arvense* L.) using machine learning algorithm method. *Agronomy* 13(11), 2835. DOI: 10.3390/agronomy13112835.

Türkoğlu, A, Haliloğlu, K., Mohammadi, S.A., Öztürk, A., Bolouri, P. *et al.* (2023b) Genetic diversity and population structure in Türkiye bread wheat genotypes revealed by simple sequence repeats (SSR) markers. *Genes* 14(6), 1182. DOI: 10.3390/genes14061182.

Uçar, C. and Şensoy, S. (2022) Validity control of markers used in molecular marker assisted selection in tomato. *Yüzüncü Yıl Üniversitesi Tarım Bilimleri Dergisi* 32(2), 300–309. DOI: 10.29133/yyutbd.1039398.

van Etten, J., de Sousa, K., Cairns, J.E., Dell'Acqua, M., Fadda, C. *et al.* (2023) Data-driven approaches can harness crop diversity to address heterogeneous needs for breeding products. *Proceedings of the National Academy of Sciences USA* 120(14), e2205771120. DOI: 10.1073/pnas.2205771120.

Vieira, M.L., Santini, L., Diniz, A.L. and Munhoz, C. deF (2016) Microsatellite markers: What they mean and why they are so useful. *Genetics and Molecular Biology* 39, 312–328. DOI: 10.1590/1678-4685-GMB-2016-0027.

Vishwakarma, M.K., Yadav, P.S., Rai, V.P., Kumar, U. and Joshi, A.K. (2022) Molecular markers and genomics assisted breeding for improving crop plants. In: Samual, J., Kumar, A. and Singh, J. (eds) *Relationship Between Microbes and the Environment for Sustainable Ecosystem Services*, Vol. 1. Microbial Products for Sustainable Ecosystem Services. Elsevier, pp. 303–334. DOI: 10.1016/B978-0-323-89938-3.00014-1.

Wang, J., Yang, J., Zhu, J., Jia, Q. and Tao, Y. (2010) Assessment of genetic diversity by simple sequence repeat markers among forty elite varieties in the germplasm for malting barley breeding. *Journal of Zhejiang University SCIENCE B* 11(10), 792–800. DOI: 10.1631/jzus.b0900414.

Wang, W., Guo, W., Le, L., Yu, J., Wu, Y. *et al.* (2023) Integration of high-throughput phenotyping, GWAS, and predictive models reveals the genetic architecture of plant height in maize. *Molecular Plant* 16(2), 354–373. DOI: 10.1016/j.molp.2022.11.016.

Wang, X., Xu, Y., Hu, Z. and Xu, C. (2018) Genomic selection methods for crop improvement: Current status and prospects. *Crop Journal* 6(4), 330–340. DOI: 10.1016/j.cj.2018.03.001.

Xu, R., Wu, Y. and Luan, J. (2016) Analysis of farmers' willingness to adopt genetically modified insect-resistant rice in China. *China Agricultural Economic Review* 8(3), 368–382. DOI: 10.1108/caer-08-2015-0102.

Xu, Y. and Jonathan H., C. (2008) Marker-Assisted selection in plant breeding: from publications to practice. *Crop Science* 48(2), 391–407. DOI: 10.2135/cropsci2007.04.0191.

Xu, Y., Ma, K., Zhao, Y., Wang, X., Zhou, K. *et al.* (2021) Genomic selection: A breakthrough technology in rice breeding. *Crop Journal* 9(3), 669–677. DOI: 10.1016/j.cj.2021.03.008.

Zampieri, E., Volante, A., Mare, C., Orasen, G. and Desiderio, F. (2023) Marker-assisted pyramiding of blast-resistance genes in a *japonica* elite rice cultivar through forward and background selection. *Plants* 12(4), 757–757. DOI: 10.3390/plants12040757.

Zargar, S.M., Mir, R.A., Ebinezer, L.B., Masi, A., Hami, A. *et al.* (2022) Physiological and multi-omics approaches for explaining drought stress tolerance and supporting sustainable production of rice. *Frontiers in Plant Science* 12, 803603. DOI: 10.3389/fpls.2021.803603.

Zehr, B.E., Dudley, J.W., Chojecki, J., Saghai Maroof, M.A. and Mowers, R.P. (1992) "Use of RFLP markers to search for alleles in a maize population for improvement of an elite hybrid." *Tag. Theoretical and Applied Genetics. Theoretische Und Angewandte Genetik* 83(6–7), 903–911. 10.1007/BF00226714 (accessed 24 May 2021).

Zommita, S., Yıldız, G.G., Zencirci, N., Özbek, Ö. and Test, K. (2023) Assessment of genetic diversity based on agro-morphological traits and ISSR molecular markers in einkorn wheat (*Triticum monococcum* ssp. monococcum) landrace populations from Turkey. *International Journal of Secondary Metabolite* 10(1), 86–105. DOI:10.21448/ijsm.1084853.

Zou, Y., Liu, Z., Chen, Y., Wang, Y. and Feng, S. (2024) Crop rotation and diversification in China: Enhancing sustainable agriculture and resilience. *Agriculture* 14(9), 1465. DOI: 10.3390/agriculture14091465.

Zurita-Silva, A., Fuentes, F., Zamora, P., Jacobsen, S. and Schwember, A. (2014) Breeding quinoa (Chenopodium quinoa willd.): Potential and perspectives. *Molecular Breeding* 34, 13–20. DOI: 10.1007/s11032-014-0023-5.

# 9 Advances in Crop Improvement through High-Throughput Phenotyping

**Jyoti Sankar Padhy[1], Ritabrata Sarkar[2], Sourav Ranjan Mohapatra[3]*, Ayesha Mohanty[4] and Kaushik Kumar Panigrahi[5]***

*[1]Department of Genetics and Plant Breeding, Bidhan Chandra Krishi Vishwavidalaya, Mohanpur, West Bengal, India; [2]Department of Genetics and Plant Breeding, ICAR-Indian Institute of Agricultural Biotechnology, Ranchi, India; [3]Department of Forest Biology and Tree Improvement, Odisha University of Agriculture and Technology, Bhubaneswar, Odisha, India; [4]AINP on Soil Biodiversity-Biofertilizers, Odisha University of Agriculture and Technology, Bhubaneswar, Odisha, India; 5AICRN on Potential crops, Odisha University of Agriculture and Technology, Bhubaneswar, Odisha, India*

**Abstract**

Plant breeding has contributed significantly to the sustenance of food security by producing crops with improved quality and yield. The selection of superior genotypes has been the core in plant breeding historically but relying only on phenotype is found to be inefficient in terms of selection accuracy and genetic gain. The development of genotyping has prompted the utilization of molecular markers in crop improvement, but this still depends on phenotypic data for preciseness. With manual phenotyping being inaccurate, destructive and technically challenging, the development of automatic, non-destructive and high-throughput plant phenotyping (HTPP) technologies can greatly complement crop improvement programmes. The current chapter briefly describes the utilization of HTPP platforms in screening breeding populations for secondary traits (highly correlated to primary traits), GS, genetic studies for QTL and GWAS approaches, and phenomic selection (PS). Even with these extensive applications, challenges are inevitable. Hence, the ways in which to deal with these challenges are discussed in this chapter.

**Keywords:** high-throughput plant phenotyping (HTP), genomic selection (GS), genome-wide association studies(GWAS), phenomic selection (PS)

**Abbreviations**

3D – three dimensional, BP – Breeding population, DR – Drought resistance, G×E×M – Genotype by environment by management, GEBV – Genomic estimated breeding value, GS – Genomic selection, GWAS – Genome-wide association studies, G×E – Genotype by environment, HTPP – High-throughput plant phenotyping, MT-RRM – Multi-trait random regression model, NDVI – normalized difference vegetation index, NGS – Next-generation sequencing, NIRS – Near-infrared spectroscopy, PEP – Phenomics estimated performance, PS – Phenomic selection, PSA – Projected shoot area, QTL – Quantitative trait locus, RGB – Red-green-blue, SNP – Single-nucleotide polymorphism, ST-RRM – Single-trait random regression model, TP – Training population, UAVs – Unmanned aerial vehicles, WU – Water usage.

*Corresponding author: mohapatrasourav23@gmail.com; kaushikpbg@gmail.com





DOI: 10.1079/9781800626638.0009

## 9.1 Introduction

Plant breeding refers to modifications made to plants by humans, ranging from inadvertent alterations in the course of agriculture to the application of molecular tools for precision and effective breeding. The basis of plant breeding has been the screening of superior genotypes among available variants in terms of the production and quality of commercial edible parts, ease of cultivation, harvest and processing, resistance against pests, and resilience to environmental challenges. Every aspect of agronomic or food value can be dissected into many distinct traits, each showcasing wide variational ranges (Breseghello and Coelho, 2013). This comprises a crucial step in phenotypic selection where the target genotype is captured by observing its phenotype. However, for various well known reasons, the genotype shows variable expression in the phenotype and there is not always a perfect relationship, which creates a gap between genotype and phenotype. One of the intricate causes, i.e. complex G×E×M interactions in which genotype (G) is exposed to the environment (E) (Blum, 2011) and also to crop management (M) (Großkinsky *et al.*, 2015), results in expression of the phenotype. An increase in breeding efficiency is the ongoing aim to meet future food grain needs. Among developed technologies, high-throughput genotyping has helped in obtaining genomic information in a rapid and low cost way. It has paved the way for the development of large mapping populations and multiple diverse recombinant inbred line (RILs) panels for phenotyping. Accurate and precise phenotypic data are still necessary for molecular breeding procedures, like genomic selection (GS) and marker-assisted recurrent selection (MARS), to make selections more effective based on genotypic information (Araus and Cairns, 2014). In this context, conventional phenotyping, which is still followed in many breeding programmes, is time-consuming, destructive and requires immense human effort to sample large populations of crop plants, making it labour-intensive and technically challenging. Additionally, it becomes more challenging to capture biochemical and physiological phenotypes at the basic level of the plant system. These phenotyping bottlenecks can be resolved by using phenotyping tools that are automatic, non-invasive and have high throughput, which means the ability to analyse a large number of plants and individual units at a particular organizational (molecular) level within plants, for multiple traits at a time (Dhondt *et al.*, 2013).

Li *et al.* (2021) defined a 'high-throughput plant phenotyping platform' as a platform capable of collecting enormous amounts of highly automated phenotypic data from hundreds of plants every day with minimal error. This new effective tool allows the monitoring and quantifying of crop progress and yield-related phenotypic characteristics in a fast, non-destructive and high-throughput way, and can assist in molecular breeding methodologies, thereby improving breeding efficiency. In this chapter, we have tried to provide brief information on recent advances in crop improvement through the application of high-throughput plant phenotyping in selected field crops. We have described how this novel platform can contribute to genetic gain and can complement various aspects of crop improvement, i.e. screening breeding populations, genetic studies, GS and phenomic selection (PS).

## 9.2 High-Throughput Phenotyping and Genetic Gain

Genetic gain may be expressed as the amount of performance advancement achieved per unit of time through artificial selection and can be denoted by $R_t$ (genetic gain) = i (selection intensity) × r (selection accuracy) × $V_A$ (genetic variance)/ y (years per cycle). Generally, breeding progress is measured by this genetic gain equation. In the context of increasing genetic gain, phenotyping can aid in enhancing the intensity and precision of selection, and detecting new genetic variants. Selection intensity is the selection rate, which indicates the percentage of individuals selected from the total population based on their performance. The development of cost-effective phenotyping tools with increased scale can facilitate the screening of larger populations, thereby increasing the possibility of identifying superior genotypes and further increasing selection intensity. One of the prime objectives of plant breeding is the estimation of the breeding value of individuals. The more accurate the breeding value, the higher the selection accuracy. Breeding programmes routinely take a record of stand establishment-related traits, abiotic and biotic stress-related traits, agronomic traits, and stage-specific observations. However, the error-prone, more subjective and less robust nature of manual measurements leads to inaccuracies in the phenotypic data. To deal with this, breeding teams conduct multi-environmental or replicated trials for phenotyping complex traits to improve the heritability of the trait. The development of low-cost, high-throughput phenotyping can assist breeding programmes by removing error and subjectivity, at reduced cost, from measurements relative to manual measurement. Moreover, these tools facilitate the quantitative measurement of relevant traits in multiple locations and improve selection precision at a lower cost (Araus *et al.*, 2018). Morota *et al.* (2022) briefly explained how increased genetic gain can be achieved by using high-throughput phenotyping data.

## 9.3 Application of High-Throughput Phenotyping in Various Aspects of Crop Improvement

Plant breeding contributes significantly to the sustenance of food security by achieving ‘steady yield gains’, making crops relatively more defensive to the continuous onset of evolved insect pests and pathogens, and providing end-use products of improved quality (Reynolds *et al.*, 2020). Historically, plant breeders have been practising phenotypic selection for genetic gain. The traditional phenotyping methods being followed are invasive, laborious and deal with a few particular plant characteristics at a time; this makes it a time-intensive process when dealing with multiple traits in large segregating generation, resulting in delayed selection and reduced efficiency of the plant breeding programme. Multi-location field trials are a vital component of plant breeding programmes in the selection of stable genotypes, since they evaluate the performance of new candidate genotypes in multiple locations and analyse the pattern of genotype by environment (G×E) interactions (Pratap *et al.*, 2019). However, less precision in traditional phenotyping makes a thorough analysis of traits to link phenotype to genotype more complex, leading to bias in selection, which may negatively affect future genetic gain (Chawade *et al.*, 2019).

### 9.3.1 Screening breeding populations

High-throughput phenotyping platforms enable the efficient screening of large populations of plants, breeding material, germplasm collections (core collections), and mapping populations. They offer enhanced accuracy and precision in acquiring phenotypic traits, while reducing the need for manual labour through remotely controlled automation; and advanced data analysis pipelines for high-throughput observation. Various mapping populations and breeding populations are being screened for a range of traits such as plant root system architecture (RSA), seedling vigour, plant architecture, leaf area and senescence, chlorophyll content, stomatal conductance, leaf water potential, canopy temperature and pollen traits. Identifying the specific characteristics linked to improved yields and enhanced stress tolerance provides clear goals for breeders to select genotypes with superior phenotypes that can be used as parental material and for the development of breeding lines for advancing crop improvement processes (Pratap *et al.*, 2019). In the initial stages of selection, plant breeding programmes often include a large number of genotypes for evaluation in small plots of approximately 1 $m^2$. The utilization of unmanned aerial vehicles (UAVs) for remote sensing, along with proximal phenotyping by way of sensors on ground-based vehicles, has proven suitable for assessing small plots. The continuous cost reduction of UAVs and the protocol development for taking and evaluating images have significantly enhanced their utility in plant breeding programmes. Conversely, satellite imaging is valuable for evaluating trial plots of moderate to large size and for multi-location yield trials, which are often distributed across large areas, making it even more challenging to conduct proximal or drone-based measurements. Mobile platforms have greater advantages compared to handheld sensors, due to their capacity to simultaneously assess multiple traits and multiple rows, thus reducing both time and cost. Manually operated platforms are cost effective and relatively easy to construct. Motorized platforms offer additional advantages, such as minimizing the need for manual labour to propel the vehicle, the capacity to accommodate multiple sensors and their loads, and the inclusion of an onboard power source to support external lighting (Chawade *et al.*, 2019). High-throughput measurements of the Normalised Difference Vegetation Index (NDVI), along with other potential spectral vegetation indices, identify variations in the physiological and structural characteristics of vegetation, including photosynthetic surface area, pigment content, hydration levels and vegetative biomass. Likewise, the use of infrared thermography to estimate canopy temperature has been employed to assess water status and stomatal conductance. Standard red-green-blue (RGB) cameras for digital imaging are being used for tasks such as the reconstruction of a 3D surface and the recognition of image patterns using advanced deep learning techniques, facilitating the estimation of traits like plant height, number of tillers, biomass and disease detection. Certain assessment protocols have been defined alongside the ongoing advancement of smaller, cheaper and more precise sensors, particularly imaging systems, thus facilitating their application across a broader array of vehicles, including lightweight UAVs. Assessing biotic stress in plants through visual observation and rating is time-consuming and lacks precision. However, utilizing imaging systems partially addresses this challenge by enabling the assessment of pathogenesis on individual leaves and facilitating comprehensive evaluations across the whole canopy. Multi- and hyperspectral imaging technology are highly effective and potential systems for identifying precise alterations in spectral reflectance of the vegetation caused by pathogenesis, ensuring accuracy and consistency in disease detection in a larger number of populations within a short timeframe. Now digital RGB imaging technology has been applied at the canopy level to identify and measure visible symptoms in individual leaves through examination of alterations in image colour, spatial patterns and variations in texture. For insect damage, remote sensing technology can effectively detect aphid damage in agricultural fields and differentiate it from other forms of plant stress. Progress in satellite imaging enhances the capability of identifying aphid damage on a large scale. The application of remote sensing for screening arthropod resistance relies significantly on how plants physiologically react to pest species. Infestations by different phloem sap feeders cause alterations in photosynthetic activity, temperature of leaf surface, gas exchange, and carotenoid concentrations that correspond to distinct spectral characteristics (Reynolds *et al.*, 2020). The advanced automated video tracking facility enables the measurement of the feeding behaviour of aphids on leaf discs and evaluates the resistance levels of plants. Additionally, magnetic resonance imaging (MRI) and X-ray computed tomography(CT) have proven effective in visualizing and measuring root growth in 3D, even in opaque soil (Mir *et al.*, 2019). Several authors have briefly outlined the utilization of high-throughput phenotyping platforms across various aspects of plant breeding programmes in their reviews. These platforms are employed to enhance the utilization of plant genetic resources (Rebetzke *et al.* 2019; Nguyen and Norton, 2020; Volk *et al.*, 2021; Kim *et al.*, 2023), and assess traits related to radiation use efficiency (RUE), photosynthesis and crop biomass (Furbank *et al.*, 2019), as well as drought response detection (Kim *et al.*, 2021) and traits associated with heat stress (Kumar *et al.*, 2015; Basavaraj and Rane, 2020). The application to selected field crops is described in Table 9.1.

### 9.3.2 Genetic studies

The prediction of various data, including location, nature, function, and interactions of genes, has been made possible by the development of genomic tools. However, prediction preciseness still requires validation by phenotyping (Pratap *et al.*, 2019). The inadequacy of traditional phenotyping methods in obtaining complete phenotypic data from individuals in large populations, the lack of trackability of every phenotypic change in the plant life cycle, and the inevitable effect of genotype by environment interaction (G × E), contribute to the low heritability of many traits. As a result, there always remains error

**Table 9.1.** Application of high-throughput phenotyping in field crops for screening populations.

| Crop | Population size | Technique | Traits | References |
|---|---|---|---|---|
| Rice | 167 accessions | DJI Phantom four unmanned aerial vehicles (UAVs (aerial platform) for RGB imaging using Pix4Dcapture | Plant height | Kawamura *et al.* (2020) |
| Rice | 150 genotypes | Conveyer type Scanalyzer 3D (LemnaTec) imaging platform installed with RGB visual camera (Prosilica GT6600 series) | Image-based traits like: leaf number per plant, canopy spread (vertical and horizontal), convex hull area, and leaf number per unit convex hull of a plant, for separating drought-tolerant and- susceptible genotypes | Vishal *et al.* (2020) |
| Durum wheat | 256 accessions | Light-induced fluorescence transient (LIFT) sensor installed in a cart | Chlorophyll fluorescence (ChlF) based canopy traits | Zendonadi dos Santos *et al.* (2021) |
| Winter wheat | 354 genotypes | DJI Matrice 600 pro UAV (unmanned aerial vehicle) platform with FLIRA65 thermal imaging camera (mounted in custom made sensor) | Canopy temperature(CT) | Perich *et al*, 2020 |
| Winter wheat | Three cultivars: Zhongmai 895, Aikang 58 and Zhoumai 18 | Mavic pro UAV platform and Sequoia 4.0 multi-spectral sensor for multi-spectral imaging | Green normalized difference vegetation index (GNDVI), red normalized difference vegetation index (RNDVI), normalized difference red-edge index (NDRE), normalized green red difference index (NGRDI), and red-edge chlorophyll index (RECI) for determining crop water and nitrogen (N) use efficiencies | Yang *et al*, 2020 |

*Continued*

**Table 9.1.** Continued

| Crop | Population size | Technique | Traits | References |
|---|---|---|---|---|
| Wheat | Two spring wheat panels | Fixed-wing (*Plat*$_{FW}$) and multi-rotor (*Plat*$_{MR}$) unmanned aerial vehicle (UAV) platforms mounted with high-resolution digital RGB camera | Plant height | Volpato *et al.* (2021) |
| Hexaploid wheat | 645 lines | 2D root phenotyping platform, Nikon D3400 DSLR camera for root imaging | Seminal root number | Shorinola *et al.* (2019) |
| Wheat | 335 varieties | Pika L System for hyperspectral imaging | Leaf chlorophyll content for identifying drought-resistant genotype | Yang *et al.* (2023) |
| Spring wheat | 64 recombinant inbred lines (RILs) | ASD FieldSpec4 standard-res portable spectroradiometer for measuring hyperspectral reflectance | Grain yield and stress tolerance indices(STIs) | El-Hendawy *et al.* (2021) |
| Cotton | 708 $F_2$ plants | FTS 3000MX FTIR spectrometer for recording Fourier transform infrared (FT-IR) spectral data, coupled with ceramic source, KBr beam splitter, and deuterated triglycine sulfate(DTGS) detector | Fibre maturity | Kim *et al.* (2019) |
| Chickpea | Moderately resistant three chickpea cultivars | Field-asymmetric ion mobility spectrometry (FAIMS) for sensing Volatile Organic Compound | Resistance to *Ascochyta* blight disease | Zhang *et al.* (2023) |

in the estimation of heritability and the assessment of phenotypic variation, even after taking all major- and minor effect quantitative trait loci (QTLs)/genes into consideration (Reynolds *et al.*, 2020). High-throughput phenotyping can overcome these drawbacks, by allowing: (i) dynamic phenotyping at various plant growth stages and recording data for multiple traits at a time in large populations; (ii) spectral or image-based characterization of traits; and (iii) the modelling based (non-destructive) assessment of biochemical parameters, which avoid time-consuming operations.

Integration of HTPP with GWAS accelerates genome analysis and investigation of complex traits for the following reasons:

1. Complex traits appear to be regulated by multiple small-effect loci. Relative to cross-sectional phenotyping, dynamic phenotyping provides more in-depth insights into the genetic architecture of these traits and facilitates detection of small-effect loci. Convenient and rapid dynamic phenotyping has been made possible by high throughput phenotyping platforms.

**2.** Some small-effect loci have a weak correlation with traits and are difficult to detect using traditional methods.

In this context, HTPP utilizes novel spectral and image based parameters that are highly correlated with primary traits, which further facilitates the identification of novel heritable loci. The ability of HTPP to take dynamic observations has made the QTL mapping approach more precise. Single-time observations identify QTLs for the phenotype of that particular time point and ignore the dependency between repeated phenotypic measurements. However, multiple time point observations discover QTLs that control the whole developmental process of the traits by detecting transient and persistent QTLs. Overall, it improves temporal resolution in QTL mapping. This approach, of considering phenotypes at multiple points, is called 'functional mapping', which is based on the assumption that for function-valued traits (any trait that varies as a function of another continuous predictor variable), phenotypic values at discrete time points are 'snapshots' of a continuous function/curve over time (Xiao *et al.*, 2022; Li and Sillanpää, 2015).

Pauli *et al.* (2016) used a field-based high-throughput plant phenotyping (HTPP) platform to record data on canopy properties (reflectance, canopy temperature and height of cotton plants) and discovered QTLs underlying the dynamic response of HTPP canopy traits in high temperature and water deficit; they also analysed the temporal patterns of QTL expression over the reproductive phase of the plant life cycle. Guo *et al.* (2018) used a non-destructive phenotyping facility to extract image-based traits (i-traits) in rice to monitor drought responses and evaluate drought resistance (DR). They performed integrated GWAS, including both i-traits and traditional DR traits, which led to identifying the previously unreported DR-related gene *OsPP15*. Likewise, Li *et al.* (2020) used novel image based traits (i-traits), known i-traits for DR evaluation in cotton, by performing integrated GWAS. They identified DR-related genes i.e. *GhRD2*, *GhNAC4*, *GhHAT22*, and *GhDREB2* for some digital traits. Gao *et al.* (2023) recorded growth traits from the seedling to the grain filling- stage using a non-invasive, high-throughput phenotyping platform and conducted integrated GWAS using image based and agronomic traits. The discovery of 277 pleiotropic QTLs revealed temporal dynamics of QTL expression controlling multiple traits at different growth stages in wheat. Further more, they demonstrated that models based on i-trait data performed well in predicting yield-related traits. Wu *et al.* (2021) performed an i-trait-based genome-wide association study and reported two new genes, *ZmcPGM2* and *ZmFAB1A*, that regulate i-traits and drought tolerance, which were validated by genetic mutation analysis. Moreover, they identified 15 i-traits as markers with high potential for breeding maize for drought tolerance. Table 9.2 provides information about the size of the population analysed, the HTPP platform used, and the traits targeted for the genetic dissection of complex phenotypes in different crop plants.

### 9.3.3 Genomic selection

Genomic selection holds significant promise in the utilization of molecular markers and phenotype data to frame innovative breeding approaches by creating statistical models for genetic assessment based on genome-wide marker profiles. This method appeared as a solution to address challenges faced in QTL mapping and GWAS. It evaluates the genetic value of individuals using extensive marker data dispersed throughout the entire genome, despite depending upon confined number of markers as in marker-assisted selection (MAS). Genomic selection uses phenotypic and genotypic information from the training population (TP) to build a prediction model. This model is then used to produce genomic estimated breeding values (GEBVs) for each member of the breeding population (BP) based on their genetic profiles. GEBVs enable prediction of individuals that will give superior performance and can be used as a parent in hybridization or for advancing breeding programmes for the next generation. This is because these individuals share a molecular marker profile similar to that of other plants in training population (TP) that have demonstrated superior performance in specific environments. However, there is a delay in the GS application in practical crop breeding due to the absence of lucrative and high-throughput genotyping platforms. The arrival

**Table 9.2.** Application of HTPP in field crops for association mapping.

| Crop | Population size | HTPP platform | Controlled/field-based HTPP | Traits targeted | No. associated loci/SNP/QTL/candidate genes | References |
|---|---|---|---|---|---|---|
| Rice | 357 accessions | Visible RGB camera–Pilot piA2400-12 gc | Greenhouse | Projected shoot area for shoot growth trajectories | 442 SNPs | Campbell *et al.* (2019) |
| Rice | 165 lines | Phantom 4 Pro UAV platform for RGB imaging | Field | Shoot biomass and vegetation fraction | 4 QTLs | Ogawa *et al.* (2021) |
| Rice | 80 accessions | ASD FieldSpec4 hi-res spectroradiometer for obtaining hyperspectral data | Laboratory | Five hyperspectral traits (normalized difference spectral index (NDSI), reflectance at wavelengths 1227 nm and 1177 nm, simple ratio index (SRI) differential spectral index (DSI)] | NDSI: 65 genes | Sun *et al.* (2019) |
| Rice | 221 accessions | Visible and near infra-red (Vis/NIR) hyperspectral imaging system: an electron-multiplying charge-coupled device (EMCCD) camera, a spectrograph, six 100 W halogen bulbs, and a translational stage | Laboratory | Grain quality | 44 chromosomal candidate regions | Barnaby *et al.* (2020) |
| Rice | 186 accessions of *Oryza sativa* and 146 accessions of *O. glaberrima* | High-throughput chlorophyll fluorescence-based screening platform | Greenhouse | Photosynthetic heat tolerance through temperature response measurements | 133 candidate genes | Robson *et al.* (2023) |

*Continued*

**Table 9.2.** Continued

| Crop | Population size | HTPP platform | Controlled/field-based HTPP | Traits targeted | No. associated loci/SNP/QTL/candidate genes | References |
|---|---|---|---|---|---|---|
| Bread wheat | 211 bread wheat cultivars | Semi automatic system for measuring canopy reflectance | Field | Three vegetation indices [the modified canopy adjusted ratio index 2 (correlated with leaf area index) (MCARI2, MERIS terrestrial chlorophyll index (correlated with chlorophyll content) (MTCI), and normalized difference vegetation index (correlated with photosynthetic activity) (NDVI)] | 105 QTLs for MTCI, 97 QTLs for NDVI, 159 QTLs for MCARI2 | Jiang *et al.* (2019) |
| Spring wheat | 1185 lines | UAS (unmanned aerial system) platform:<br>1. IRIS + quadcopter installed with a 3-channel Canon S100 digital camera for imaging<br>2. DJI Matrice100 quadracopter installed with 5-channel multispectral RedEdge camera for imaging | Field | Crop lodging assessment based on image-derived digital elevation model | A key genomic region on chromosome 2A | Singh *et al.* (2019) |

| | | | | | | |
|---|---|---|---|---|---|---|
| Winter wheat | 330 winter wheat accessions | Light detection and ranging (LIDAR) mounted on field phenotyping platform | Field | Canopy height, and average daily stem elongation rates (SERs) to investigate short-term growth response to temperature | Three temperature-responsive and four temperature-irresponsive QTLs | Kronenberg *et al*. (2021) |
| Winter wheat | 335 cultivars | Automated image analysis using 1200 dpi flatbed scanners | Field | *Septoria tritici* blotch (STB) resistance (e.g. the lesions coverage percentage of leaf area, average density of pycnidia within lesions, pycnidia size, average pycnidia darkness) | 26 chromosomal segments | Yates *et al*. (2019) |
| Wheat | 288 accessions | Automated and high-throughput phenotyping platform (HTP) | Greenhouse | 31 image based traits and 17 yield related agronomic traits | 1605 QTLs including a number of known genes or QTLs; 277 pleiotropic QTLs | Gao *et al*. (2023) |
| Cotton | 200 accessions | Automated phenotyping Platform: RGB imaging with high-throughput rice phenotyping platform | Greenhouse | 119 image based digital traits (56 morphological traits and 63 texture traits) | 390 genetic loci | Li *et al*. (2021) |

*Continued*

**Table 9.2.** Continued

| Crop | Population size | HTPP platform | Controlled/field-based HTPP | Traits targeted | No. associated loci/SNP/QTL/ candidate genes | References |
|---|---|---|---|---|---|---|
| Maize | 369 inbred lines | Core root feature extraction platform, EOS 5D Mark III camera with an EF 24–105 mm f/4L IS USM lens (Canon) used to capture images | Field | Root area, maximum width, median width, convex hull area, width-profile angle, adjusted depth | 77 genes | Zheng *et al*. (2020) |
| Maize | 252 inbred lines | UAV platform: octocopter (DJI Spreading Wings S1000) with a 20.2 megapixel digital camera (Cyber shot DSC-QX100) | Field | Plant height | 68 unique QTLs | Wang *et al*. (2019) |
| Maize | 765 plants of 255 maize hybrids | RGB imaging: Grasshopper3 RGB camera, 12.5–75 mm TV zoom lens, LED illumination | Laboratory in the PhenoArch phenotyping platform | Plant leaf area, biomass and 3D architecture, radiation use efficiency | – | Chen *et al*. (2019) |
| Maize | 468 inbred lines | Zeiss Z1/ApoTome stereo-microscope in bright field with a 1× objective lens + CNN algorithms: U-net, to quantify patterning in phenotype | Laboratory | Bulliform cell pattern | Five candidate genes | Qiao *et al*. (2019) |
| Maize | 480 inbred lines | Micro-CT scanning platform: Skyscan 1172 | Laboratory | Vascular bundle traits of stems including geometry related, morphology related and distribution related traits | 1562 SNPs | Zhang *et al*. (2020) |

| | | | | | | |
|---|---|---|---|---|---|---|
| Soybean | 189 genotypes | Chlorophyll fluorescence phenoyping platform: Fluorpen Z995-PAR fluorometers | Field | Several fluorescence phenotypes | 288 SNPs | Herritt *et al*. (2018) |
| Soybean | 200 accessions | DJI Phantom 4 Pro unmanned aerial system (UAS) platform, with factory-installed camera (2.54 cm, 20 megapixel CMOS sensor) | Field | Dark green colour index (DGCI) | 45 SNPs for three environments and 16 significant SNPs for all environments | Kaler *et al*. (2020) |
| Soybean | 341 accessions | Photosynthetic System II imaging system: CCD (charge-coupled device) camera, spectral LEDs, 70 cm × 70 cm *illuminated area*, and spectral range 350–1000 nm | Greenhouse | Normalized difference vegetation index(NDVI), chlorophyll index (CHL) | 38 QTLs for NDVI, 32 QTLs for CHL | Wang *et al*. (2021) |

of next-generation sequencing (NGS) in recent years has significantly decreased both the cost and duration of sequencing procedures, along with facilitating the discovery of single nucleotide polymorphisms (SNPs). Advancement led to the progress of genome-wide SNP genotyping platforms with high throughput capabilities, in particular genotype-by-sequencing (GBS), which has made available a cost-effective means of utilizing SNPs that are well suited for GS applications. Moreover, advancements in genotyping platforms does not limit accurate prediction in GS. Instead, now the reliability of phenotypic data becomes technically challenging for the implementation of GS in crop plants, which gives rise to the genotype phenotype gap(GP gap). The GS prediction model generates GEBV for every genotyped individual in the observed set and relies on on the accuracy and precision of the phenotypic data of the TP, and therefore in every generation of selection there is a genetic gain. Hence, phenotypic data play a crucial role in training the GS model to accurately predict the GEBV of the BP. The invasive, manual and destructive techniques used for plant phenotyping were labour intensive, expensive and lacked precision, resulting in inaccuracies in genomic-assisted breeding (GAB) and were not very suitable for large population sizes. Automatic, non-invasive high throughput phenotyping facilities can precisely record accurate data for a large number of plants in a quick manner by making use of image capturing and analysis, spectroscopy, robotics, and high-performance computing facilities. According to Bhat *et al.* (2019), the combination of HTPP and GS will revolutionize the field of plant breeding and result in a substantial enhancement of genetic gain for complex traits.

Sun *et al.* (2019b) categorized utilization of HTPP platforms in GS into two groups. The first group involves directly utilizing phenotypic data generated by HTPP platforms as the main trait in training genomic prediction models. As evidence of this, Watanabe *et al.* (2017) utilized a UAV based remote sensing platform to collect plant height indicators in sorghum. Their study revealed that the predictive capability of the GS model based on UAV measured phenotypic data closely resembled that of traditional measurements, while significantly reducing labour costs. The second group enhances prediction accuracy by first using HTPP platforms to measure the traits that exhibit genetic correlation with the primary trait (also called secondary traits). Subsequently, these secondary trait data with primary traits are incorporated into a multi-trait genomic prediction model. For example, Baba *et al.* (2020) utilized a greenhouse based HTPP platform and assessed the efficacy of both a single-trait random regression model (ST-RRM) and a multi-trait random regression model (MT-RRM) in predicting water usage (WU), the primary trait, by borrowing information from projected shoot area (PSA), a secondary trait. WU, being a trait with low heritability and challenging to measure, was improved in prediction accuracy through a joint analysis with PSA, which demonstrates higher heritability and is more easily measurable. This study concluded that joint analysis of WU and PSA with the MT-RRM enhances the capability to predict future genetic values for WU over a single trait model. Multi-trait approaches have advantages when one low heritable target trait is correlated to a highly heritable secondary non-target trait. Using an MT model for joint analysis has the potential to enhance the prediction of genetic values for traits with low heritability, consequently enhancing the efficacy of selection in plant breeding programmes. In breeding programmes, the agronomic values of breeding materials are assessed through the examination of multiple traits, which frequently exhibit some genetic correlations. The breeding values can be predicted by fitting all phenotypes jointly in a single model using a multi-trait (MT) approach. The genetic covariances between the traits can be captured by these approaches and have been shown to enhance accuracy in predicting breeding values as compared to single-trait approaches, particularly for limitedly recorded or low heritable phenotypes. Thus, in the context of HTPP, these platforms are used to generate high dimensional multi-trait data that are handled by MT genomic prediction approaches.

Moreover, the prediction framework can be employed by secondary phenotypes obtained from HTPP to improve the precision of foretelling a desired characteristic such as yield. The MT data acquired using HTPP frequently correspond with an added level of a complexity–time axis. When different phenotypes are continuously recorded at regular intervals for the course of the experiment or the entire growing season, the usual MT framework might not be appropriate. Thus, the validity of employing this conventional MT model is questioned based on its assumptions.

For instance, one presumption is that every phenotype included in the MT model has unique characteristics. While this holds true for certain phenotypes, such as yield protein content or yield, assumptions regarding temporal phenotypes, which are infinite-dimensional features, imply that, although records are only available for specific time intervals, they are expected to vary continuously with time between the two intervals. With these data, a more suitable approach is to consider the temporal phenotypes as continuous traits and conduct genetic analyses using random regression models (RRM). Therefore, the RRM framework for temporal traits has been integrated into an MT model. Application of high throughput phenotyping in GS has been presented in Table 9.3.

### 9.3.4 Phenomic selection

Phenomic selection may be referred to as the use of high-throughput phenotyping to obtain numerous variables that can be used as regressors or to estimate kinship in the statistical models classically used in GS. It relies on routinely used NIRS reflectance instead of molecular markers (as in GS) to make predictions (phenomics) in any other environment without specific NIRS data at the genotypic level rather than the individual level. It also uses a prediction model as in GS but uses NIRS reflectance for prediction instead of a molecular marker. NIRS reflectance can replace molecular markers as it is related to the molecular interactions of the analysed biological sample as a result, projected to be linked with endophenotypes which are transcripts, small RNAs, metabolites and various molecular levels between genotype and phenotype. Phenomic selection can be applied in breeding programmes of orphan crops that still lack cost-reliable genotyping tools to screen closely fixed material or clones, as PS captures non-additive genetic effects which are often difficult to capture with molecular markers, as well as to screen out a specified portion of selected candidates. The effectiveness of screening and selection should be based on PS accuracy and reliability: higher accuracy means the breeder is confident in selecting many individuals without losing the finest candidates (Rincent *et al.*, 2018). Robert *et al.* (2022a) reported the at PS can be a potential substitute or complement to GS in plant breeding programmes.

#### 9.3.4.1 Approaches in phenomic prediction

1. In the first scenario, NIRS data of the training set (TS) and prediction set (PS) individuals are recorded in a uniform environment (Env 1). Training set trait data (of Env 1) and NIRS data are employed to predict the phenomics estimated performance (PEP) of the prediction set (untested individuals); environment specifies a location–year combination.
2. In the second scenario, NIRS data (of the training and prediction sets) from the central NIRS location (Env 1) are used to forecast the PEP (of untested individuals) at another environment (Env 2) for which trait data for test genotypes (training set) are recorded in Env 2.
3. In the third scenario, NIRS and trait data for the training set are taken in Env 1 to predict the PEP of untested individuals (prediction set) for the same environment (Env 1). NIRS data of these new (untested) candidates in the prediction set are generated for another year, which is another environment (Env 2) than the training set. In this case, NIRS or spectral data are taken from different environments. Spectral data are unlike molecular markers and may contradict due to the year (environmental) effect. To deal with such effects, connecting genotypes are used, which are genotypes selected from the training set (of Env 1) and grown with the prediction set (in Env 2), and NIRS or spectral data are recorded from the connecting genotypes in this environment (Env 2). These data are used to estimate the adjusted mean across environments and reduce environmental or year effects.
4. In the fourth scenario, the NIRS and trait data are recorded for the training set in one environment (Env 1), and NIRS data of the prediction set are recorded from another

**Table 9.3.** Application of high throughput phenotyping in genomic selection.

| Crop | Technique | Environmental condition | Traits targeted | Prediction model | Prediction accuracies | Reference |
|---|---|---|---|---|---|---|
| Wheat | Near-infrared (NIR) spectroscopy, XDS Rapid Content Analyser (FOSS) for NIR prediction | Field | Six end-product quality traits (grain protein, water absorption, crumb yellowness, texture hardness, flour swelling volume, flour yield) | Multi-trait model | Increased 30% | Azizinia *et al.* (2023) |
| Wheat | Aerial platform, with hyperspectral and thermal cameras | Field | Green normalized difference vegetation index and canopy temperature, and for improving primary trait, i.e. yield | Multi-trait random regression model | Increased 146% | Sun *et al.* (2019b) |
| Wheat | Unmanned aerial platform with Micasense RedEdge M and DJI Phantom four multispectral camera | Field | Raw canopy reflectance values (blue, green, red, NIR, and Red Edge) for improving primary trait, i.e. yield | Genetic (G) single-environment genomic prediction (GP), G multi-environment GP, G + E multi-environment GP with environment covariance (E) matrix, G+M+E-multi-environment GP supplemented with M and E matrices | – | Mróz *et al.* (2024) |
| Soybean | Mul-tispectral (MS) imaging by using four-eye MS camera mounted on a quadcopter unmanned aerial vehicle UAV | Field | Vegetation indices at the different stages of growth for aboveground biomass prediction | Multi-trait random regression model | – | Sakurai *et al.* (2022) |

environment (Env 2) to predict PEP there (in Env 2).

Zhu *et al.* (2021) reported that the RR-BLUP model performed well in phenomic prediction. There were only minor differences in predicting the performance at a single location using NIRS data from different locations. Overall, this implies that RR-BLUP performed well for prediction using NIRS data from the different locations. Unlike genomic prediction, phenomic prediction is less dependent on kinship between training and prediction set individuals and less affected by population structure. In context to phenomic prediction training set size appears to be a less important factor for predictive ability, and even small training sets that are insufficient for genomic prediction, show better results with phenomic data. Further, Weiß *et al.* (2022) reported similar results concerning kinship between training and prediction sets, and population structure. In the context of sample material from which NIRS reflectance data would be obtained, seed based NIR spectra show better results in predicting grain yield, which is considered an important trait for breeders. Therefore, phenomic selection acts as a promising tool for practical breeding, to improve selection gain for complex traits in plant breeding (Zhu *et al.*, 2022) (Table 9.4).

## 9.4 Challenges and Future Perspectives

While rapid advancements in genetic analysis have increased crop population size, the constrained pace of phenotyping has become the bottleneck confining crop breeding. Breaking through this bottleneck, high-throughout phenotyping has emerged as a boon in crop improvement. However, there are numerous challenges in the advancement of high-throughput plant phenotyping systems. Ground based plant phenotyping platforms face difficulty in navigation, especially in scenarios with restricted plant row spacing or wet soil conditions, leading to extended time requirements for completing entire experiments (McCarthy *et al.*, 2010; Araus and Cairns, 2014; Andrade-Sanchez *et al.*, 2014; Condorelli *et al.*, 2018; Xu *et al.* 2018). On the other hand, aerial-based plant phenotyping platforms have a shorter flight duration, high sensitivity to weather conditions, lower payload capacity, and less resistance to high wind speed. The spatial resolution is contingent on both the speed and altitude of the platform (Herwitz *et al.*, 2004; Zhang and Kovacs, 2012; Shi *et al.*, 2016; Han *et al.*, 2018; Wang *et al.*, 2018; Ziliani *et al.*, 2018; Xu *et al.*, 2019).

The ability to conduct precise phenotyping, especially under repeatable and representative field conditions, is significantly trailing behind the capability to generate genotypic data. This limitation is impeding progress in breeding programmes. Besides, it remains unclear for a plant breeder how to effectively integrate and make use of the vast amount of data produced by HTPP in their breeding programmes (Awada *et al.*, 2018). While breeders require a unified pipeline or tool, the majority of existing HTPP technologies are segmented, with even data management being compartmentalized. In essence, we can say that current HTPP technologies are designed with a focus on technology rather than being focused on meeting the specific needs and preferences of breeders. To address these challenges, it is necessary to employ user friendly phenotyping tools capable of managing these processes. Subsequently, there is a need to integrate these tools into a streamlined pipeline on a common data exchange platform equipped with standardized application programming interfaces (APIs) and access to genotypic data (Ninomiya, 2022). In recent years, HTPP has experienced advancements due to the swift development of various technologies, encompassing novel sensors, robotics, image analysis and modelling, remote control, and data mining. Two primary constraints that require enhancement include the management of data and the application of bioinformatics for extracting valuable insights from extensive phenotypic datasets (Araus and Cairns, 2014). The incorporation of low-cost and user friendly tools for the effective phenotyping of relevant traits should be a crucial element in the breeding process. This is especially important for national and international agricultural research platforms and micro to medium sized seed producing sectors in developing countries, which may face financial constraints in establishing and maintaining expensive HTPP platforms. The utilization of technological advancements

**Table 9.4.** Application of phenomics prediction in selected field crops.

| Crop | HTPP platform and sensors | Traits predicted | Predictor/regressor | Phenomic model | Prediction accuracy of PS | References |
|---|---|---|---|---|---|---|
| Bread wheat | Near infrared spectrophotometer:<br>NIR 6500<br>FOSS spectrometer<br>XDS NIR Analysers, FOSS Spectrometer<br>MPA II FT-NIR analyser | Heading date<br>Grain yield | Near Infrared (NIR) spectral data | Single NIRS model showed lower prediction accuracy than muilti NIRS model | PS showed slightly better PA for both traits relative to genomic prediction | Robert *et al.*, 2022b |
| Winter wheat | (UAS) platform, PhenoFlyf or RGB imaging | Protein content of grain<br>Yield performance<br>Yield stability | UAS (Unmanned Aerial System) based RGB images | Partial least squares regression (PLS) model was favoured over the random forest (RF) regression model | The phenomic selection(PS) showed comparable prediction accuracy relative to genomic selection (GS) and selection efficiency is increased by 28% | Roth *et al.*, 2023 |

in phenotyping instruments should be complemented by strategies to understand and manage variations in field sites to enhance repeatability. This involves implementing suitable experimental designs, choosing relevant traits and, effectively, integrating diverse datasets, analyses and applications (Araus *et al.*, 2008; Prasanna *et al.*, 2013). To facilitate the upcoming adoption of field HTPPs, all system components must be integrated efficiently, which involves combining data collection and processing with a more user-friendly data management approach.

## References

Andrade-Sanchez, P., Gore, M.A., Heun, J.T., Thorp, K.R., Carmo-Silva, A.E. *et al.* (2014) Development and evaluation of a field-based high-throughput phenotyping platform. *Functional Plant Biology* 41(1), 68. DOI: 10.1071/fp13126.

Araus, J.L. and Cairns, J.E. (2014) Field high-throughput phenotyping: The new crop breeding frontier. *Trends in Plant Science* 19(1), 52–61. DOI: 10.1016/j.tplants.2013.09.008.

Araus, J.L., Slafer, G.A., Royo, C. and Serret, M.D. (2008) Breeding for yield potential and stress adaptation in cereals. *Critical Reviews in Plant Sciences* 27(6), 377–412. DOI: 10.1080/07352680802467736.

Araus, J.L., Kefauver, S.C., Zaman-Allah, M., Olsen, M.S. and Cairns, J.E. (2018) Translating high-throughput phenotyping into genetic gain. *Trends in Plant Science* 23(5), 451–466. DOI: 10.1016/j.tplants.2018.02.001.

Awada, L., Phillips, P.W.B. and Smyth, S.J. (2018) The adoption of automated phenotyping by plant breeders. *Euphytica* 214, 148. DOI: 10.1007/s10681-018-2226-z.

Azizinia, S., Mullan, D., Rattey, A., Godoy, J., Robinson, H. *et al.* (2023) Improved multi-trait prediction of wheat end-product quality traits by integrating NIR-predicted phenotypes. *Frontiers in Plant Science* 14, 1167221. DOI: 10.3389/fpls.2023.1167221.

Baba, T., Momen, M., Campbell, M.T., Walia, H. and Morota, G. (2020) Multi-trait random regression models increase genomic prediction accuracy for a temporal physiological trait derived from high-throughput phenotyping. *PLOS ONE* 15(2), e0228118. DOI: 10.1371/journal.pone.0228118.

Barnaby, J.Y., Huggins, T.D., Lee, H., McClung, A.M., Pinson, S.R.M. *et al.* (2020) Vis/NIR hyperspectral imaging distinguishes sub-population, production environment, and physicochemical grain properties in rice. *Scientific Reports* 10(1), 9284. DOI: 10.1038/s41598-020-65999-7.

Basavaraj, P.S. and Rane, J. (2020) Avenues to realize potential of phenomics to accelerate crop breeding for heat tolerance. *Plant Physiology Reports* 25, 594–610. DOI: 10.1007/s40502-020-00552-2.

Blum, A. (2011) Phenotyping and selection. In: *Plant Breeding for Water-Limited Environments*. Springer, pp. 153–216.

Breseghello, F. and Coelho, A.S.G. (2013) Traditional and modern plant breeding methods with examples in rice (*Oryza sativa* L.). *Journal of Agricultural and Food Chemistry* 61(35), 8277–8286. DOI: 10.1021/jf305531j.

Campbell, M., Momen, M., Walia, H. and Morota, G. (2019) Leveraging breeding values obtained from random regression models for genetic inference of longitudinal traits. *Plant Genome* 12(2), 180075. DOI: 10.3835/plantgenome2018.10.0075.

Chawade, A., van Ham, J., Blomquist, H., Bagge, O., Alexandersson, E. *et al.* (2019) High-throughput field-phenotyping tools for plant breeding and precision agriculture. *Agronomy* 9(5), 258. DOI: 10.3390/agronomy9050258.

Chen, T.-W., Cabrera-Bosquet, L., Alvarez Prado, S., Perez, R., Artzet, S. *et al.* (2019) Genetic and environmental dissection of biomass accumulation in multi-genotype maize canopies. *Journal of Experimental Botany* 70(9), 2523–2534. DOI: 10.1093/jxb/ery309.

Condorelli, G.E., Maccaferri, M., Newcomb, M., Andrade-Sanchez, P., White, J.W. *et al.* (2018) Comparative aerial and ground based high throughput phenotyping for the genetic dissection of NDVI as a proxy for drought adaptive traits in durum wheat. *Frontiers in Plant Science* 9, 893. DOI: 10.3389/fpls.2018.00893.

Dhondt, S., Wuyts, N. and Inzé, D. (2013) Cell to whole-plant phenotyping: The best is yet to come. *Trends in Plant Science* 18(8), 428–439. DOI: 10.1016/j.tplants.2013.04.008.

El-Hendawy, S., Al-Suhaibani, N., Mubushar, M., Tahir, M.U., Refay, Y. *et al.* (2021) Potential use of hyperspectral reflectance as a high-throughput nondestructive phenotyping tool for assessing salt tolerance in advanced spring wheat lines under field conditions. *Plants* 10(11), 2512. DOI: 10.3390/plants10112512.

Furbank, R.T., Jimenez-Berni, J.A., George-Jaeggli, B., Potgieter, A.B. and Deery, D.M. (2019) Field crop phenomics: Enabling breeding for radiation use efficiency and biomass in cereal crops. *New Phytologist* 223(4), 1714–1727.

Gao, J., Hu, X., Gao, C., Chen, G., Feng, H. *et al.* (2023) Deciphering genetic basis of developmental and agronomic traits by integrating high-throughput optical phenotyping and genome-wide association studies in wheat. *Plant Biotechnology Journal* 21(10), 1966–1977. DOI: 10.1111/pbi.14104.

Großkinsky, D.K., Svensgaard, J., Christensen, S. and Roitsch, T. (2015) Plant phenomics and the need for physiological phenotyping across scales to narrow the genotype-to-phenotype knowledge gap. *Journal of Experimental Botany* 66(18), 5429–5440. DOI: 10.1093/jxb/erv345.

Guo, Z., Yang, W., Chang, Y., Ma, X., Tu, H. *et al.* (2018) Genome-wide association studies of image traits reveal genetic architecture of drought resistance in rice. *Molecular Plant* 11(6), 789–805. DOI: 10.1016/j.molp.2018.03.018.

Han, X., Thomasson, J.A., Bagnall, G.C., Pugh, N.A., Horne, D.W. *et al.* (2018) Measurement and calibration of plant-height from fixed-wing UAV images. *Sensors* 18(12), 4092. DOI: 10.3390/s18124092.

Herritt, M., Dhanapal, A.P., Purcell, L.C. and Fritschi, F.B. (2018) Identification of genomic loci associated with 21 chlorophyll fluorescence phenotypes by genome-wide association analysis in soybean. *BMC Plant Biology* 18(1), 213. DOI: 10.1186/s12870-018-1517-9.

Herwitz, S., Johnson, L., Dunagan, S., Higgins, R., Sullivan, D. *et al.* (2004) Imaging from an unmanned aerial vehicle: Agricultural surveillance and decision support. *Computers and Electronics in Agriculture* 44(1), 49–61. DOI: 10.1016/j.compag.2004.02.006.

Jiang, L., Sun, L., Ye, M., Wang, J., Wang, Y. *et al.* (2019) Functional mapping of N deficiency-induced response in wheat yield-component traits by implementing high-throughput phenotyping. *Plant Journal* 97(6), 1105–1119. DOI: 10.1111/tpj.14186.

Kaler, A.S., Abdel-Haleem, H., Fritschi, F.B., Gillman, J.D., Ray, J.D. *et al.* (2020) Genome-wide association mapping of dark green color index using a diverse panel of soybean accessions. *Scientific Reports* 10(1), 5166. DOI: 10.1038/s41598-020-62034-7.

Kawamura, K., Asai, H., Yasuda, T., Khanthavong, P., Soisouvanh, P. *et al.* (2020) Field phenotyping of plant height in an upland rice field in Laos using low-cost small unmanned aerial vehicles (UAVs). *Plant Production Science* 23(4), 452–465. DOI: 10.1080/1343943x.2020.1766362.

Kim, H.J., Liu, Y., Fang, D.D. and Delhom, C.D. (2019) Feasibility assessment of phenotyping cotton fiber maturity using infrared spectroscopy and algorithms for genotyping analyses. *Journal of Cotton Research* 2(1), 8. DOI: 10.1186/s42397-019-0027-0.

Kim, M., Lee, C., Hong, S., Kim, S.L., Baek, J.H. *et al.* (2021) High-throughput phenotyping methods for breeding drought-tolerant crops. *International Journal of Molecular Sciences* 22(15), 8266. DOI: 10.3390/ijms22158266.

Kim, S.H., Subramanian, P., Na, Y.W., Hahn, B.S. and Kim, Y. (2023) RDA-genebank and digital phenotyping for next-generation research on plant genetic resources. *Plants* 12(15), 2825. DOI: 10.3390/plants12152825.

Kronenberg, L., Yates, S., Boer, M.P., Kirchgessner, N., Walter, A. *et al.* (2021) Temperature response of wheat affects final height and the timing of stem elongation under field conditions. *Journal of Experimental Botany* 72(2), 700–717. DOI: 10.1093/jxb/eraa471.

Kumar, S., Gupta, P., Kumar, J. and Pratap, A. (2015) Phenotyping crop plants for drought and heat-related traits. In: Kumar, J., Pratap, A. and Kumar, S. (eds) *Phenomics in Crop Plants: Trends, Options and Limitations*. Springer, pp. 89–100.

Li, Z. and Sillanpää, M.J. (2015) Dynamic quantitative trait locus analysis of plant phenomic data. *Trends in Plant Science* 20(12), 822–833.

Li, B., Chen, L., Sun, W., Wu, D., Wang, M. *et al.* (2020) Phenomics-based GWAS analysis reveals the genetic architecture for drought resistance in cotton. *Plant Biotechnology Journal* 18(12), 2533–2544. DOI: 10.1111/pbi.13431.

Li, D., Quan, C., Song, Z., Li, X., Yu, G. *et al.* (2021) High-throughput plant phenotyping platform (HT3P) as a novel tool for estimating agronomic traits from the lab to the field. *Frontiers in Bioengineering and Biotechnology* 8, 623705. DOI: 10.3389/fbioe.2020.623705.

McCarthy, C., Hancock, N. and Raine, S.R. (2010) Apparatus and infield evaluations of a prototype machine vision system for cotton plant internode length measurement. *Journal of Cotton Science* 14(4), 221–232.

Mir, R.R., Reynolds, M., Pinto, F., Khan, M.A. and Bhat, M.A. (2019) High-throughput phenotyping for crop improvement in the genomics era. *Plant Science* 282, 60–72. DOI: 10.1016/j.plantsci.2019.01.007.

Morota, G., Jarquin, D., Campbell, M.T. and Iwata, H. (2022) Statistical methods for the quantitative genetic analysis of high-throughput phenotyping data. In: Lorence, A. and Jimenez, K.M. (eds) *High-Throughput Plant Phenotyping: Methods and Protocols*. Springer US, New York, USA, pp. 269–296.

Mróz, T., Shafiee, S., Crossa, J., Montesinos-Lopez, O.A. and Lillemo, M. (2024) Multispectral-derived genotypic similarities from budget cameras allow grain yield prediction and genomic selection augmentation in single and multi-environment scenarios in spring wheat. *Molecular Breeding* 44(1), 5. DOI: 10.1007/s11032-024-01449-w.

Nguyen, G.N. and Norton, S.L. (2020) Genebank phenomics: A strategic approach to enhance value and utilization of crop germplasm. *Plants* 9(7), 817. DOI: 10.3390/plants9070817.

Ninomiya, S. (2022) High-throughput field crop phenotyping: Current status and challenges. *Breeding Science* 72(1), 3–18. DOI: 10.1270/jsbbs.21069.

Ogawa, D., Sakamoto, T., Tsunematsu, H., Kanno, N., Nonoue, Y. *et al.* (2021) Haplotype analysis from unmanned aerial vehicle imagery of rice MAGIC population for the trait dissection of biomass and plant architecture. *Journal of Experimental Botany* 72(7), 2371–2382. DOI: 10.1093/jxb/eraa605.

Pauli, D., Andrade-Sanchez, P., Carmo-Silva, A.E., Gazave, E., French, A.N. *et al.* (2016) Field-based high-throughput plant phenotyping reveals the temporal patterns of quantitative trait loci associated with stress-responsive traits in cotton. *G3: Genes, Genomes, Genetics* 6(4), 865–879. DOI: 10.1534/g3.115.023515.

Perich, G., Hund, A., Anderegg, J., Roth, L., Boer, M.P. *et al.* (2020) Assessment of multi-image unmanned aerial vehicle based high-throughput field phenotyping of canopy temperature. *Frontiers in Plant Science* 11, 150. DOI: 10.3389/fpls.2020.00150.

Prasanna, B.M., Araus, J.L., Crossa, J., Cairns, J.E., Palacios, N. *et al.* (2013) High-throughput and precision phenotyping for cereal breeding programs. In: Gupta, P. and Varshney, R. (eds) *Cereal Genomics II*. Springer, Dordrecht, Netherlands, pp. 341–374. DOI: 10.1007/978-94-007-6401-9_13.

Pratap, A., Gupta, S., Nair, R.M., Gupta, S.K., Schafleitner, R. *et al.* (2019) Using plant phenomics to exploit the gains of genomics. *Agronomy* 9(3), 126. DOI: 10.3390/agronomy9030126.

Qiao, P., Lin, M., Vasquez, M., Matschi, S., Chamness, J. *et al.* (2019) Machine learning enables high-throughput phenotyping for analyses of the genetic architecture of bulliform cell patterning in maize. *G3: Genes, Genomes, Genetics* 9(12), 4235–4243. DOI: 10.1534/g3.119.400757.

Rebetzke, G.J., Jimenez-Berni, J., Fischer, R.A., Deery, D.M. and Smith, D.J. (2019) High-throughput phenotyping to enhance the use of crop genetic resources. *Plant Science* 282, 40–48. DOI: 10.1016/j.plantsci.2018.06.017.

Reynolds, M., Chapman, S., Crespo-Herrera, L., Molero, G., Mondal, S. *et al.* (2020a) Breeder friendly phenotyping. *Plant Science* 295, 110396. DOI: 10.1016/j.plantsci.2019.110396.

Rincent, R., Charpentier, J.P., Faivre-Rampant, P., Paux, E., Le Gouis, J. *et al.* (2018) Phenomic selection is a low-cost and high-throughput method based on indirect predictions: Proof of concept on wheat and poplar. *G3: Genes, Genomes, Genetics* 8(12), 3961–3972. DOI: 10.1534/g3.118.200760.

Robert, P., Auzanneau, J., Goudemand, E., Oury, F.X., Rolland, B. *et al.* (2022) Phenomic selection in wheat breeding: Identification and optimisation of factors influencing prediction accuracy and comparison to genomic selection. *Theoretical and Applied Genetics* 135(3), 895–914. DOI: 10.1007/s00122-021-04005-8.

Robert, P., Goudemand, E., Auzanneau, J., Oury, F.X., Rolland, B. *et al.* (2022b) Phenomic selection in wheat breeding: Prediction of the genotype-by-environment interaction in multi-environment breeding trials. *Theoretical and Applied Genetics* 135(10), 3337–3356. DOI: 10.1007/s00122-022-04170-4.

Robson, J.K., Ferguson, J.N., McAusland, L., Atkinson, J.A., Tranchant-Dubreuil, C. *et al.* (2023) Chlorophyll fluorescence-based high-throughput phenotyping facilitates the genetic dissection of photosynthetic heat tolerance in African (*Oryza glaberrima*) and Asian (*Oryza sativa*) rice. *Journal of Experimental Botany* 74(17), 5181–5197. DOI: 10.1093/jxb/erad239.

Roth, L., Fossati, D., Krähenbühl, P., Walter, A. and Hund, A. (2023) Image-based phenomic prediction can provide valuable decision support in wheat breeding. *Theoretical and Applied Genetics* 136(7), 162. DOI: 10.1007/s00122-023-04395-x.

Sakurai, K., Toda, Y., Kajiya-Kanegae, H., Ohmori, Y., Yamasaki, Y. *et al.* (2022) Time-series multispectral imaging in soybean for improving biomass and genomic prediction accuracy. *Plant Genome* 15(4), e20244. DOI: 10.1002/tpg2.20244.

Shi, Y., Thomasson, J.A., Murray, S.C., Pugh, N.A., Rooney, W.L. *et al.* (2016) Unmanned aerial vehicles for high-throughput phenotyping and agronomic research. *PLOS ONE* 11(7), e0159781. DOI: 10.1371/journal.pone.0159781.

Shorinola, O., Kaye, R., Golan, G., Peleg, Z., Kepinski, S. *et al.* (2019) Genetic screening for mutants with altered seminal root numbers in hexaploid wheat using a high-throughput root phenotyping platform. *G3: Genes, Genomes, Genetics* 9(9), 2799–2809. DOI: 10.1534/g3.119.400537.

Singh, D., Wang, X., Kumar, U., Gao, L., Noor, M. *et al.* (2019) High-throughput phenotyping enabled genetic dissection of crop lodging in wheat. *Frontiers in Plant Science* 10, 394. DOI: 10.3389/fpls.2019.00394.

Sun, D., Cen, H., Weng, H., Wan, L., Abdalla, A. *et al.* (2019a) Using hyperspectral analysis as a potential high throughput phenotyping tool in GWAS for protein content of rice quality. *Plant Methods* 15(1), 54. DOI: 10.1186/s13007-019-0432-x.

Sun, J., Poland, J.A., Mondal, S., Crossa, J., Juliana, P. *et al.* (2019b) High-throughput phenotyping platforms enhance genomic selection for wheat grain yield across populations and cycles in early stage. *Theoretical and Applied Genetics* 132(6), 1705–1720. DOI: 10.1007/s00122-019-03309-0.

Vishal, M.K., Tamboli, D., Patil, A., Saluja, R., Banerjee, B. *et al.* (2020) Image-based phenotyping of diverse rice (*Oryza sativa* L.) genotypes. *arXiv:2004.02498*. DOI: 10.48550/arXiv.2004.02498.

Volk, G.M., Byrne, P.F., Coyne, C.J., Flint-Garcia, S., Reeves, P.A. *et al.* (2021) Integrating genomic and phenomic approaches to support plant genetic resources conservation and use. *Plants* 10(11), 2260. DOI: 10.3390/plants10112260.

Volpato, L., Pinto, F., González-Pérez, L., Thompson, I.G., Borém, A. *et al.* (2021) High throughput field phenotyping for plant height using UAV-based RGB imagery in wheat breeding lines: Feasibility and validation. *Frontiers in Plant Science* 12, 591587. DOI: 10.3389/fpls.2021.591587.

Wang, L., Liu, F., Hao, X., Wang, W., Xing, G. *et al.* (2021) Identification of the QTL-allele system underlying two high-throughput physiological traits in the Chinese soybean germplasm population. *Frontiers in Genetics* 12, 600444. DOI: 10.3389/fgene.2021.600444.

Wang, X., Singh, D., Marla, S., Morris, G. and Poland, J. (2018) Field-based high-throughput phenotyping of plant height in sorghum using different sensing technologies. *Plant Methods* 14, 53. DOI: 10.1186/s13007-018-0324-5.

Wang, X., Zhang, R., Song, W., Han, L., Liu, X. *et al.* (2019) Dynamic plant height QTL revealed in maize through remote sensing phenotyping using a high-throughput unmanned aerial vehicle (UAV). *Scientific Reports* 9(1), 3458. DOI: 10.1038/s41598-019-39448-z.

Watanabe, K., Guo, W., Arai, K., Takanashi, H., Kajiya-Kanegae, H. *et al.* (2017) High-throughput phenotyping of sorghum plant height using an unmanned aerial vehicle and its application to genomic prediction modeling. *Frontiers in Plant Science* 8, 421. DOI: 10.3389/fpls.2017.00421.

Weiß, T.M., Zhu, X., Leiser, W.L., Li, D., Liu, W. *et al.* (2022) Unraveling the potential of phenomic selection within and among diverse breeding material of maize (*Zea mays* L.). *G3: Genes, Genomes, Genetics* 12(3), jkab445. DOI: 10.1093/g3journal/jkab445.

Wu, X., Feng, H., Wu, D., Yan, S., Zhang, P. *et al.* (2021) Using high-throughput multiple optical phenotyping to decipher the genetic architecture of maize drought tolerance. *Genome Biology* 22(1), 185. DOI: 10.1186/s13059-021-02377-0.

Xiao, Q., Bai, X., Zhang, C. and He, Y. (2022) Advanced high-throughput plant phenotyping techniques for genome-wide association studies: A review. *Journal of Advanced Research* 35, 215–230. DOI: 10.1016/j.jare.2021.05.002.

Xu, R., Li, C. and Velni, J.M. (2018) Development of an autonomous ground robot for field high throughput phenotyping. *IFAC-PapersOnLine* 51(17), 70–74. DOI: 10.1016/j.ifacol.2018.08.063.

Xu, R., Li, C. and Paterson, A.H. (2019) Multispectral imaging and unmanned aerial systems for cotton plant phenotyping. *PLOS ONE* 14(2), e0205083. DOI: 10.1371/journal.pone.0205083.

Yang, M., Hassan, M.A., Xu, K., Zheng, C., Rasheed, A. *et al.* (2020) Assessment of water and nitrogen use efficiencies through UAV-based multispectral phenotyping in winter wheat. *Frontiers in Plant Science* 11, 927. DOI: 10.3389/fpls.2020.00927.

Yang, Y., Nan, R., Mi, T., Song, Y., Shi, F. *et al.* (2023) Rapid and nondestructive evaluation of wheat chlorophyll under drought stress using hyperspectral imaging. *International Journal of Molecular Sciences* 24(6), 5825. DOI: 10.3390/ijms24065825.

Yates, S., Mikaberidze, A., Krattinger, S.G., Abrouk, M., Hund, A. *et al.* (2019) Precision reveals novel loci for quantitative resistance to *Septoria tritici* blotch . *Plant Phenomics* 2019, 3285904. DOI: 10.34133/2019/3285904.

Zendonadi dos Santos, N., Piepho, H., Condorelli, G.E., Licieri Groli, E., Newcomb, M. *et al.* (2021) High-throughput field phenotyping reveals genetic variation in photosynthetic traits in durum wheat under drought. *Plant, Cell & Environment* 44(9), 2858–2878. DOI: 10.1111/pce.14136.

Zhang, C. and Kovacs, J.M. (2012) The application of small unmanned aerial systems for precision agriculture: A review. *Precision Agriculture* 13(6), 693–712. DOI: 10.1007/s11119-012-9274-5.

Zhang, Y., Wang, J., Du, J., Zhao, Y., Lu, X. *et al.* (2021) Dissecting the phenotypic components and genetic architecture of maize stem vascular bundles using high-throughput phenotypic analysis. *Plant Biotechnology Journal* 19(1), 35–50. DOI: 10.1111/pbi.13437.

Zheng, Z., Hey, S., Jubery, T., Liu, H., Yang, Y. *et al.* (2020) Shared genetic control of root system architecture between *Zea mays* and *Sorghum bicolor*. *Plant Physiology* 182(2), 977–991. DOI: 10.1104/pp.19.00752.

Zhu, X., Leiser, W. L., Hahn, V., & Würschum, T. (2021) Phenomic selection is competitive with genomic selection for breeding of complex traits. *Plant Phenome Journal*, 4(1), e20027. DOI: 10.1002/ppj2.20027.

Zhu, X., Maurer, H.P., Jenz, M., Hahn, V., Ruckelshausen, A. *et al.* (2022) The performance of phenomic selection depends on the genetic architecture of the target trait. *Theoretical and Applied Genetics* 135(2), 653–665. DOI: 10.1007/s00122-021-03997-7.

Ziliani, M., Parkes, S., Hoteit, I. and McCvabe, M. (2018) Intra-season crop height variability at commercial farm scales using a fixed-wing UAV. *Remote Sensing* 10, 12. DOI: 10.3390/rs10122007.

# 10 Climate-Resilient Agricultural Interventions for Crop Production: From a Global Perspective to Local Solutions

**Govinda Rizal[1] and Shanta Karki[2*]**
[1]*Institute of Agriculture and Animal Science, Tribhuvan University, Kirtipur, Nepal;*
[2]*National Centre for Fruit Development, Kirtipur, Nepal*

**Abstract**

Climate change often disrupts established agricultural systems, thereby threatening food security. To prevent or minimize such threats to global and national food security, climate-resilient agricultural (CRA) interventions are necessary. CRA encompasses diverse practices and technologies to enhance productivity, reduce vulnerability, and ensure food security amid climatic challenges. These interventions focus on crop improvement, soil and water management, and climate-informed decision making. Key strategies include managing local resources, developing crop varieties that withstand, and even thrive in, stressful conditions, promoting conservation agriculture, expanding precision resource management, and enhancing weather forecasting tools. The Food and Agriculture Organization (FAO) has developed the CRA framework to integrate climate considerations into agricultural policies. This underscores the vital role of CRA in sustainable agricultural development. Aligning global perspectives with local solutions is essential for mitigating climate risks and safeguarding food production. Therefore, prioritizing CRA interventions is crucial for securing resilient agricultural systems and ensuring food security amidst climate uncertainties.

**Keywords:** climate-resilient agriculture, climate-smart agriculture, maladaptation, sustainability, technology transfer

**Abbreviations**

CRA – Climate-resilient agriculture; CSA – Climate-smart agriculture; DRE – Dehydration-responsive element.

## 10.1 Introduction

The adverse effects of climate change are global phenomena in which every nation is prone to extreme weather events, leading to deadly and expensive natural disasters (TrapBag, 2022). In South Asia and South-east Asia, where agriculture is the main source of the economy (ADB and UNCTAD, 2015), large populations rely on nature based traditional agriculture practices. Such populations are particularly vulnerable to climate change events (Aryal *et al.*, 2020). Fostering the growth of the agricultural sector is an effective strategy for poverty reduction, agricultural growth and economic development (Janvry and Sadoulet, 2010). Temperature

*Corresponding author: shantakyoto@gmail.com


DOI: 10.1079/9781800626638.0010

fluctuations, and climate-sensitive agriculture, such as rain-fed farming, along with frequent and unpredictable heatwaves or cold waves stress crops at any stage, resulting in reduced yields (Bryan *et al.*, 2013).

Agriculture in developing countries needs momentous change to produce enough food and tackle climate change. By 2050, the world needs to produce 70% more food because of a growing population and changing diets. But climate change could make it harder to grow food in some already struggling areas. The answer is climate-smart agriculture (CSA). It focuses on ways to make farming more resilient and productive while also cutting emissions. This approach suggests ways governments and organizations can help small farmers switch to climate-smart practices and discusses how to find the money to support these changes, suggesting new ideas for funding (FAO, 2010). However, CSA is not a silver bullet. There are trade-offs to consider (FAO, 2010). While some practices might improve yields, they might also need more water or other costs. Some of the practices might reduce greenhouse gas emissions but have a lower overall output.

To address the increasing challenges brought by climate change, mitigate the vulnerability of agricultural systems to climate related risks and uncertainties, and bolster the resilience of food production systems considering shifting climatic conditions, stakeholders developed CRA. This initiative aims to ensure food security and sustainable livelihoods by adapting agricultural practices to withstand climate change impacts, including extreme weather events, shifting precipitation patterns, and temperature fluctuations. The CRA initiative, a transformative farming approach, prioritizes sustainable food security amidst climate uncertainties. It emphasizes efficient resource management through practices such as crop rotation and integrated pest management, alongside the development and cultivation of stress-tolerant crop varieties and livestock breeds. Addressing climate risks through strategies such as drought–flood and high-temperature tolerant crop varieties, integrated soil health management, and the use of precision agriculture technologies, CRA aims for consistent, higher yields. The benefits include enhanced food security, improved farm income and environmental sustainability, which are crucial in the face of global climate change threats. Despite challenges like limited resources, technological advancement and knowledge gaps, CRA's success lies in education, research, access to technology and fair-trade initiatives, marking a requirement for a comprehensive strategy for a resilient and sustainable agricultural system.

The trend of escalating rural–urban migration, growing urban populations and increasing unplanned urbanization causes elevated evaporation rates, reduced groundwater recharging and diminishing water availability for drinking and irrigation, thereby affecting communities dependent on agriculture.

The impact of climate change extends beyond crop yields and directly, and indirectly, affects livestock production. Fluctuations in heat and temperature stress animals, setting back dairy and meat products, as well as the overall health of animals.

Across South and South-east Asia, climate change threatens agriculture, often jeopardizing food security for millions. Rising temperatures and altered rainfall patterns are already disrupting staple crops like rice in Vietnam's Mekong Delta, where crops suffer from both heat stress and saltwater intrusion due to rising sea levels (Kontgis *et al.*, 2019). Frequent typhoons wreak havoc in the Philippines, a major producer of coconuts and mangoes, causing substantial crop damage and disrupting agricultural activities (Israel and Briones, 2013). Water scarcity is another growing concern, particularly in India, a major grain producer (Amarasinghe *et al.*, 2007). Changing monsoon patterns worsen water stress, jeopardizing crop yields. Warmer temperatures and shifting weather patterns also create ideal conditions for pests and diseases, leading to significant crop losses, as seen in Indonesia's coffee plantations (Djufry *et al.*, 2022). Finally, sea level rise and salinization threaten low lying agricultural areas in Thailand, a major producer of rice and fruits, disrupting freshwater supplies and inundating farmland (Rujivanarom, 2023). These stark examples underscore the urgent need for adaptation strategies and the implementation of climate-resilient agriculture (CRA) practices to mitigate the devastating effects of climate change on the region's agriculture.

Erratic rainfall patterns bring about unpredicted floods in many regions, and droughts in others, significantly affecting the sustainability of crop production, which has a direct impact on national food security. Such climatic variabilities often lead to cascading hazards, such as the Melamchi flood disaster of 2021, which was initiated by intense precipitation in the upstream areas but triggered cascading hazards along the river corridors, causing loss of life as well as damaged roads and bridges, dams, houses, shops and schools; this led to disrupted settlements and forcing the displacement of communities (Maharjan *et al.*, 2021). Such hazards require preparedness to reduce loss and damage, as well as post-disaster rebuilding resilience.

Addressing these climate induced vulnerabilities calls for comprehensive integrated strategies that consider the interconnected nature of agriculture, livelihoods and environmental well-being.

While significant portions of Asia enjoy economic stability, there are still impoverished regions facing vulnerability. In these areas, natural disasters pose the greatest threat, as there are limited resources for protection and rebuilding. Ensuring affordable protection becomes critically important in such contexts.

This chapter reviews the major hazards relating to climate change that have been reported as occurring in countries in South and South-east Asia. It provides strategic recommendations for preventing and mitigating these hazards in the future to minimize the adverse impact of climate change related disasters.

## 10.2 Methodology

The authors obtained the information and data presented in this chapter from secondary sources and discourse with experts. To gain a thorough understanding of the impact of climate change and associated vulnerabilities in South and South-east Asia, particularly in the context of climate-resilient agriculture, insights have been gathered from publications authored by experts that are well versed in the field. Secondary data have been collected through a systematic review of research articles, available online and in print, along with books and newspaper articles. The authors referred to review articles to ensure a comprehensive overview and to find and address any gaps in the existing literature. This dual sourced approach enabled a comprehensive examination of the subject matter, encompassing both primary insights and a thorough review of existing scholarly and non-scholarly materials. Based on these findings, the authors have recommended strategies for the implementation of CRA.

## 10.3 Findings

The CRA can enhance agriculture to confront and adapt to climate related challenges. The findings include endeavours to address specific obstacles in agriculture in the face of climate change, encompassing factors such as erratic rainfall patterns, escalating temperatures, occurrences of extreme weather phenomena like droughts and floods, and alterations in pest and disease profiles. Various CRA practices are examined, as potential solutions available to farmers. These include the cultivation of drought-tolerant crop varieties; adoption of water management techniques, such as mulching, drip irrigation and rainwater harvesting to conserve water and bolster soil moisture retention; implementation of climate-smart cropping systems involving crop rotation, cover cropping, and diverse crop varieties to fortify soil health, fertility and resilience to shifting conditions; and the integration of integrated pest management strategies emphasizing non-chemical pest control methods to mitigate environmental harm and adapt to evolving pest resistance. The findings highlight the impact of these CRA practices on agricultural resilience, analysing outcomes such as enhanced crop yields amidst adverse conditions, diminished vulnerability to crop failure in the face of extreme weather events, bolstered soil health and fertility, reduced dependency on external inputs like water and fertilizers, and the promotion of farm profitability and sustainability amidst the challenges posed by climate change.

### 10.3.1 Recent natural disasters

There have been frequent reports of major climate change associated disasters in countries in South and South-east Asia, reported in the media (Table 10.1).

The list of disasters included in the table highlights disasters with stronger evidence of potential climate change influence. The availability of data and attribution to climate change varies across disaster types and countries. However, this is not an exhaustive list. It is based on information available online at the time of drafting this chapter. Human casualties, crop damage and livestock deaths are difficult to quantify precisely and might have gone unreported.

It is clear from the aforementioned list that the region is frequently hit by devastating disasters related to or aggravated by climate change.

### 10.3.2 Effects of disasters

Disasters aggravated by climate change vary depending on location, vulnerability and adaptation measures in place. The effects in separate locations are not comparable as the ways of reporting disasters vary. Climate change can worsen the severity and frequency of natural disasters, leading to devastating impacts on human agriculture and livestock.

Climate change related disasters like floods, droughts, heatwaves and cyclones are increasingly affecting human agriculture and livestock in South and South-east Asia. These disasters have disrupted agricultural production, threatened food security, and repeatedly caused economic losses.

Prolonged dry spells and water scarcity can severely stress crops, leading to stunted growth, wilting and reduced yields.

Excessive rainfall and flooding can inundate fields, destroy standing crops and wash away fertile topsoil. The 2021 floods in Thailand (AFP and Al Jazeera, 2023) submerged more than 13,600 km$^2$ of, mostly, rice paddies, and affected more than 1.3 million people (Pratt, 2021).

Rising temperatures create favourable conditions for pests and diseases, which are less endemic to the places previously, allowing them to thrive and spread more rapidly. For example, warmer days in temperate regions could trigger the growth and spread of brown planthopper, a destructive rice pest, leading to significant yield losses.

Droughts and changing rainfall patterns can stress plants, making them more susceptible to pest and disease attacks. Extreme heatwaves can cause heat stress in animals, leading to decreased milk production, weight loss and even death. In 2019, Myanmar experienced severe heatwaves that resulted in livestock deaths and economic losses for farmers. Myanmar reports around 7000 livestock deaths per year (WEF, 2019).

These disasters disrupt access to food and water for livestock, leading to malnutrition, disease and death.

Floods can damage irrigation systems, storage facilities, and transportation networks, hindering transportation and market access.

Frequent and intense storms disrupt agricultural activities, damage infrastructure and reduce crop yields. High winds and heavy rains associated with storms can damage greenhouses, barns and other agricultural structures. Rising sea levels threaten coastal agricultural lands with saltwater intrusion, making them unsuitable for crops. This is a growing concern in countries like Vietnam and Bangladesh, and other countries surrounded by seas and oceans. Bangladesh faces frequent floods. In 2020, monsoon flooding affected 5.4 million people and inundated 1,059,295 households (Begum *et al.*, 2021).

Reports have mentioned climate change-driven disasters forcing farmers and herders in Russia (Amstislavski *et al.*, 2013) and Nigeria (Akinyemi and Olaniyan, 2017) to abandon their land and livelihoods, resulting in significant social and economic disruptions. However, cases in South and South-east Asia have not received similar attention. The absence of reporting, or difficulty in finding such reports, does not negate the possibility of climate change impacted migration occurring in these regions.

The disasters between 2005 and 2015 in Asia incurred a loss of just over US$6 billion in fruit and nut production, a loss of US$7.3 billion in livestock production, and a loss of about US$5 billion in vegetable production (FAO, 2017).

**Table 10.1.** List of major climate change-associated disasters in South and South-east Asian countries reported in news media.

| Country | Year | Disaster | Reference |
|---|---|---|---|
| Afghanistan | 2016 | Floods & Landslides | (Davies, 2016) |
| Afghanistan | 2022 | Drought | (FAO, 2023) |
| Afghanistan | 2023 | Flash Floods | (Popalzai and Tawfeeq, 2023) |
| Bangladesh | 2017 | Flash Floods | (Chakraborty *et al*., 2021) |
| Bangladesh | 2022 | Flash Floods | (IFRC, 2023) |
| Bhutan | 2019 | Glacial Lake Outburst Flood(GLOF) | (Anon., 2020) |
| Cambodia | 2011 | Floods | (IFRC and DREF, 2011) |
| Cambodia | 2015 | Drought | (ReliefWeb, 2016) |
| India | 2015 | Heatwave | (Ghatak *et al*., 2017) |
| India | 2018 | Droughts | (Jain, 2021) |
| India | 2020 | Locust outbreak | (Mukhopadhyay, 2020) |
| India | 2020 | Heatwave | (WWA, 2022) |
| India | 2023 | Himalayan Glacial Outburst Flood | (Anon., 2023) |
| Indonesia | 2015 | Drought | (ReliefWeb, 2016) |
| Laos | 2015 | Drought | (ReliefWeb, 2016) |
| Laos | 2016 | Floods | (AHA Centre, 2016) |
| Laos | 2023 | Heatwave | (Dotto *et al*., 2023) |
| Malaysia | 2015 | Drought | (ReliefWeb, 2016) |
| Malaysia | 2020 | Drought | (ReliefWeb, 2020) |
| Malaysia | 2023 | Heatwave | (Dotto *et al*., 2023) |
| Maldives | 2018 | Heatwave & Coral mortality | (Bessell-Browne *et al*., 2021) |
| Myanmar | 2023 | Heatwave | (Dotto *et al*., 2023) |
| Nepal | 2017 | Floods & Landslides | (National Planning Commission, 2017) |
| Nepal | 2021 | Floods & Landslides | (ReliefWeb, 2021) |
| Pakistan | 2015 | Heatwave | (Ghatak *et al*., 2017) |
| Pakistan | 2018 | Droughts | (Jain, 2021) |
| Philippines | 2015 | Drought | (ReliefWeb, 2016) |
| Philippines | 2017 | Typhoon Haiyan | (Disaster Emergency Committee, 2023) |
| Sri Lanka | 2020 | Floods and landslides | (ReliefWeb, 2023) |
| Sri Lanka | 2023 | Drought | (IWMI, 2023) |
| Thailand | 2015 | Drought | (ReliefWeb, 2016) |
| Thailand | 2021 | Floods | (AFP and Al Jazeera, 2023) |
| Thailand | 2023 | Heatwave | (Dotto *et al*., 2023) |
| Vietnam | 2015–16 | Drought | (ReliefWeb, 2016) |
| Vietnam | 2020 | Drought | (ReliefWeb, 2020) |
| Vietnam | 2023 | Floods | (Miller, 2023) |
| Vietnam | 2023 | Heatwave | (Dotto *et al*., 2023) |

### 10.3.3 Limitations on coping with disasters

South and South-east Asia are at the forefront in experiencing the harsh realities of climate change, with frequent and intense disasters like floods, droughts, cyclones and heatwaves becoming the 'new normal'. While efforts are underway to adapt and cope, several limitations hinder achieving true resilience. The 2023 heatwave in India and Pakistan caused widespread crop losses and health problems, emphasizing the need for heat action plans and climate-resilient agriculture practices (WWA, 2022).

Developing nations in these regions lack the financial means to invest in robust disaster preparedness, early warning systems, and infrastructure upgrades for resilience. The inadequacy of irrigation systems leads to difficulties in coping with changing weather patterns.

Poverty, healthcare and other pressing needs often take precedence over long-term climate adaptation projects. Poverty worsens the impact of climate change related disasters. Impoverished communities have fewer resources with which to prepare and recover from such events. The 2015 earthquake in Nepal showed the challenges faced by impoverished communities in rebuilding; this event exposed the vulnerability of smallholder farmers. The limitations affecting resilience to climate change disasters are outlined here:

- Funds for disaster response tend to favour urban areas, leaving rural communities, often the most vulnerable, with inadequate support.
- Limited coordination between government agencies impedes effective disaster management and the implementation of adaptation plans.
- Marginalized communities, women and children, who face greater risks, have limited access to resources during disasters, worsening their vulnerability.
- Competition for land and water resources, compounded by climate change impacts, can lead to social unrest and hinder adaptation efforts.
- Rapid urbanization and high population density in cities make it challenging to evacuate and provide shelter to affected populations during disasters. Disasters often transcend national borders, requiring coordinated efforts among neighbouring countries. Political tensions and disputes can impede collaborative disaster management strategies. The 2004 Indian Ocean tsunami (Kanayev *et al.*, 2024) highlighted the difficulties in managing densely populated coastal areas.
- Lack of comprehensive data on climate risks and vulnerabilities makes it difficult to prioritize adaptation needs and target resources effectively.
- Insufficient understanding of climate change and its impacts among communities hinders preparedness and participation in adaptation initiatives. In 2021, the Melamchi flood disaster of Nepal caused loss of lives and agricultural lands, and destroyed enterprises (Maharjan *et al.*, 2021).

Bridging the gap between traditional coping mechanisms and modern scientific knowledge is crucial for effective adaptation strategies. Despite early warnings, the sheer scale of the floods can overwhelm existing infrastructure and resources, highlighting the need for more robust systems. For instance, the 2013 Typhoon Haiyan in the Philippines, and the 2010 floods in Pakistan highlighted the vulnerability of these regions to extreme weather events. The 2008 Cyclone Nargis in Myanmar highlighted the impact of deforestation on storm surges. Rural and remote areas lack effective early warning systems, reducing the ability to provide prompt alerts to at-risk populations. Improved early warning systems can save lives, as seen in the case of the 2007 Cyclone Sidr in Bangladesh.

## 10.4 Adoption of CSA

The CSA strives to achieve a balance between environmental sustainability, resilience to climate change and a steady food supply. For the adoption of technologies and practices in farming, CSA emphasizes the 'triple win' that is characterized as follows. Firstly, the CSA focuses on enhancing adaptation that helps farmers and ecosystems adjust to the changing climatic conditions; secondly, it aims to reduce greenhouse gas emissions to mitigate climate change; and lastly, CSA endeavours to bolster food security, ensuring a stable and reliable food supply even in the face of climate-related challenges.

However, CSA has often unintentionally led to 'maladaptation', reinforcing existing vulnerabilities by helping a few powerful elites or transferring risks between groups. This may result from overly technical programming that overlooks socio-political dynamics. By maladaptation, experts refer to actions, strategies or policies implemented in response to climate change that, instead of reducing vulnerability and enhancing resilience, unintentionally exacerbate existing vulnerabilities or create new ones (Schipper, 2022). Maladaptation, in the context of climate change, occurs when interventions to cope with the negative impacts of climate change lead to unintended consequences, and undermine the overall goal of increasing adaptive ability. This concept emphasizes the importance of carefully planning and implementing adaptation measures to avoid inadvertently making communities or systems more susceptible to the adverse effects of climate change. Maladaptation can result from an inadequate understanding of local conditions, misjudgement of risks, or the unintended negative consequences of well-intentioned interventions. Effective adaptation strategies aim to minimize maladaptation and contribute positively to the resilience of communities and ecosystems in the face of climate change.

This compels us to find an alternative contextualization of CSA with socio-economic dynamics that can address structural contributors to vulnerability and differential exposure to climate risk, thereby enhancing CSA's impact on reducing rural poverty and increasing equity. This understanding has led to the concept of transformative adaptation from climate-smart to climate-resilient agriculture (Hellin *et al.*, 2023).

## 10.5 Concept of CRA

To make agriculture transformative and align it with broader development goals, it is crucial to focus on the diversity of farmers, avoid maladaptation risks and prioritize social equity. This includes acknowledging that resilience involves both agri- and socio-ecological aspects. Practitioners should be cautious about helping wealthier farmers at the expense of the most vulnerable, or focusing on those for whom agriculture does not alleviate poverty. Success depends on genuine transdisciplinary partnerships and systems approaches, ensuring adaptation and mitigation goals, fair incomes, food security, and development. The emphasis on social equity and human well-being distinguishes climate-resilient from climate-smart agriculture (Hellin *et al.*, 2023).

While CSA and CRA share some common goals to reach sustainable and climate-resilient farming practices, and there are enough overlaps in their goals and practices, they are not synonymous with each other. While CSA integrates a range of practices, technologies and policies for sustainable agriculture, including both adaptation and mitigation efforts, CRA focuses on building resilience to climate-related challenges within the agricultural system itself (Table 10.2).

## 10.6 Strategies

### 10.6.1 Use of molecular mechanisms to regulate plant resilience to climate change

Genetic engineering approaches can use inherent plant molecular mechanisms to create crop varieties with specific traits for climate change elements, thereby developing a climate-resilient agriculture system. Plants respond to climate change by sensing the extreme environmental conditions at cell level, which start significant changes in their physiology, metabolism and gene expression. At the cell membrane, plants activate certain genes like arabinogalactan proteins (AGPs) and glycine-rich proteins(GRPs), which provide strengthening to the cell wall (Wu *et al.*, 2019). Expression of these genes is developmentally regulated and has tissue specific expression. Their expression is also modulated by biotic and abiotic factors, thus providing resilience to plants to overcome such climatic adversities.

**Table 10.2.** Some basic differences between climate-smart agriculture (CSA) and climate-resilient agriculture (CRA).

| Climate-smart agriculture (CSA) | Aspect | Climate-resilient agriculture (CRA) |
|---|---|---|
| Focuses on sustainable practices to enhance productivity, adaptation, and mitigation | Definition | Emphasizes the ability of the agricultural system to recover from shocks and to sustain productivity |
| Aims to optimize farm production while minimizing environmental impact and enhancing resilience | Goal | Aims at building resilience to climate variability and extreme events |
| Integrates technology, practices, and policies to achieve efficient and sustainable farming systems | Approach | Focuses on building natural and ecological resilience by incorporating diverse crops and sustaining ecosystems |
| Broad approach addressing the entire agricultural value chain and associated sectors | Scope | Specific approach building the capacity of agricultural systems to withstand climate challenges |
| Includes strategies to reduce greenhouse gas emissions from agricultural activities | Mitigation focus | Focuses on adapting to climate change, while also emphasizing mitigation efforts |
| Encourages adoption of innovative technologies for precision farming and resource management | Innovation and technology | Values indigenous, traditional, and local knowledge, with an emphasis on low-tech and nature-based solutions |
| Incorporates risk reduction strategies through diversification and sustainable practices | Risk management | Directly addresses risks by developing strategies to cope with and recover from climate-related disturbances |
| Supports adaptive measures to live with changing climate patterns and weather events | Adaptation measures | Supports building adaptive capacity by enhancing the overall resilience of the farming system |
| Strives to ensure economic viability for farmers while balancing environmental sustainability | Economic considerations | Focuses on economic viability of farming in the face of unpredictable climate conditions and risks |
| Encourages the development and implementation of policies supporting climate-smart practices | Policy integration | Advocates for policies that integrate climate resilience into agricultural planning and decision making |
| Agroforestry, biotechnology for stress-tolerant crops, data-driven decision making, digital platforms for information dissemination, greenhouse farming, ICT for precision agriculture, precision farming, precision irrigation, and weather forecasting apps | Examples | Adaptive management, climate-resilient livestock breeding, conservation agriculture, abiotic stress-tolerant and disease - and pest-resistant crop varieties, agroecological practices, early warning systems, ICT for real-time monitoring, water management technologies, and weather forecasting |

Similarly, drought and salinity stress tolerance are obtained by osmotic adjustments, activation of transcriptional factors like abscisic acid-responsive elements (ABREs: AREB, ABF and DREB2), and regulation of $Na^+$ homeostasis via transporters (like NSCC, NHX1, SOS1, HKT1 and LTC1). The transcription factor CAMTA6 significantly influences the transcriptome of germinating seeds under salt stress, impacting genes responsible for $Na^+$ transporters, such as AtHKT1, which are crucial for maintaining $Na^+$ balance (Shkolnik *et al.*, 2019).

For adaptations to chilling and frost stress, plants use hydrophobic barriers (waxes/cuticles) and antinuclear (cryoprotective glycoprotein) and antifreeze proteins (Sharma and Deswal, 2014).

Higher expression of heat-shock proteins such as HSP70, HSP100, HSP90 and HSP60 is important for thermal tolerance (Wang *et al.*, 2014).

Tolerance to heavy-metal stress can be achieved via vacuolar sequestration and production of phytochelatin, organic acids and metallothionein. Reactive oxygen species generated due to abiotic stresses can be alleviated through enzymatic (APX, CAT, POD, SOD, GR and GST) and non-enzymatic (ascorbate, glutathione, carotenoids and flavonoids) antioxidants (Hasanuzzaman *et al.*, 2020).

Genetic manipulation of these genes in transgenic plants resulted in better tolerance to various abiotic stresses. Genetic engineering of plants through various genome editing tools, such as clustered regularly interspaced short palindromic repeats/CRISPR-associated protein 9 (CRISPR-Cas9), improves abiotic stress tolerance and enhances crop quality, texture and shelf life (Liu *et al.*, 2021).

Mitogen-activated protein kinases convert extracellular stimuli into a wide range of cellular responses and are involved in phosphorylating other protein molecules that start a downstream oxidative signalling cascade (Cargnello and Roux, 2011).

The *NPK1* gene, isolated from tobacco, has been inserted in maize to increase drought tolerance of transgenic maize by protecting photosynthetic machinery under drought stress (Shou *et al.*, 2004).

Temperature variations across the world have a direct effect on plant productivity. There is a prominent change in the growth and survival patterns of plants under temperature stress. Temperature stress is divided into two categories: low-temperature stress (chilling injury) and high-temperature stress (heat stress). Induction of dehydration-responsive element (DRE) through the interaction of transcription factors, such as dehydration response element binding (DREB1A) with DRE, can be used to induce the expression of cold-responsive genes (COR). At low-temperature, the inducer of CBF expression (ICE) transcription factor turns on the expression of CBF/DREB genes, which induces freezing tolerance (Gilmour *et al.*, 1998).

Heat stress means temperatures above a threshold level, which causes irreversible damage to plant growth and development. The scorching of leaves, leaf senescence, abscission, fruit discoloration, reflective leaf hair, leaf curling, and vertical leaf orientation are the main morphological effects of higher temperatures. Heat stress induces the production of ammonia within plant tissues. This leads to ammonia toxicity. The crassulacean acid metabolism (CAM) pathway handles the high production of organic acids, such as pyruvate, citrate, malate, phosphoenolpyruvate and oxaloacetic acid. These organic acids prevent ammonia toxicity within the cytoplasm under heat stress.

### 10.6.2 Developing CRA interventions

Resilience involves agri- and socio-ecological dimensions and can be improved by incorporating digital interventions for forecasting services and social equity into agricultural interventions. Based on examples from Africa (Glover *et al.*, 2016) and Vietnam (Kyeyune and Turner, 2016), it is understood that CRA requires transdisciplinary partnerships and systems approaches to meet adaptation and mitigation goals, ensuring equitable incomes, food security and development. To target agricultural interventions effectively, researchers, extension workers, and farmers must apply a social equity lens to agricultural research and development (Hellin *et al.*, 2023).

To cope with rapid and unpredictable climatic change, all the agri-ecosystem stakeholders are encouraged to develop and implement new agricultural practices, such as early maturing crop varieties, drought tolerant varieties, crop diversification, conservation agriculture, small livestock (which require less feeding), bee keeping for efficient pollination services and multi-layered cultivation, as well as digital interventions to automate environmental conditions within a protected structure to produce off-season fruits and vegetables, and to forecast and disseminate weather conditions for the early warning and preparedness of farming households. Similarly, integrated soil fertility management practices, integrated pest management practices, and increasing resource use efficiencies (land, water and inputs) are to be practised developing a more resilient agriculture system as a whole.

### 10.6.3 Grasping the opportunity of climate change

While we have focused on the adverse effects of climate change, there are some positive aspects associated with it. For example, the nymphs of the brown planthopper, *Nilaparvata lugens* (Stål), are already living at temperatures close to their upper thermal limits. The climate warming in tropical regions and certain extreme elevated temperature events could limit their survival and distribution (Piyaphongkul *et al.*, 2012).

The growing season and region for crops increase. The length of growing season is calculated as the time between the last spring frost and the first frost of the following autumn or winter (EPA, 2021). Thus, fewer frost days mean a longer growing season. As temperatures rise across seasons, and the growing seasons become longer (USDA, 2003), growers can take advantage of the change, if other associated changes such as water availability and pest infestation are managed properly.

There will be new opportunities for crop diversification as individual crops can be grown in previously untapped areas, creating new venues of niches (KC *et al.*, 2022).

Dragon fruit (pitaya), originally from Central America, has found its way to cultivation in various countries, including Vietnam, Thailand and the Philippines. It is now extensively grown in countries in South and South-east Asia. This resilient crop is particularly well suited to arid and semi-arid regions, because of its ability to thrive in dry conditions, thereby making it a valuable choice for areas grappling with water scarcity (Oliveira-Aparecido *et al.*, 2022).

Kiwi, with its origins in China, has been introduced and cultivated in nations such as New Zealand, Italy and Chile. It has now become a prominent crop in the rest of Asia. Kiwi plants show adaptability to various climates and marginal lands, although they thrive in temperate regions. The cultivation of kiwi in Asian countries has added opportunities for diversifying agriculture in these regions (Gurbuz *et al.*, 2024).

Avocado, native to south-central Mexico, is now cultivated in countries worldwide, including those in South and South-east Asia. Avocado trees flourish in warm and subtropical climates. The introduction of avocados in Asia contributes significantly to enhancing agricultural diversity and resilience in the region (FAO, 2023).

Drumsticks, or moringa, have their origin in Africa and Asia but have been introduced to numerous tropical and subtropical regions. Recognized for its resilience in arid and semi-arid conditions, moringa's adaptability to diverse climates makes it an asset for regions grappling with climate-related challenges (The Hindu, 2017).

As both snow line and tree line move upwards in the Himalayas and towards the Arctic due to environmental changes, particularly global warming, there is an increment in area for new vegetation. As the line advances or if the forest beneath it becomes denser, this could significantly impact the global carbon cycle by increasing the amount of carbon stored on land (Subedi, 2009).

### 10.6.4 Addressing the negative impact of climate change

Mitigating the negative impacts of climate change requires a concerted effort in various sectors across ecosystems, economies and livelihoods. One key strategy involves implementing renewable energy initiatives to reduce greenhouse gas emissions and mitigate further global warming. Transitioning from fossil fuels to solar and wind energy decreases carbon emissions, as well as creating job opportunities and fostering sustainable development. Additionally, enhancing climate resilience in vulnerable communities is crucial. This includes investing in infrastructure that can withstand extreme weather events such as floods and storms, as well as implementing early warning systems and disaster preparedness measures to mitigate risks.

Furthermore, promoting sustainable land management practices like afforestation, reforestation and soil conservation helps sequester carbon dioxide from the atmosphere, mitigating climate change while also preserving biodiversity and enhancing ecosystem services.

Ultimately, addressing the negative impacts of climate change needs a multifaceted approach that integrates mitigation, adaptation and sustainable development strategies to build a more resilient and equitable future for all.

## 10.7 Recommendations

Every individual is a stakeholder in climate issues, either contributing to their causes or experiencing the consequences as victims. However, more than the masses, the vulnerable and the victims, the people with greater access to policy making and financial resources have more significant roles. These stakeholders, upon whom higher expectations are placed, are expected to invest in knowledge sharing and capacity building, which includes social networking, peer learning, information and communication technologies, group and individual training, demonstration, and agriculture extension services. The authors' recommendations are as follows:

- Enhance or modify field schools to meet local needs and, thus, strengthen knowledge and building capacities.
- Put in place, and support, initiatives for financial approaches, encompassing credit, subsidies like cash transfers, matching grants, insurance against loss, and risk management strategies.
- Establish effective organizational structures with institutional arrangements for farmer cooperatives and federations, contract farming and land titling.
- Develop community infrastructure like dams for irrigation, canals, artificial ponds and reservoirs and water retention mechanisms.
- Put structures and mechanisms in place to prevent disasters, establish early warning systems, and institute rescue and rehabilitation protocols for all scenarios.
- Increase crop water use efficiency through precision agriculture technologies, such as micro-irrigation, drip irrigation, rainwater and snow harvesting, and use of cover crops. Mulching with organic mulching materials can reduce adverse effects of climate variations and improve crop yields. Choosing suitable crop varieties, adjusting planting dates, proper irrigation scheduling considering water sensitive stages of crops, groundwater recharging techniques, proper soil nutrient management practices, and integrated disease/pest management help in ensuring sustainable crop yields despite erratic rainfalls, an increase in the number of warmer days, and other climatic variations.
- Advocate for, or increase, funding in agriculture and climate research. This initiative will empower and uplift all, including marginalized communities, by expanding their access to improved agricultural practices and technologies.

## 10.8 Conclusion

Impoverished farmers in developing and underdeveloped countries are particularly vulnerable to the ill impacts of climate change. It pushes them further into poverty and hunger. The concept and practices of CRA focus on fostering resilience against these adverse effects through the adoption of sustainable agricultural practices. However, the absence of comprehensive climate change adaptation policies and strategies can compromise the effective implementation of these practices. Addressing this gap requires robust trans-border policies, tackling the root causes of climate change, reducing vulnerability and ensuring long-term resilience. Thus, there is an immediate need to develop proper coping mechanisms and mitigation measures in affected regions, while intensively developing and implementing climate resilient agricultural strategies.

## References


ADB and UNCTAD (2015) *Food security in South Asia: Developing regional supply chains for the food processing industry* [Online]. Available at: www.adb.org/sites/default/files/publication/189217/south-asia-food-processing.pdf (accessed 20 February 2024).

AFP and Al Jazeera (2023) *Tens of thousands affected as severe flooding hits Thailand's south* [Online]. Available at: www.aljazeera.com/news/2023/12/26/tens-of-thousands-affected-as-severe-flooding-hits-thailands-south (accessed 10 February 2024).

AHA Centre (2016) *Situation update No. 1: Flood and flashflood in Lao PDR* [Online]. Available at: https://reliefweb.int/disaster/ff-2016-000093-lao (accessed 10 February 2024).

Akinyemi, T. and Olaniyan, A. (2017) Nigeria: Climate war. Migratory adaptation and farmer–herder conflicts. *Conflict Studies Quarterly* (21), 3–21. DOI: 10.24193/csq.21.1.

Amarasinghe, U., Shah, T., Turral, H. and Anand, B. (2007) *India's water future to 2025-2050: Business as usual scenario and deviations*. International Water Management Institute, Colombo, Sri Lanka.

Amstislavski, P., Zubov, L., Chen, H., Ceccato, P., Pekel, J.F. *et al*. (2013) Effects of increase in temperature and open water on transmigration and access to health care by the Nenets reindeer herders in northern Russia. *International Journal of Circumpolar Health* 72(1), 21183. DOI: 10.3402/ijch.v72i0.21183.

Anon. (2020) *Detailed assessment report on GLOF hazard from Thorthormi Glacial Lake and associated glaciers* [Online]. Available at: www.nchm.gov.bt/attachment/ckfinder/userfiles/files/Report_on_Thorthormi_assessment_2019.pdf (accessed 10 February 2024).

Anon. (2023) *Glacial Lakes Threaten Millions in a Warming World* [Online]. Available at: www.wired.com/story/glacial-lake-outburst-flood-glof-india-sikkim/ (accessed 10 February 2024).

Aryal, J.P., Sapkota, T.B., Khurana, R., Khatri-Chhetri, R., Rahut, D.B. *et al*. (2020) Climate change and agriculture in South Asia: Adaptation options in smallholder production systems. *Environment, Development and Sustainability* 22, 5045–5075. DOI: 10.1007/s10668-019-00414-4.

Begum, A., Dutta, S., Norton, R. and Venkateswaran, K. (2021) *Post Event Review Capability (PERC) Study. Learning from the 2020 Floods in Faridpur District, Bangladesh to build resilience*. Institute for Social and Environmental Transition-International, Boulder, CO, USA.

Bessell-Browne, P., Epstein, H.E., Hall, N., Buerger, P. and Berry, K. (2021) Severe heat stress resulted in high coral mortality on Maldivian reefs following the 2015–2016 El Niño Event. *Oceans* 2(1), 233–245. DOI: 10.3390/oceans2010014.

Bryan, E., Ringler, C., Okoba, B., Roncoli, C., Silvestri, S. *et al*. (2013) Adapting agriculture to climate change in Kenya: Household strategies and determinants. *Journal of Environmental Management* 114, 26–35. DOI: 10.1016/j.jenvman.2012.10.036.

Cargnello, M. and Roux, P. (2011) Activation and function of the MAPKs and their substrates, the MAPK-activated protein kinases. *Microbiology and Molecular Biology Reviews* 75(1), 50–83. DOI: 10.1128/MMBR.00031-10.

Chakraborty, D., Mondal, K., Islam, S. and Roy, J. (2021) 2017 flash flood in Bangladesh: Lessons learnt. In: Pal, I., Shaw, R., Djalante, R. and Shrestha, S. (eds) *Disaster Resilience and Sustainabilit*. Elsevier, pp. 591–610. DOI: 10.1016/B978-0-323-85195-4.00007-X.

Davies, R. (2016) *Afghanistan – Deadly Flash Floods in Khost Province* [Online]. Available at: https://floodlist.com/asia/afghanistan-floods-khost-province-july-2016 (accessed 10 February 2024).

Disaster Emergency Committee (2023) *Philippines typhoon facts and figures* [Online]. Available at: https://www.dec.org.uk/article/philippines-typhoon-facts-and-figures (accessed 10 February 2024).

Djufry, F., Wulandari, S. and Villano, R. (2022) Climate smart agriculture implementation on coffee smallholders in Indonesia and strategy to accelerate. *Land* 11(7), 1112. DOI: 10.3390/land11071112.

Dotto, C., Shveda, K. and Robinson, L. (2023) *A 'once-in-200 years' heat wave caught Southeast Asia off guard. Climate change will make them more common* [Online]. Available at: https://edition.cnn

.com/2023/06/06/asia/southeast-asia-heat-wave-humidity-climate-intl-hnk-dst-scn-dg/index.html (accessed 10 February 2024).

EPA (2021) *Climate change indicators: Length of growing season* [Online]. Available at: www.epa.gov/climate-indicators/climate-change-indicators-length-growing-season (accessed 10 February 2024).

FAO (2010) *Climate-smart agriculture: Policies, practices and financing for food security, adaptation and mitigation* [Online]. Available at: www.fao.org/3/i1881e/i1881e00.pdf (accessed 20 February 2024).

FAO (2017) *The impact of disasters and crises on agriculture and food security* [Online]. Available at: www.fao.org/3/I8656EN/i8656en.pdf (accessed 20 February 2024).

FAO (2023) *Assessment of agricultural drought severity in Afghanistan for 2022* [Online]. Available at: www.fao.org/geospatial/resources/detail/en/c/1661713/ (accessed 20 February 2024).

FAO (2023) *Ripe for change: Adapting avocado production to a changing climate* [Online]. Available at: www.fao.org/3/cc7119en/cc7119en.pdf (accessed 20 February 2024).

Ghatak, D., Zaitchik, B., Hain, C. and Anderson, M. (2017) The role of local heating in the 2015 Indian Heat Wave. *Scientific Reports* 7, 7707. DOI: 10.1038/s41598-017-07956-5.

Gilmour, S., Zarka, D.G., Stockinger, E.J., Salazar, M.P., Houghton, J.M. *et al.* (1998) Low temperature regulation of the Arabidopsis CBF family of AP2 transcriptional activators as an early step in cold-induced COR gene expression. *Plant Journal* 16(4), 433–442. DOI: 10.1046/j.1365-313x.1998.00310.x.

Glover, D., Sumberg, J. and Andersson, J. (2016) The adoption problem; or why we still understand so little about technological change in African agriculture. *Outlook on Agriculture* 45, 3–6. DOI:10.5367/oa.2016.0235.

Gurbuz, I., Ozkan, G. and Er, S. (2024) Exploring kiwi fruit producers' climate change perceptions. *Applied Fruit Science* 66, 475–483. DOI: 10.1007/s10341-023-01021-4.

Hasanuzzaman, M., Bhuyan, M.B., Zulfiqar, F., Raza, A., Mohsin, S.M. *et al.* (2020) Reactive oxygen species and antioxidant defense inplants under abiotic stress: Revisiting the crucial role of a universal defense regulator. *Antioxidants* 9(8), 681. DOI: 10.3390/antiox9080681.

Hellin, J., Fisher, E. and Taylor, M. (2023) Transformative adaptation: From climate-smart to climate-resilient agriculture. *CABI Agriculture and Bioscience* 4, 30. DOI: 10.1186/s43170-023-00172-4.

IFRC (2023) *Bangladesh: Floods - Final Report (MDRBD028)* [Online]. Available at: https://reliefweb.int/report/bangladesh/bangladesh-floods-final-report-mdrbd028 (accessed 10 February 2024).

IFRC and DREF (2011) *DREF operation update, Cambodia: Floods 2011* [Online]. Available at: www.ifrc.org/docs/appeals/11/MDRKH003drefou1.pdf (accessed 10 February 2024).

Israel, D. and Briones, R. (2013) *Impacts of natural disasters on agriculture, food security, and natural resources and environment in the Philippines.* ERIA Discussion Paper Series, ERIA-DP-2013-15. Available at: www.eria.org/ERIA-DP-2013-15.pdf (accessed 25 November 2024).

IWMI (2023) *Sri Lanka Drought Outlook* [Online]. Available at: www.iwmi.cgiar.org/tools/drought-bulletin/2023/sri-lanka-drought-outlook-issue-2.pdf (accessed 10 February 2024).

Jain, N. (2021) *Southern India's 2016-2018 drought was the worst in 150 years* [Online]. Available at: https://india.mongabay.com/2021/05/southern-indias-2016-2018-drought-was-the-worst-in-150-years/ (accessed 10 February 2024).

Janvry, A. and Sadoulet, E. (2010) Agricultural growth and poverty reduction: Additional evidence. *World Bank Research Observer* 25(1), 1–20.

Kanayev, V.F., Verlaan, P.A. and Morgan, J.R. (2024) Indian Ocean. In: *Encyclopedia Britannica*. Available at: www.britannica.com/place/Indian-Ocean (accessed 24 September 2014).

KC, K., Fraser, E., Wassmansdorf, D. and Tzadok, E. (2022) Climate change creates opportunities to expand agriculture in the Hindu Kush Himalaya but will cause considerable ecosystem trade-offs. *Environmental Research Communications* 4(11), 111001. DOI: 10.1088/2515-7620/ac9aea.

Kontgis, C., Schneider, A., Ozdogan, M., Kucharik, C., Duc, N.H. *et al.* (2019) Climate change impacts on rice productivity in the Mekong River Delta. *Applied Geography* 102, 71–83. DOI: 10.1016/j.apgeog.2018.12.004.

Kyeyune, V. and Turner, S. (2016) Yielding to high yields? Critiquing food security defnitions and policy implications for ethnic minority livelihoods in upland Vietnam. *Geoforum* 71, 33–34. DOI: 10.1016/j.geoforum.2016.03.001.

Liu, Q., Yang, F., Zhang, J., Liu, H., Rahman, S. *et al.* (2021) Application of CRISPR/Cas9 in crop quality improvement. *International Journal of Molecualr Sciences* 22(8), 4206. DOI: 10.3390/ijms22084206.

Maharjan, S.B., Steiner, J.F., Shrestha, A.B., Maharjan, A., Nepal, S. *et al.* (2021) *The Melamchi flood disaster: Cascading hazard and the need for multihazard risk management* [Online]. Available at: https://lib.icimod.org/record/35284 (accessed 10 February 2024).

Miller, L. (2023) *Up close and personal in Viet Nam's flooded central region* [Online]. Available at: https://www.unicef.org/vietnam/close-and-personal-viet-nams-flooded-central-region (accessed 10 February 2024).

Mukhopadhyay, A. (2020) *India swarmed by locust invasion* [Online]. Available at: www.dw.com/en/india-locusts/a-53579409 (accessed 10 February 2024).

National Planning Commission (2017) *Nepal Flood 2017: Post Flood Recovery Needs Assessment* [Online]. Available at: https://un.org.np/resource/nepal-flood-2017-post-flood-recovery-needs-assessment (accessed 25 November 2024).

Oliveira-Aparecido, L.E., Dutra, A.F., de Lima, R.F., de Alcântara Neto, F., Botega Torsoni, G. *et al.* (2022) Climate change scenarios and the dragon fruit climatic zoning in Brazil. *Theoretical and Applied Climatology* 149, 897–913. DOI: 10.1007/s00704-022-04090-9.

Piyaphongkul, J., Pritchard, P. and Bale, J. (2012) Can tropical insects stand the heat? A case study with the brown planthopper *Nilaparvata lugens* (Stål). *PLOS ONE* 7(1), e29409. DOI: 10.1371/journal.pone.0029409.

Popalzai, E. and Tawfeeq, M. (2023) *Dozens swept away in flash floods in Afghanistan* [Online]. Available at: https://edition.cnn.com/2023/07/23/middleeast/afghanistan-flash-flood-deaths-intl/index.html (accessed 10 February 2024).

Pratt, S. (2021) *Thailand Swamped by Monsoon Flooding* [Online]. Available at: https://earthobservatory.nasa.gov/images/148983/thailand-swamped-by-monsoon-flooding (accessed 10 February 2024).

ReliefWeb (2016) *South-East Asia: Drought - 2015-2017* [Online]. Available at: https://reliefweb.int/disaster/dr-2015-000180-vnm (accessed 10 February 2024).

ReliefWeb (2020) *Southeast Asia: Drought – 2019–2020* [Online]. Available at: https://reliefweb.int/disaster/dr-2019-000113-phl (accessed 10 February 2024).

ReliefWeb (2021) *Nepal: Floods and Landslides - Jun 2021* [Online]. Available at: https://reliefweb.int/disaster/fl-2021-000134-npl (accessed 10 February 2024).

ReliefWeb (2023) *Sri Lanka: Floods and Landslides - Oct 2023* [Online]. Available at: https://reliefweb.int/disaster/fl-2023-000196-lka (accessed 10 February 2024).

Rujivanarom, P. (2023) Rising sea levels forcing residents in Thailand's low-lying areas to adapt – or leave. *Mekong Eye* 21, 8. Available at: www.mekongeye.com/2023/08/21/rising-sea-samut-songkhram

Schipper, E. (2022) Catching maladaptation before it happens. *Nature Climate Change* 12, 617–618. DOI: 10.1038/s41558-022-01409-2.

Sharma, B. and Deswal, R. (2014) Antifreeze proteins in plants: An overview with an insight into the detection techniques including nanobiotechnology. *Journal of Proteins and Proteomics* 5(2), 89–107.

Shkolnik, D., Finkler, A., Pasmanik-Chor, M. and Fromm, H. (2019) CALMODULIN-BINDING TRANSCRIPTION ACTIVATOR 6: A key regulator of $Na^+$ homeostasis during germination. *Plant Physiology* 180(2), 1101–1118. DOI: 10.1104/pp.19.00119.

Shou, H., Bordallo, P. and Wang, K. (2004) Expression of the *Nicotiana* protein kinase (NPK1) enhanced drought tolerance in transgenic maize. *Journal of Experimental Botany* 55, 1013–1019. DOI: 10.1093/jxb/erh129.

Subedi, M. (2009) Climate change and its potential effects on tree line position: An introduction and analysis. *The Greenery* 7(1), 17–21.

The Hindu (2017) *How the drumstick tree is fighting climate change in the Maghreb* [Online]. Available at: www.thehindu.com/specials/impact-journalism-day-2017/how-an-indian-tree-is-fighting-climate-change-in-the-maghreb/article19126345.ece (accessed 20 February 2024).

TrapBag (2022) *Natural Disasters in Asia: Types and Prevention*. TrapBag, Fort Myers, FL, USA.

USDA (2003) *Growing seasons in a changing climate* [Online]. Available at: www.climatehubs.usda.gov/growing-seasons-changing-climate (accessed 10 February 2024).

Wang, K., Zhang, X., Goatley, M. and Ervin, E. (2014) Heat shock proteins in relation to heat stress tolerance of creeping bentgrass at different N levels. *PLOS ONE* 9(7), e102914. DOI: 10.1371/journal.pone.0102914.

WEF (2019) *This chart shows where extreme weather is causing the most fatalities* [Online]. Available at: www.weforum.org/agenda/2019/12/extreme-weather-environment-climate-change/ (accessed 10 February 2024).

Wu, W., Zhu, S., Chen, Q., Lin, Y., Tian, J. *et al.* (2019) Cell wall proteins play critical roles in plant adaptation to phosphorus deficiency. *International Journal of Molecular Sciences* 20(21), 5259. DOI: 10.3390/ijms20215259.

WWA (2022) *Climate change made devastating early heat in India and Pakistan 30 times more likely* [Online]. Available at: www.worldweatherattribution.org/climate-change-made-devastating-early-heat-in-india-and-pakistan-30-times-more-likely/ (accessed 10 February 2024).

# Index

Note: Page numbers in **bold** type refer to **figures;** page numbers in *italic* type refer to *tables*

Printed and bound by CPI Group (UK) Ltd, Croydon, CR0 4YY

13/09/2026

14961116-0001